中传学者文库编委会

主　任： 廖祥忠　张树庭
副主任： 蔺海波　李　众　刘守训　李新军　王　晖
　　　　　杨　懿　柴剑平

成　员（按姓氏笔画排序）：
　　　　王廷信　王栋晗　王晓红　王　雷　文春英
　　　　龙小农　付　龙　叶　龙　刘东建　刘剑波
　　　　任孟山　李怀亮　李　舒　张绍华　张　晶
　　　　张根兴　张毓强　林卫国　郑　月　金　炜
　　　　金雪涛　周建新　庞　亮　赵新利　徐红梅
　　　　贾秀清　高晓虹　隋　岩　喻　梅　熊澄宇

中传学者文库

主编／柴剑平
执行主编／龙小农
副主编／张毓强　周建新

洞观智能音视频

张勤团队学术文集

张勤　等著

中国传媒大学出版社
·北京·

图书在版编目（CIP）数据

洞观智能音视频：张勤团队学术文集 / 张勤等著 .-- 北京：中国传媒大学出版社，2024.8.

（中传学者文库 / 柴剑平主编）.

ISBN 978-7-5657-3774-9

Ⅰ.TN912.2-53；TN948.65-53

中国国家版本馆 CIP 数据核字第 2024R9L890 号

洞观智能音视频：张勤团队学术文集
DONGGUAN ZHINENG YINSHIPIN: ZHANG QIN TUANDUI XUESHU WENJI

著　　者	张勤 等
责任编辑	杨小薇
封面设计	锋尚设计
责任印制	李志鹏
出版发行	中国传媒大学出版社
社　　址	北京市朝阳区定福庄东街 1 号　　邮　编　100024
电　　话	86-10-65450528　65450532　　传　真　65779405
网　　址	http://cucp.cuc.edu.cn
经　　销	全国新华书店
印　　刷	北京中科印刷有限公司
开　　本	710mm×1000mm　1/16
印　　张	25.25
字　　数	303 千字
版　　次	2024 年 8 月第 1 版
印　　次	2024 年 8 月第 1 次印刷
书　　号	ISBN 978-7-5657-3774-9/TN・3774　　定　价　126.00 元

本社法律顾问：北京嘉润律师事务所　郭建平

总 序

媒介是人类社会交流和传播的基本工具。从口语时代到印刷时代，再经电子时代至今天的数智时代，媒介形态加速演变、融合程度深入发展，媒介已然成为现代社会运行的基础设施和操作系统。今天，人类已经迈入媒介社会，万物皆媒、人人皆媒，无媒介不社会、无传播不治理。今天，无论我们怎么用力于信息传播的研究、怎么重视信息传播人才的培养都不为过。

中国传媒大学（其前身为北京广播学院）作为新中国第一所信息传播类院校，自1954年创建伊始，即与媒介形态演变合律同拍、与国家发展同频共振，努力探索中国特色信息传播人才培养模式、构建中国信息传播类学科自主知识体系，执信息传播人才培养之牛耳、发信息传播研究之先声，被誉为"中国广播电视及传媒人才摇篮""信息传播领域知名学府"。

追溯中传肇始发轫之起源、瞩望中传砥砺跨越之未来，可谓创业维艰而其命维新。昔日中传因广播而起，因电视而兴，因网络而盛，今天和未来必乘风破浪、蓄势而上，因人工智能而强。在这期间，每一种媒介兴起，中传均吸引一批志于学、问于道、勤于术的

学者汇聚于此，切磋学术、传道授业，立时代之潮头，回应社会需求，成为学界翘楚、行业中坚，遂有今日中传学术研究之森然气象，已历七秩而弦歌不断，将传百世亦风华正茂。

自新时代以来，中传坚守为党育人、为国育才初心，励精图治、勠力前行，秉承"系统治理、创新图强、交叉融合、特色发展"的办学理念，牢牢把握高等教育发展大势、传媒业态发展趋势，瞄准"智能传媒"和"国际一流"两大主攻方向，以世界为坐标、以未来为向度，完成了全面布局和系统升级，正在蹄疾步稳、高质量推动学校从传统高等教育向未来高等教育跨越、从传统传媒教育向智能传媒教育跨越、从国内一流向世界一流跨越，全力建设中国特色、世界一流传媒大学。

中国特色、世界一流，在于有大先生扎根中国大地，汇聚古今、融通中外；在于有大先生执教黉门，学高为师、身正为范；在于有大先生躬耕杏坛，敦品积学、启智润心。习近平总书记更强调，高校教师要立志成为大先生，在教书育人和科研创新上不断创造新业绩。中传广大教师素来以做大先生为毕生职志，努力成为新时代"经师"与"人师"的统一者，做真学问、立高品行，践履"立德树人"使命。

2024岁在甲辰，欣逢中传建校70华诞，学校特邀约部分学者钩玄勒要、增删批阅，遴选已公开刊发的论文汇编成集，出版"中传学者文库"，意在呈现学校在学科建设、科学研究、服务行业实践等方面的最新成果，赓续中传文脉，谱写时代新声。

文库汇聚老中青三代学者，资深学者渊渟岳峙、阐幽抉微；中年学者沉潜蓄势、厚积薄发；青年学者踌躇满志、未来可期。文库与五十周年校庆所出版的"北广学者文库"相承接，大致可勾勒中传知识生产薪火相传、三代辉映之概貌，反映中传在构建中国特色新闻传播类、传媒艺术类、传媒技术类学科体系、学术体系和话语体系方面的耕耘与收获，窥见中国特色信息传播类学科知识体系构建的发展脉络与轨迹。

这一构建过程，虽筚路蓝缕，却步履铿锵；虽垦荒拓野，亦四方辐辏。一批肇始于中传，交叉融合、具有中国特色的学科，如播音主持艺术学、广播电视艺术学、传媒艺术学、数字媒体艺术学、政治传播学等，从涓涓细流汇入滔滔江河，从中传走向全国，展现了中传学者构建中国自主知识体系的学术想象力和创新力。文库展示的虽然是历史，实则是呈现今天；看似是总结过去，实则是召唤未来。与其说这套文库的出版，是对既有学术成果的展示，毋宁说是对未来学术创新的邀约。

回首过往，七秩芳华。我们深知，唯有将马克思主义基本原理与中华优秀传统文化相结合，才能推动中华学术创造性转化和创新性发展，推动中国自主知识体系的构建。我们深知，唯有准确把握媒介形态演变的脉动、深刻认知媒介形态变革所产生的影响，才能推动中国信息传播类学科自主知识体系的构建与时俱进。

展望未来，星辰大海。我们深知，以人工智能为代表的产业和科技革命正迅疾而来，媒介生态正在加速重构，教育形态正在全面重塑，大学之使命与价值正在被重新定义；我们深知，唯有"胸怀国之大者"、面向世界科技前沿、面向经济主战场、面向国家重大需求，才能确保中传始终屹立于中国乃至世界传媒教育发展之潮头。

如何应对人工智能带来的深刻变革，对中传而言是一场要么"冲顶"、要么"灭顶"的"兴亡之战"。我们坚信，不管前方是雄关漫道，还是荆棘满途，唯有勇敢直面"教育强国，中传何为？"这一核

心命题，奋力书写"智能传媒教育，中传师生有为！"的精彩答卷，才能化危为机，奋力开创人工智能时代中传智能传媒教育新纪元。

功不唐捐，芳华七秩；风帆正举，赓续创新。

是为序。

第十四届全国政协委员，中国传媒大学党委书记、教授、博士生导师

序 言

在中国传媒大学建校70周年之际，我们精选了张勤教授在过去17年间（2008-2024）的22篇具有代表性的期刊和会议论文，汇编成本书。

张勤教授于1990-1995年任加拿大英属哥伦比亚大学图像处理实验室研究工程师，1996-2000年任美国摩托罗拉公司DNS前端工程高级技术顾问，开发了第一代和第二代数字卫星电视和交互式有线电视网络分配系统，获得授权发明专利十余项。为了投身祖国的科研事业，张勤教授于2000年返回祖国，全职任教于中国传媒大学，担任通信与信息系统专业博士生导师、新创信息技术研究所所长。自回国任教以来，张勤教授一直致力于下一代数字广播电视技术中的音视频理论与系统研究，主持并完成了多项国家自然科学基金重点项目、面上项目和国家科技支撑计划项目，培养和指导青年教师获批20余项国家级项目。此外，张勤教授还致力于研究成果的产业化，其主持研发的IPTV系统于2003年开始运营，于2007年由TCL公司生产，并被荷兰数字家庭标准采用，其中软件机顶盒部分成为我国信息产业部研发规划的第三代机顶盒。2010年，由张勤教授发起筹建的媒介音视频教育部重点实验室通过教育部建设计划论证，并于2016年正式开放运行，张勤教授被聘任为实验室主任。实验室面向国家文化传播战略需求与音视频处理领域的重大科学问题，经过不断建设与发展，形成了沉浸式计算与视觉艺术呈现、有声文化与仿声计算、云边端沉浸式内容处理、沉浸交互与空间艺术评价四个研究方向，并取得了一系列的研究成果。

本书由四个部分组成，第一部分为"情智信息"，第二部分为"媒介音频"，第三部分为"视觉处理"，第四部分为"人工智能"。本书全面展示了张勤教授在智能音视频领域的理论探索和实际应用成果。

随着科技的迅猛发展，智能音视频技术不断迎来新的挑战和机遇，我们期待与各界同仁携手并肩，共同推动智能音视频技术迈向更高的台阶，谨以此为序。

Contents

第一部分 情智信息

情智信息的建模与应用 ·· 003

Interaction Between Dynamic Affection and Arithmetic Cognitive Ability: a Practical Investigation with EEG Measurement ·· 016

SSTM-IS: Simplified STM Method Based on Instance Selection for Real-Time EEG Emotion Recognition ·· 041

Multi-Source Information-Shared Domain Adaptation for EEG Emotion Recognition ··· 067

Emotional Quality Evaluation for Generated Music Based on Emotion Recognition Model ·· 080

第二部分 媒介音频

Design of Linear-Phase Nonsubsampled Nonuniform Directional Filter Bank with Arbitrary Directional Partitioning ··· 093

Multi-Source Separation Using over Iterative Empirical Mode Decomposition ········· 108

A Two-Stage Complex Network Using Cycle-Consistent Generative Adversarial Networks for Speech Enhancement ··· 114

Analysis of Music Rhythm Based on Bayesian Theory ·································· 144

Learning to Generate Emotional Music Correlated with Music Structure Features ········ 152

Visually Aligned Sound Generation via Sound-Producing Motion Parsing ············· 167

MovieREP: a New Movie Reproduction Framework for Film Soundtrack ············· 201

第三部分　视觉处理

Distributed Markov Chain Monte Carlo Kernel Based Particle Filtering for Object Tracking ... 209

Use Hierarchical Genetic Particle Filter to Figure Articulated Human Tracking ... 222

Human Action Recognition Using Multi-Velocity STIPs and Motion Energy Orientation Histogram ... 231

Semantic Based Autoencoder-Attention 3D Reconstruction Network ... 251

Flexible Light Field Angular Superresolution via a Deep Coarse-to-Fine Framework · 270

Cross-Domain Feature Similarity Guided Blind Image Quality Assessment ... 287

A Dataset and Benchmark for 3D Scene Plausibility Assessment ... 306

第四部分　人工智能

Interpretability Diversity for Decision-Tree-Initialized Dendritic Neuron Model Ensemble ... 333

Pruning of Dendritic Neuron Model with Significance Constraints for Classification ... 362

A General Paradigm of Knowledge-Driven and Data-Driven Fusion ... 379

第一部分
情智信息

情智信息的建模与应用*

一、引言

在经过几十年发展而逐步健全的人工智能技术体系支持下,针对文本、音频、视频等数据的分析与处理技术有效地解决了人类在体力劳动过程中的诸多问题:工业生产中的操作问题、医疗手术中的控制问题、驾驶与航行中的导航问题、图像识别与知识搜索问题等。然而,在信息技术飞速发展的今天,我们不能将目光停留在仅通过人工智能技术来实现对人类体力的解放上,如何将其应用于服务人类智力发展的新型媒体上将是未来信息技术领域研究的核心问题之一。

信息技术如何促进智力水平的提高对人类社会的发展具有重要的意义。从情感与智力关联性角度的研究发展来看,人类智力是以脑活动为基础的,而智商的高低取决于皮层神经元的工作效率与相关信息的存储。早在 1937 年,Kluver 和 Bucy 就曾通过实验发现,猴子边缘性区域(尤其是颞叶内侧杏仁核区域)的损伤会带来"精神失明"——尽管动物保留了正常视力,但是视觉刺激失去了它们的情感意义,影响动物对外界事物的认知。此后的大量研究也证实了大脑边缘区域在刺激信息加工过程中的主观作用。Hanying 通过实验验证注意力对视觉刺激诱发响应具有调制作用,在不同背景脑电与刺激强度条件下注意力的调制作用是不同的。在特定脑电节律振荡与初始相位及视觉刺激对比度高时,注意力对视觉刺激响应有显著调节作用。因此,在视听刺激下,人体的生理信号(主要包括外周生理信号和脑信号等)会产生一系列的变化,对应不同的情感状态。良好的情境会使人产生愉悦的情感,有助于集中注意力,进而提高大脑的活动效率,由情感变化导致的注意力改变会直接影响大脑对视听刺激的接收。情感与智力表现之间存在一定的相互影响作用,智力是情感的基础,并引导情感的发展。只有通过对客观事物的反映,主体才能确定客观事物是否满足自身的需求,从而产生相应的态度体验,进而引发不同的情感,而情感反过来对认知过程也起调节作用。

从音视频等信息与情感关联性角度的研究发展来看,情感与音乐有着紧密的关系,

---

* 本文原载于《中国传媒大学学报(自然科学版)》,作者系叶龙、段丹婷、钟微、胡飞、张勤。收入本书时有改动。

情感对人的思维与行为一直存在着重要的调控作用，情感是音乐影响智力表现的重要桥梁。心理学实验表明，当一个人欣赏喜爱的音乐时，更容易激发积极的情感，从而有效地促进其智力水平的正常发挥。音乐认知可以在很大程度上活化脑力，锻炼、提高大脑的工作效率，从而起到促进智力正常或者超常发挥的作用。虽然音乐和智力之间的关联为人所知，但还是处于知其然不知其所以然的阶段，对这种关联的模型和关联的度量都缺少科学的研究与本质的认知。因此，本文从情感角度出发，针对信息、情感与智力发展的耦合问题进行研究，给出了情智模型的定义及其研究范畴，提出了六觉（听觉、视觉、味觉、嗅觉、触觉和意境知觉）交叉感知模型以建立起情感与智力互通的桥梁，并在此基础之上构建了服务于人类智力发展的情智模型。此外，在构建的情智模型指导下，实现了脑波信号驱动的情感音乐生成系统，从而达到促进人类智力水平提升的目的。

二、情智模型的提出

（一）情智模型的定义

情感与智力是人类进化的结晶，情感是人类各种机械能力之上的核心抽象概念，提供了人类行为和思维管理与调制的机能。智力被认为与一系列认知任务的表现有密切联系，智力发展意为在这些认知任务中智力的表现情况。人的信息接收、情感生成与智力表现是一个相互耦合的过程。如果把用于影响人的智力表现的情感信息称为情智信息的话，对人类自我认知的主要科学问题之一就是信息、情感与智力发展的关系与机理，本文将其定义为情智模型。如图1所示，未来的智能是具有完善类人功能的智能，我们认为，在类人方面，"信息+情感"的非理性调节与"训练+记忆"的理性调节同样重要。情智模型的研究目的是要通过实现理性调节与非理性调节的耦合，达到完善的类人智能。

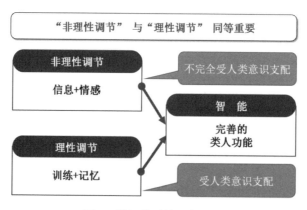

图1　情智模型的研究目的

（二）与现有研究的关联

近年来，基于深度神经网络的特定机器学习方法在人工智能领域占据了主流，从而使得计算机视觉、人机交互等技术能够更好地体现出服务于人类体力发展的价值。然而，针对服务于人类智力发展的情感表征方式与处理方法的研究，目前国内外还基本处于空白阶段。因此，本节主要从以下两个方面体现情智模型与现有研究的关联。

1. 情感与智力表现的交互作用

情感是人类一种重要的本能，在人们的日常生活、工作、交流、处理事务和决策中扮演着重要的角色。从美国哲学家 James 提出"情感是什么"这个问题到现在，"情感"始终是心理学及哲学领域里引人注目的研究对象。一百三十年后的今天，"情感"已经变成了神经科学、生物学、工学、社会学、经济学等诸多领域的重要研究对象。在已有的研究中，离散表示和连续（维度）表示是两种最基本的情感描述形式。以 Ekman 为代表的基本情感模型将情感状态分类成离散的情感类别，根据情感的纯度和原始度，可以将情感划分为基本情感（主要情感或原始情感）和复合情感（次要情感）两大类。生气、害怕、难过、惊讶、高兴和厌恶，Ekman 提出的这六种基本情感在情感计算领域认可程度较高。与此类似的理论还有"调色板理论"。Buck 从生理情感、认知状态情感、社会情感和精神状态情感等方面分别给出了基于不同角度的基本情感类型。

近年来，随着交叉领域研究的兴起，对于"情感"的研究正朝着一个通过跨领域的研究来探索的新方向前进。情感影响智力表现的研究往往聚焦于情感对认知能力的影响，尤其是注意力和情感的相互作用过程。考虑到事件流进入认知系统的过程及认知系统处理能力的限制，情感和注意力的产生均包含一个关键过程——相对于中性或一般事件，会优先处理相关事件，这就导致了知觉分析的增强、记忆和运动行为的激活。因此，面部表情和声音中社会信号的情感处理可能与注意力相关机制密切关联。有结果表明，具有特定情感相关性的视觉事件，如面部表情或情感图片，也能比中性事件更容易吸引人们的注意力。Sander 等人通过功能磁共振实验探究了声音情感刺激与个体注意力之间的关系，根据实验中受试者呈现出的愤怒和中性听力范式下的表现，确定了人类情感的产生受个体内部注意力的调节，以及听觉区域存在对情感信号加工的调节机制。Blair 等人用情感 Stroop 范式进行实验，指出至少在一些患者中，额叶外侧及中间皮层的能力下降，会导致情感和焦虑障碍患者抑制情感对目标产生注意力的能力受损。

另外，情感与记忆之间也存在一定的联系。早在 1981 年 2 月，美国斯坦福大学心理学家 Pourtois 在《美国心理学家》杂志上介绍了他的一项研究成果，认为记忆力与人的情感状态密切相关，并提出了记忆与情感相关效应假说。Phelps 研究发现，在不同的情感状态下，杏仁核可以调节依赖于海马体的记忆编码和存储功能，而海马复合物通过形成情感意义和事件解释的情节表征，可以影响遇到情感刺激时的杏仁核响应，这进一步说明了情感会对人的记忆产生一定的影响。

目前大多数关于情感与智力表现相互作用的研究都是基于单一感觉刺激的。然而已有研究表明，大多的情感都是多种刺激叠加诱发而成的，即听觉、视觉、味觉、触觉、嗅觉等多感觉刺激下所产生的复杂的情智信息。因此需要开展基于多种感觉刺激的情感影响智力表现的基础理论研究，分析多感觉刺激下不同感知系统的信息交互和作用、系统之间信息的不同表示形式和信息传递的方式。通过研究多感觉刺激下情感对包含感觉、知觉、注意力、记忆等在内的认知的影响，构建多感觉刺激下情感促进智力表现的多媒体模型，才能够进一步地开发人的大脑，丰富精神世界、培养创新意识，从而促进人的智力发展。

2. 情感模型

情感分析被证明是一个心理学、社会学、神经科学、计算机科学等多学科高度交叉融合的研究领域。2012 年，Koelstra 等人提出了一个用于分析人类情感状态的多模态数据库，并研究了各个生理信号与主观情感分析之间的相关性。这些生理反应信号包括心率变异性（Heart Rate Variability, HRV）、脑电图（Electroencephalogram, EEG）、皮肤电导（Skin Conductance, SC）、血容量脉搏（Blood Volume Pulse, BVP）和呼吸率（Respiration Rate, RR）。悉尼大学的 Daniel 等人收集了 65 名志愿者休息状态下的 HRV，实验表明 HRV 能够为识别人类情感提供新的标记。Fabien 等人通过提取 RECOLA 数据库中包括 HRV 在内的生理信号以及音视频特征，采用时序神经网络预测回归模型，并对比特征级融合与决策级融合的结果，确定每段音视频的最优唤醒度——效价值。Anuharshini 等人通过受试者的 HRV 和 RR 这两个生理参数的有效变化以及行为反应来研究两种印度古典音乐的情感反应。Cabredo 等人通过提取音乐切片的特征以及听众对应的 BVP 指数，采用生物学的 motif 发现算法构建出音乐特征和生理特征的映射关系。Takahashi 等人使用音频内容进行心理学实验以激发受试者的情感，来收集生物电位信号。测量实验采集了两种生命体征 BVP 和 SC 以评估三种情感：积极情感（放松和愉悦），消极情感（压力和不愉快）和中性情感，并采用支持向量机（Support Vector Machines, SVM）设计情感识别系统，对于三种情感实现了 41.2% 的识别率。

EEG 信号是一种非侵入性的脑机接口，它允许外部机器在没有手术的情况下感知来自大脑的神经生理信号。从中枢神经系统捕获的非侵入性 EEG 信号已被用于探索情感，这种与个性化差异关联性较小的特征能更好地用于情感分析的研究。TanuSharma 等人通过实时提取各个参与者在不同情感下的生理信号（包括 HRV、BR、BVP、EEG），探究各种生理信号与情感之间的关联性。日本大阪大学产业科学研究所的 Cabredo 等人针对个人的音乐感受独立性，提出了基于 EEG 信号的音乐模型，通过记录多人听取音乐的 EEG 信号以及相应的个人情感标签，对 EEG 信号的谱特征和情感标签做了关联性分析。中国台湾辅仁大学的 Hsu 等人提取音乐特征和个人 EEG 信号特征，通过人工神经网络融合两种特征完成了相应的情感分类任务。Park 等人通过生理信号识别主体的负面情感，其中皮肤电活动（Electrodermal Activity, EDA）、皮肤温度（Skin Temperature, SKT）、心电图（Electrocardiogram, ECG）和体积描记（photoplethysmography, PPG）被记录为情感的生理信号，使用机器学习算法分析提取 28 个特征用于情感识别，并通过比较每个算法的识别结果来确定优选算法，结果显示支持向量机（SVM）取得了最高的训练精度，而线性判断分析（Linear Discriminant Analysis, LDA）获得了最高的测试精度。

从上面的分析可知，现有的情感模型，无论是特征训练的神经网络模型，还是基于生理反应信号处理的模型，都没有明确地揭示出激励信号与人类情感之间的关系。因此，如何构建能够真正反映出激励与人类情感之间关系的模型，仍然是情感计算领域的基础问题和难点问题之一。

三、情智信息的建模

目前，关于情智模型的研究工作主要集中在情感计算阶段。关于情感的产生，至今还没有一个合理的模型解释。詹姆斯—朗格学说、丘脑情感学说、激活学说是目前主要的三种学说，而现有的关于脑科学的一些研究已经逐步验证了激活学说的正确性，即情感的来源是对情景的评估，情景向脑提供感觉信息，引起皮层皮下的整合活动，产生情感体验，整个过程是一种"场景—评价—情感"的激活过程。由这个模型导出的情感计算方式已经成为目前情智模型问题研究的基础，但是无论在定性还是定量的研究上，现有方式都存在很多不足，导致情感计算只有在特定的数据库中准确率较高（作为一个数据处理问题），距离实际应用还有一定的差距。其主要原因是在研究复杂的情感问题时，采用了过于简化的情感模型，缺少各种影响情感反应的调制因子，比如文化水平、经济基础、社会地位等。

众多不同场景因素的人群会有不同的情感反应。另一方面，目前情智模型的研究也缺少各场景类型（听觉、视觉、嗅觉等）交叉感知的桥梁。在情感产生的过程中，皮层下神经过程的作用处于情感形成的显著地位，大脑皮层神经中枢、脑垂体、下丘脑、肾上腺等部位和腺体对情感起着调节作用。为了建立起情感智力互通的桥梁，本文把人的感觉场景按照听觉、视觉、味觉、嗅觉、触觉和意境知觉分别表述，提出了基于上述六觉的交叉感知模型，如图 2 所示。该模型将情感产生过程中的神经类别分为感知神经和处理神经两种神经元，各种感觉在情感层面耦合形成联觉。

在上述六觉情感联觉模型的基础之上，图 3 给出了情智模型的构建过程。如图 3 所示，情智模型是融合脑科学、生物学、心理学、人工智能、数字媒体等的交叉领域的研究，其以脑科学中情感影响智力的神经反应机理为基础，通过采集多模态生理信号，在心理学领域寻求情感联觉模型，使用人工智能领域的模型计算方法，结合数字媒体领域的数据生成手段，力求建立情感与智力耦合的桥梁。这里，情智模型是一个复杂的变参数的概率问题，可以表示为式（1）：

$$p(A,E,I,\theta_1,\theta_2) = p(A|E,\theta_2)p(E|I,\theta_1)p(I)p(\theta_2)p(\theta_1) \quad (1)$$

其中，A 表示智力反应，E 表示情感联觉，I 表示信息，θ_1、θ_2 分别表示附加因素。

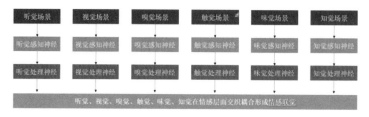

图 2　情感联觉模型

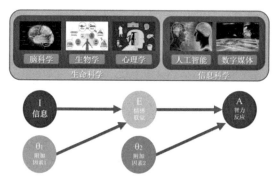

图 3　情智模型

从科学领域的层面来看，随着信息科学的发展和人工智能研究的不断深入，科学家们对人类智能的认识也不断深入，由人脑表现出来的心智现象不仅体现在单纯智能的方面，还体现在知觉、情感方面。智能并不是一个单独或者割裂的人脑功能，智能和感知、情感等有着密切的联系。在情感影响智力表现的研究中，情感对人脑认知能力、记忆能力的影响已经有了很多的测试数据作为佐证，因此人的信息接收、情感生成与智力表现是一个相互耦合的过程。情智模型的研究突破了以往人工智能研究单纯从智能出发实现智能计算，在激发情感影响智能方向上独树一帜，使机器智能与情智模型相结合，真正朝着类人智能、类脑智能的方向开展原创性工作。已有的脑认知和心理学的模型可以作为情智模型研究的基石，但还远远不够，通过多种方式进行情智模型建模，能够推动情感和智力耦合的脑认知研究，促进大脑神经认知学和心理学的发展。

四、情智模型在音乐生成系统中的应用

音乐是人类传递情感的一种载体，也是抒发和表达人类情感的最佳工具，听众渴望从音乐中产生情感的共鸣。因此，为听众生成符合心境的个性化音乐成为艺术创作领域的关键问题之一。然而，传统的人工作曲方法不仅要求创作者具备扎实的乐理知识，而且创作过程耗时耗力。近年来，随着人工智能与艺术融合技术的迅猛发展，如何借助计算机实现音乐生成已成为一个炙手可热的课题。

脑波音乐是将 EEG 信号按照特定的规则转化而成的乐曲，它是一种兼具音乐性与生理性的新颖的音乐形式。EEG 信号自 1924 年首次被发现并记录以来，各个领域的学者围绕 EEG 信号开展了大量的研究。1934 年，Andian 和 Mattews 实现了人类脑波的发声，从而开创了脑波音乐的先河。遗憾的是，由于早期的研究局限于单一的 EEG 信号处理方式，脑波音乐效果并不理想。经过几十年的发展，尤其是 20 世纪 90 年代以来，对脑波音乐的探索逐步深入，脑波音乐的实用价值受到越来越多学者的青睐。现今，大量的脑波音乐涌入音乐市场，极大程度地提高了音乐创作的工作效率。然而，以往的脑波音乐生成研究大多忽视了对其情感表达的分析。对音乐作品的评价标准不应该仅停留在提升可听性这个层面，一首好的音乐作品必定是会打动人心的。音乐作品的精髓就是情感产生共鸣进而激发人脑智力的其他表现。我们认为，在人工智能技术的加持下，对情智模型的探索将加速服务于智力发展的新媒体系统的开发，从而达到真正的类人智能。为了探索情智模型在音乐生成中的应用，本文设计并实现了一个 EEG 信号驱动的情感音乐生成系统，我们将此系统取名为"人人都是贝多芬"。顾名思义，通过使用此系统，任何人都可以像著名音乐家贝多芬一样创作出优美动听又情感饱满的音乐。

（一）"人人都是贝多芬"系统

"人人都是贝多芬"系统如图 4 所示，该系统主要包括三个模块，即 EEG 信号采集及预处理、EEG 信号情感识别模块和情感音乐生成模块。首先，被试佩戴脑电帽观看诱发情感的短视频（短视频时长 24-177s，分为积极、中性、消极三种情感类型），采集被试观看短视频时的 EEG 信号。然后，对 EEG 信号进行预处理、提取差分熵特征。接着，将提取到的特征输入分类器进行实时情感识别，得到情感分类结果。而后，由计算得到的情感类型驱动音乐生成网络，生成带有情感的个人专属音乐。最后，将生成的情感脑波音乐作为新的情感诱发材料刺激被试产生情感，不断循环上述步骤直到评价模型的评价指标收敛，从而得到最终反映被试真实情感变化的音乐。

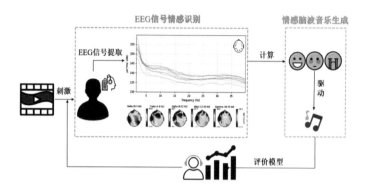

扫二维码
查看彩图

图 4 "人人都是贝多芬"系统流程图

（二）EEG 信号采集及预处理

实验环境由被试室和主试室组成。如图 5 所示，EEG 信号采集实验在安静、明亮的被试室进行，通过一台 24.5 寸、刷新率为 165Hz 的显示屏呈现刺激。EEG 信号数据通过无线便携式脑电仪 Emotiv EPOC X 设备进行采集而得，其主要参数为 14 导（包括国际通用 10-20 系统中的 AF3、F7、F3、FC5、T7、P7、O1、O2、P8、T8、FC6、F4、F8、AF4，共 14 个通道），0.2—43Hz 带宽，256Hz 采样频率。被试的 EEG 信号数据通过 EmotivPRO 软件记录。同时，本系统还可使用 MER-502-79U3C 工业相机、Tobiipro 眼动仪、GSR 皮肤电传感器采集被试的微表情、眼动、皮肤电阻等多种生理信号。为了避免电磁设备对 EEG 信号质量的干扰，被试室显示屏与主机采用有线的方式连接。主机放置于主试室，主试通过主试室显示器实时监控设备连接情况及被试状态。

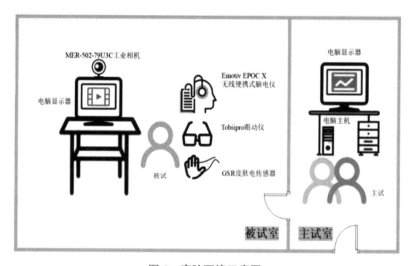

图 5 实验环境示意图

实验开始前,由1名主试向被试详细说明实验流程。当了解实验流程后,被试被带进被试室,由2名主试帮助他佩戴脑电帽,并提醒被试调整座椅至观看舒适位置。正式采集数据前,被试先进行一次练习来熟悉这个系统。在15s基线记录后,播放一段短视频,然后被试填写主观评价。接下来,确保被试完全理解实验流程、音量调整至合适大小后,主试提醒被试EEG信号采集实验正式开始并离开房间,之后被试点击屏幕上的"开始"键开始实验。被试参与EEG信号采集实验如图6所示。

图6 脑电信号采集情景图

EEG信号预处理包括人工去除伪迹、滤波、特征提取。人工去除伪迹是为了避免实验过程中由于补充脑电液而产生的干扰信号,以及由于被试头部肌肉收缩、眨眼或眼球移动产生的明显的肌电、眼电信号。滤波过程使用带通滤波器保留了0.1—50Hz频率的EEG数据,陷波滤波器去噪50Hz,采样频率设置为256Hz。特征提取过程以2秒为一个片段进行切分,提取EEG信号在δ(1-4Hz)、θ(4-8Hz)、α(8-12Hz)、β(13-30Hz)、γ(31-45Hz)五个频段的差分熵特征。

(三)EEG信号情感识别模块

实验共收集了40名被试(男性20人、女性20人,平均年龄27.6岁、标准差为3.92)的EEG信号数据用于情感识别的分类器选择。分类器性能的优劣直接影响了脑波音乐生成过程中输入情感音乐生成网络的情感类型是否准确。本文比较了机器学习方法中常用的11种分类器在EEG信号情感识别中的分类性能。这些分类器使用前面提到的差分熵特征作为输入,将采集到的EEG信号数据按照8:2的比例划分为训练集和测试集,输出为情感三分类结果(积极、中性、消极)。图7给出了不同分类器对EEG信号的情感识别结果。实验结果表明,Gradient Boosting算法在分类准确率和泛化能力上表现最好,准确率达到了95%。因此,本系统选用Gradient Boosting算法作为EEG信号情感识别模块的分类器。

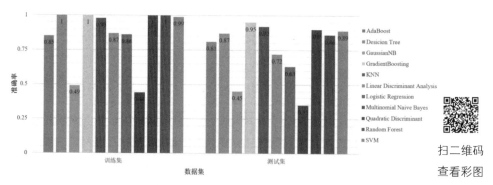

图7　11种常见的机器学习算法情感识别结果

(四)情感音乐生成模块

在情感音乐生成模块,将嵌入情感标签和音乐结构特征作为条件输入生成网络,同时添加感知优化的情感分类损失函数,生成带有情感的脑波音乐。该网络模型包括情感音乐生成器和情感音乐分类器两部分,如图8所示。具体来说,情感音乐生成器采用自监督训练模式来构建模型,将EEG信号的情感识别结果和音乐结构特征作为条件输入,输出为MIDI事件序列。在情感音乐分类器中,为了缩短生成的特征分布与真实特征分布之间的距离,对情感音乐分类模型进行预训练。网络的输入是MIDI序列,输出是情感分类的结果。

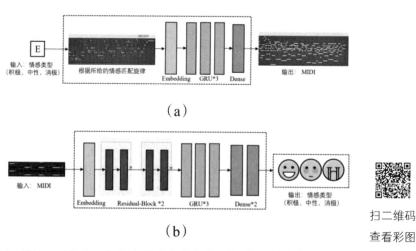

图8　脑波音乐生成网络(a)情感音乐生成网络,(b)情感音乐分类网络

在系统中，使用 VGMIDI 数据集进行模型训练。VGMIDI 的原始标签分为积极和消极两类，经主观评价可将情感不突出的音乐片段标注为中性，得到情感标签为积极、中性和消极三类的数据集。随后，将 EEG 实时情感识别的结果输入训练好的音乐生成网络，以控制网络生成相应情感的音乐。

五、结束语

本文创新性地提出了"信息、情感与智力的耦合模型构建"这一科学前沿问题，创建了情智模型理论体系，并在此理论研究的基础上，提出并实现了基于情智模型的脑波情感音乐生成系统，从而在"情智信息的感知、建模与生成"研究中迈出了坚实的一步。

情智模型问题涉及未来信息技术、工业技术、国防科学、社会科学、医疗健康等多个关乎国计民生的重大领域，在现实应用中也将服务于我国新旧动能转换、社会安全、城乡均衡发展、"一带一路"等国家策略需求，具有很强的研究意义与应用价值。情智模型的研究突破了以往人工智能单纯研究机器智能的局限，在情感影响智能方向上独树一帜，使机器智能与情智模型相结合，有助于真正实现类人智能、类脑智能。未来工作中，我们将针对"信息、情感与智力的耦合模型构建"这一重大科学前沿问题，围绕情智信息的感知提取、建模表征、生成验证三个层面，开展相关研究工作以创建情智信息理论体系，并在此理论研究的基础上，探索基于情智耦合模型的音频、视频与场景的生成技术，构建情智信息理论验证的音频、视频与场景数据测试平台，从而建立服务于智力发展的新媒体理解架构。

参考文献

[1] KLÜVER H, BUCY P C. An analysis of certain effects of bilateral temporal lobectomy in the rhesus monkey, with special reference to "psychic blindness"[J]. The journal of psychology, 1938, 5(1): 33-54.

[2] SHARMA T, BHARDWAJ S, MARINGANTI H B. Emotion estimation using physiological signals[C]. Tencon IEEE Region 10 Conference, 2008: 1-5.

[3] CABREDO R, LEGASPI R S, INVENTADO P S, et al. An emotion model for music using brain waves[C]. Proc. 13th international society for music information retrieval conference, 2012: 265-270.

[4] HSU J L, ZHEN Y L, LIN T C, et al. Affective content analysis of music emotion through EEG[J]. Multimedia systems, 2018, 24(2): 195-210.

[5] PARK B J, YOON C, JANG E H, et al. Physiological signals and recognition of negative emotions[C]. IEEE international conference on information and communication technology convergence, 2017: 1074-1076.

[6] HAN Y, XU M, KE Y, et al. Modulation of attention in visual stimulation response[J]. Nanotechnology and precision engineering, 2015, 13(05): 353-358.

[7] HUSAIN G, THOMPSON W F, SCHELLENBERG E G. Effects of musical tempo and mode on arousal, mood, and spatial abilities[J]. Music perception, 2002, 20(2): 151-171.

[8] KALANTHROFF E, COHEN N, HENIK A. Stop feeling: inhibition of emotional interference following stop-signal trials[J]. Frontiers in human neuroscience, 2013, 7: 78.

[9] BUSCHKUEHL M, JAEGGI S M. Improving intelligence: a literature review[J]. Swiss medical weekly, 2010, 140(19-20): 266.

[10] ARNOLD M B. Emotion and personality[J]. The American journal of psychology, 1960, 76(3): 516-519.

[11] JAMES W. What is an emotion?[J]. Mind, 1884, 9: 188-205.

[12] EKMAN P, FRIESEN W V. Unmasking the face: a guide to recognizing emotions from facial clues[M]. Cambridge, MA: Malor Books, 2003.

[13] EKMAN P. An argument for basic emotions[J]. Cognition & emotion, 1992, 6(3-4): 169-200.

[14] RUSSELL J A, WEISS A, MENDELSOHN G A. Affect grid: a single-item scale of pleasure and arousal[J]. Journal of personality and social psychology, 1989, 57(3): 493-502.

[15] BUCK R. The biological affects: a typology[J]. Psychological review, 1999, 106(2): 301-336.

[16] SANDER D, GRANDJEAN D, POURTOIS G, et al. Emotion and attention interactions in social cognition: brain regions involved in processing anger prosody[J]. Neuroimage, 2005, 28(4): 848-858.

[17] BLAIR K S, SMITH B W, MITCHELL D G V, et al. Modulation of emotion by cognition and cognition by emotion[J]. Neuroimage, 2007, 35(1): 430-440.

[18] POURTOIS G, GRANDJEAN D, SANDER D, et al. Electrophysiological correlates of rapid spatial orienting towards fearful faces[J]. Cerebral cortex, 2004, 14(6): 619-633.

[19] CHUN M M, PHELPS E A. Memory deficits for implicit contextual information in amnesic subjects with hippocampal damage[J]. Nature neuroscience, 1999, 2(9): 844-847.

[20] KOELSTRA S, MUHL C, SOLEYMANI M, et al. Deap: a database for emotion analysis; using physiological signals[J]. IEEE transactions on affective computing, 2011, 3(1): 18-31.

[21] QUINTANA D S, GUASTELLA A J, OUTHRED T, et al. Heart rate variability is associated with emotion recognition: direct evidence for a relationship between the autonomic nervous system and social cognition[J]. International journal of psychophysiology, 2012, 86(2): 168-172.

[22] RINGEVAL F, EYBEN F, KROUPI E, et al. Prediction of asynchronous dimensional emotion ratings from audiovisual and physiological data[J]. Pattern recognition letters, 2015, 66: 22-30.

[23] ANUHARSHINI K, SIVARANJANI M, SOWMIYA M, et al. Analyzing the music perception based on physiological signals[C]. IEEE 5th International Conference on Advanced Computing & Communication Systems, 2019: 411-416.

[24] CABREDO R, LEGASPI R S, NUMAO M. Identifying emotion segments in music by discovering motifs in physiological data[C]. Proc. 13th International Society for Music

Information Retrieval Conference, 2011: 753-758.
[25] TAKAHASHI K. Remarks on computational emotion recognition from vital information[C]. IEEE Proceedings of 6th International Symposium on Image and Signal Processing and Analysis, 2009: 299-304.
[26] MINSKY M. The emotion machine: commonsense thinking, artificial intelligence, and the future of the human mind[M]. New York: Simon and Schuster, 2007.
[27] SANDER D, GRANDJEAN D, POURTOIS G, et al. Emotion and attention interactions in social cognition: brain regions involved in processing anger prosody[J]. Neuroimage, 2005, 28(4): 848-858.
[28] HERREMANS D, CHUAN C H, CHEW E. A functional taxonomy of music generation systems[J]. ACM computing surveys, 2017, 50(5): 1-30.
[29] BRIOT J P, PACHET F. Deep learning for music generation: challenges and directions[J]. Neural computing and applications, 2020, 32(4): 981-993.
[30] GAO Z, DANG W, WANG X, et al. Complex networks and deep learning for EEG signal analysis[J]. Cognitive neurodynamics, 2021, 15(3): 369-388.
[31] ADRIAN E D, MATTHEWS B H C. The Berger rhythm: potential changes from the occipital lobes in man[J]. Brain, 1934, 57(4): 355-385.
[32] ROSENBOOM D. Extended musical interface with the human nervous system: assessment and prospectus[J]. Leonardo, 1999, 32(4): 257-257.
[33] HUANG R, WANG J, WU D, et al. The effects of customised brainwave music on orofacial pain induced by orthodontic tooth movement[J]. Oral diseases, 2016, 22(8): 766-774.
[34] ZHENG W L, LU B L. Investigating critical frequency bands and channels for EEG-based emotion recognition with deep neural networks[J]. IEEE transactions on autonomous mental development, 2015, 7(3): 162-175.
[35] FERREIRA L, LELIS L, WHITEHEAD J. Computer-generated music for tabletop role-playing games[C]. Proceedings of the AAAI Conference on Artificial Intelligence and Interactive Digital Entertainment, 2020, 16(1): 59-65.

Interaction Between Dynamic Affection and Arithmetic Cognitive Ability: a Practical Investigation with EEG Measurement[*]

1. Introduction

AFFECT is a feeling, emotion, or mood, which has been found to have a substantial influence on cognitive processes, including attention, working memory, and decision-making. Recent advances in affective neuroscience have demonstrated the interconnection between affection and cognition, both of which may emerge from the same general cortical system. This implies that the performance of cognitive abilities is influenced by the different emotional states of individuals, and the emotion itself may also be much more likely to change when carrying out cognitive activities. Although a growing number of research studies have focused on the interaction between emotional state and cognitive ability (e.g., [7], [8]), experimental validation is still in the preliminary stages. Nonetheless, the interaction study between emotion and cognition plays an important role in both learning and working scenarios. It may help people better understand the relationship between emotion and cognition, provide guidance for education and human resource management and is of great importance for understanding human behavior and psychological mechanisms.

Many studies in the literature (e.g., [12]–[18]) have investigated the relationship between emotion and cognition. Among the current studies, most researchers have only focused on the effects of emotion on cognitive performance, where little study has been seen to explore the interaction between these two. Initial studies focused on negative emotions, such as test anxiety and math anxiety. However, this approach measures trait anxiety rather than real-time affective experience. Therefore, recent work first used positive, neutral and

---

[*] The paper was originally published in *IEEE Transactions on Affective Computing*, 2023, and has since been revised with new information. It was co-authored by Xiaonan Yang, Yilu Peng, Yuyang Han, Fangyi Li, Qin Zhang, Shuo Wu and Xia Wu.

negative pictures as emotional stimuli to induce emotions of different valences. It was found that the reaction time became faster for solving difficult problems with negative emotions and for solving simple problems with positive emotions. Since then, there have been conflicting findings in studies related to emotions and reaction time. Reference [17] explored the effect of picture emotional stimulus material on the performance of young and old subjects and found that the reaction time of subjects became longer with negative emotions compared to neutral emotions, and this negative effect was more evident in the young subjects' group. However, reference [18] obtained the opposite conclusion, with shorter reaction times for subjects under negative emotions compared to positive and neutral emotions. However, all of the above studies only utilized behavioral data to investigate the impact of emotional state on cognitive ability, suffering the adverse impact of being too time-sensitive and easily falsifying, which may have caused inconsistent conclusions.

As one of the physiological signals commonly used in neuroscientific research, Electroencephalogram (EEG) can rapidly reflect electrophysiological activities throughout the cortex and scalp surface by directly measuring the activities of neurons. It has high temporal resolution and, more importantly, can provide more reliable information due to its specific feature of difficulty in disguising. In addition, EEG acquisition devices are noninvasive to subjects and have relatively low costs. It has therefore become more popular for identifying the neurophysiological responses and cognitive abilities of subjects. Nevertheless, few studies have used EEG to analyze brain states during emotions that influence cognitive abilities. To the best of our knowledge, only one research group has analyzed amplitude changes in the P1, N170 or CNV (contingent negative variation) as a way to represent cognitive abilities during cognitive tasks under different emotional face evocations based on several similar experimental paradigms. However, these studies did not focus on the subjects' own emotional experiences. Moreover, ERP analysis requires considerable paradigm and temporal synchronization, and it is difficult to measure brain activity extending beyond a few seconds, making it difficult to perform continuous change analysis using ERP.

The accurate determination of emotional state is the first crucial step in analyzing and revealing the interaction between emotion and cognition. Most previous studies share a common issue in that the evoked emotions are simply presumed to be the real emotions, without taking into account the observation that emotions may change when carrying out cognitive activities. This may lead to the inaccurate detection of true emotions, which further adversely impacts the investigation of the interactions between emotion and cognition. For instance, some studies have resolved this through a scale approach, such as the validity-arousal scale. Nevertheless, participants may repeatedly think about their emotions and thus potentially exacerbate them. Additionally, participants may not be able to realize or describe the emotions they experience or mask their true emotions due to factors such as social expectations, social reframing, and self-esteem. Recently, EEG-based emotion recognition has become very popular. As an external manifestation of the brain's cognitive mental activity, EEG can be used as an effective measure to study emotional psychology and thus to make real-time monitoring of emotions a reality.

Because the true labels of the emotions are not known in the cognitive task, it is unrealistic to quantitatively measure whether the predicted emotions are accurate or not. Therefore, it is crucial for obtaining more accurate prediction that the performance of the proposed dynamic emotion monitoring model can achieve desired accuracies. Many deep learning networks have achieved desirable results in emotion recognition, such as the three-dimensional convolutional neural network CNN (3D-CNN), EmotionNet, and four-dimensional convolutional recurrent neural network (4D-CRNN), etc. Amongst them, 4D-CRNN can integrate frequency, spatial and temporal information of multichannel EEG signals to improve emotion recognition accuracy. Single-subject binary classification accuracy in the DEAP dataset (i.e., a well-known emotion recognition data repository with physiological signals) achieved 93%, which can meet our monitoring needs. These models expanded the dataset by segmenting the signal; however, they do not take into account that the emotional state of the subject is undergoing constant evolution. Therefore, at the early stage of video elicitation, the subjects' emotional states might not be elicited, which may cause a decrease in accuracy during the training process due to labeling errors.

This work attempts to address a *critical question*, that is, to investigate the interaction between emotional state and cognitive ability by the use of EEG measurement. In particular, an innovative EEG-based study is carried out for developing a continuous cognitive task of a realistic scenario, where different emotional states are evoked. As indicated above, the current study did not consider the influence of cognitive activity on emotion during analysis. To further examine the underlying relationships, it is necessary to make two hypotheses in the first place: 1) emotions would change during the cognitive task; and 2) the subjects' emotional states might not be elicited at the early stage of video elicitation. For this purpose, this work designs an experimental paradigm, collecting a dataset of affection and cognition interactions in a simulated continuous cognitive task scenario in a laboratory situation with EEG measurement during the entire process. Based on this, we construct a real-time monitoring model by 4D-CRNN using EEG time series data, to dynamically monitor the current emotional state. Given the observed emotion, the analysis of the interaction between cognitive abilities and dynamic emotions is undertaken from the perspectives of both behavioral performance and brain mechanisms. On this basis, this paper aims to explore the interactions between emotion and cognitive ability through EEG and to explain the analysis in terms of brain mechanisms. Therefore, the objectives of this paper are as follows:

- To construct a dynamic emotional monitoring model that can reflect emotional changes in real time (for verifying Hypotheses 1 and 2);

- To explore how emotions and current cognitive abilities interact and to explain the potential explanatory brain mechanisms (for addressing Critical Question).

The rest of this paper is structured as follows. Section II presents the details of the proposed methodology for investigating the interaction between emotion and cognition. Section III reports on the experimental results on the dynamic monitoring model of emotion and the analysis of interactive relationships. Section IV discusses the experimental outcomes and limitations of this study. Finally, Section V summarizes the contributions of this paper.

2. Methodologies

This section describes the main research process and research methods of this paper. Among them, subsections II−A, II−B, II−C and II−D introduce the design and acquisition details of the dataset, focusing on the protocol of investigating the interaction between dynamic emotion and cognition. Subsection II−E presents the process of constructing a dynamic monitoring model for affective detection (Objective 1), and II−F describes the specific interaction analysis from the perspectives of behavior and brain (Objective 2).

2.1 Participants

Forty-two undergraduate students and graduate students (22 men and 20 women) participated in this experiment from Beijing Normal University. All of them were right-handed and had no mental disease or color vision disorders. Participants were between the ages of 19 and 23 ($Mean = 22, STD = 1.83$). None of them had experience in psychology.

2.2 Materials

Experimental materials include videos used to elicit emotions and multi-digit additions to measure cognitive performance.

1) Emotional video: There are several popular datasets currently for emotion analysis using EEG which adopt videos as emotion-evoked material, such as DEAP and SEED. It has also been proved empirically that videos have more sensory stimulation and better emotion-rendering ability than pictures and audios do. Therefore, we constructed a novel emotion-evoking video dataset to be used as emotion-evoking material.

We recruited fifteen undergraduate students ($Mean = 23.27, STD = 1.98$) from Beijing Normal University to take part in the video material scoring experiments. None of the subjects had a history of neurological conditions and did not participate in the formal experiment. In reference [40], the authors found that signals with video stimulation times longer than 60s achieved better classification results. Thus, 24 video fragments (2-4 mins, $Mean = 188.2s, STD = 53.05$) were selected from the online video website, which were divided into three categories according to their meaning, i.e., positive, neutral, and negative emotional videos. Each category contained eight video fragments. To evaluate the emotion induction effect, we invited 15 participants to sit in a quiet, dimly lit room. They were asked to watch all 24 video fragments and finish a valence-arousal-domination questionnaire and PANAS after each fragment. The fragments were displayed randomly, and this experiment lasted approximately 1.5 hours.

Participants' emotional state scores including arousal, valence, and dominance scores were calculated to select videos by internal consistency reliability. The results showed that the internal consistency reliability was 0.969, 0.841, and 0.709 for arousal, valence, and dominance, respectively, indicating high consistency and high rating reliability. Additionally, a variance analysis of valence and arousal was performed to judge whether there were significant differences in the scores of different emotions. The results of the P value between valence scores were 2.9×10^{-5} (positive and neutral), 3.2×10^{-6} (positive and negative) and 6×10^{-6} (neutral and negative). The P values were all far less than 0.01, indicating that there were significant differences in the valence scores of the subjects under different emotional videos. Thus, all three types of videos could induce corresponding emotional states. Finally, to select the video materials with the best induction effect, principal component analysis (PCA) was conducted on all PANAS scores to find the factors with the main influence. According to the importance of the component factors, 15 videos (5 videos of each emotion type) were finally selected to construct the new emotion-induced video dataset.

2) Cognitive performance test: There is a correlation between mathematical arithmetic ability and working memory or cognitive ability. In order to avoid the bias caused by different specialties over the experimental results, the cognitive task chooses simple multi-digit addition operations to characterize the current cognitive ability, which are mainly related to working memory but essentially do not require much specialized knowledge.

Additional problems were presented in a standard form (i.e., $a+b$) with 7 difficulty levels, and the detailed design rules are shown in Table I. Subjects needed to watch formula (2s) first, and then a result was displayed. They needed to judge whether it was correct (within 2s). There were 14 questions after each emotional video. Finally, subjects' answer, the completion time and the difficulty level of the question were recorded.

Table I The design rules for addition formula of different difficulty levels

Difficulty Levels	Design Rules	Example
Very low(L1)	1&2 digit numbers	74+2
Low(L2)	1&2 digit numbers with 1 carry	53+8
Medium(L3)	2 digit numbers with 1 carry	67+42

Table I Continued

Difficulty Levels	Design Rules	Example
Medium–high(L4)	2 digit numbers with 2 carries	39+65
High(L5)	2&3 digit numbers with 1 carry	337+32
Very high(L6)	2&3 digit numbers with 2 carries	76+347
Extremely high(L7)	3 digit numbers with 3 carries	983+748

2.3 Protocol: Investigating the Interaction Between Dynamic Emotion and Cognition

The total duration of the experiment was approximately 100 mins, including 15 mins preparation (5 mins rest+10 mins practice), 80 mins formal experiment, and 5 mins rest. EEG data were recorded throughout the whole experiment. In the preparation stage, participants were introduced to the procedure of the task and some considerations during the EEG measurement and then signed informed consent forms. During the practice stage, participants could perform cognitive tests similar to the formal experiments in order to get familiar with the procedure. During the formal experiment, the external environments were kept consistent and participants were asked to complete the experiment alone in a quiet room to rule out factors affecting emotions other than emotion videos and mathematical problems.

The formal experimental procedure is described in Fig. 1. A 2 (Group: Emotion Induction task vs. Intelligence task)×3 (Emotion: Positive vs. Neutral vs. Negative) repeated measures design was employed. The formal experiment contained three sessions. Each session consisted of 5 experiment blocks and a 5 mins break for participants to rest and for experimenters to adjust the EEG electrodes. An experiment block includes the emotion induction stage and cognitive performance stage. Participants were asked to focus on the emotional video first. After watching an attention cross (1s), participants needed to complete 14 trials (63s) in the cognitive performance stage. Each trial displayed a calculation formula without the answer (2s, look), an attention cross (0.5s), and a true/wrong answer (2s, decision). During this stage, participants needed to judge whether the answer given was correct within 2s and respond as quickly as possible on the keyboard. Moreover, video fragments and calculation formulas were displayed randomly, but to keep the data balanced, the number of different emotional videos and difficulty level of each experiment were equally distributed. That is, each participant needed to watch 15 (5 positive+5 neutral+5 negative) video fragments and complete 210 (30×7, 7 levels) calculation formulas. This experimental protocol was implemented using PsychoPy software.

2.4 EEG Data Recordings and Preprocessing

EEG activity was recorded continuously via a high-density 128-channel EGI system of Net Amps 300 amplifiers. The EEG was recorded continuously with a 128-channel GSN using the vertex sensor (Cz) as the reference electrode. Direct current acquisition was used, and the data were sampled at 500 Hz during recording. The impedances of all electrodes were kept below 50 $k\Omega$, as recommended for this type of amplifier by EGI guidelines.

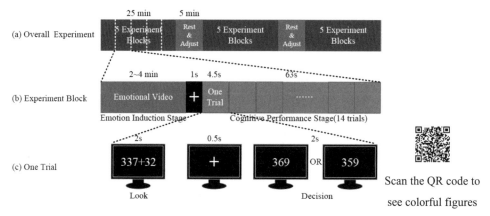

Scan the QR code to see colorful figures

Fig. 1 The formal experimental procedure of the emotion and cognition experiment. (a) The overall experiment contains three sessions, and each session consists of 5 experiment blocks and a 5 mins break. (b) An experiment block includes the emotion induction stage and cognitive performance stage (14 trials). (c) A trial displays a calculation formula without the answer (2s, look), an attention cross (0.5s), and a true/wrong answer (2s, decision).

Owing to the weakness and low signal-to-noise ratio of the EEG signal, it is necessary to preprocess the collected EEG data. In this study, the offline EEG data were preprocessed by MATLAB R2020b and the plug-in EEGLAB toolbox (version 2021.0)[①]. The preprocessing procedure of this study was as follows: First, data for the two 5 mins breaks, that is, REST & ADJUST between two sessions in Fig. 1(a), were removed because of distinct artifacts. Then, the data were filtered by a low-pass filter of 50 Hz, a high-pass filter of 0.1 Hz and a band-stop filter of 49-51 Hz. Next, the data were referenced to the average reference and resampled to 200 Hz. Then, bad channels were removed and interpolated. Additionally, eye movements and blink artifacts were corrected by using the independent component analysis (ICA) algorithm and ADJUST function[②] implemented in EEGLAB.

After finishing the preprocessing, each participant's EEG data were divided into 15 movie fragments data long epochs and 210×2 (look & decision) cognitive performance data short epochs.

---

① *https://eeglab.org/download/*

② *https://www.nitrc.org/projects/adjust/*

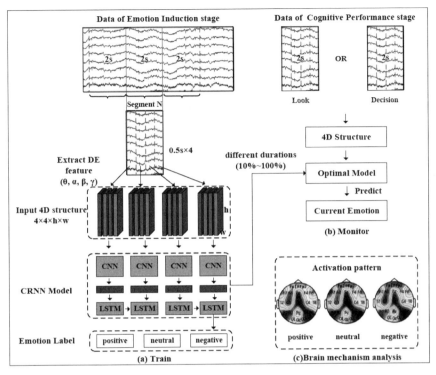

Fig. 2 Processing procedures for interactive analysis. (a) 4D-CRNN model construction during the emotion induction stage and using different durations to find the optimal model. (b) Monitoring real-time emotion by the optimal model during the cognitive performance stage. (c) Brain mechanism analysis by activation pattern.

2.5 Construction of Dynamic Monitoring Model for Affective Detection

1) Differential entropy feature: The differential entropy (DE) feature has been proven to be the most stable feature for emotion recognition, which is used to measure the complexity of EEG signals. The DE feature is defined as:

$$h(Z) = -\int_Z f(z)\log(f(z))dz \qquad (1)$$

where Z is a random variable, and $f(z)$ is the probability density function of Z. Furthermore, if $f(z)$ is from the Gaussian distribution, the DE feature can be calculated as follows:

$$h(Z) = -\int_{-\infty}^{+\infty} \frac{1}{\sqrt{2\pi\sigma^2}} e^{-\frac{(z-\mu)^2}{2\sigma^2}} \log\left(\frac{1}{\sqrt{2\pi\sigma^2}} e^{-\frac{(z-\mu)^2}{2\sigma^2}}\right) dz = \frac{1}{2}\log(2\pi e\sigma^2) \qquad (2)$$

where Z follows the Gaussian distribution $N(\mu, \sigma^2)$, and e is the Euler's constant.

2) 4D-CRNN model: 4D-CRNN was used as the monitoring model during the emotion

induction stage. 4D-CRNN integrates frequency, spatial and temporal information of multichannel EEG signals to improve emotion recognition accuracy. Specifically speaking, CNN is used to learn the frequency and spatial representation for each temporal slice in 4D structures, and RNN with LSTM cells takes the outputs of CNN as input to extract the temporal dependency between slices.

3) Procedure: The procedure includes model construction during the emotion induction stage and monitoring real-time emotion during the cognitive performance stage, which is shown in Fig. 2(a)(b). During the model construction stage, each subject had 15 2~4 mins EEG signals. According to [35], we divided the signals with 2s. Then, DE features were extracted from four frequency bands (theta (4-8 Hz), alpha (8-13 Hz), beta (13-30 Hz), and gamma (30-45 Hz)) of each segment ($0.5s \times 4$). Next, we converted the DE features of different channels to the 2D map ($h \times w$) and then concatenated four frequency bands to obtain a 3D structure ($4 \times h \times w$). The 2D map we used is consistent with [35], that is $h = 8, w = 9$, which is shown in Fig. 3. Finally, we took one sample per 2s ($T = 2$) without overlapping to obtain the final input structure of the 4D-CRNN model ($4 \times 4 \times h \times w$).

0	0	AF3	FP1	FPZ	FP2	AF4	0	0
F7	F5	F3	F1	FZ	F2	F4	F6	F8
FT7	FC5	FC3	FC1	FCZ	FC2	FC4	FC6	FT8
T7	C5	C3	C1	CZ	C2	C4	C6	T8
TP7	CP5	CP3	CP1	CPZ	CP2	CP4	CP6	TP8
P7	P5	P3	P1	PZ	P2	P4	P6	P8
0	PO7	PO5	PO3	POZ	PO4	PO6	PO8	0
0	0	CB1	O1	OZ	O2	CB2	0	0

Fig. 3 The 2D map of 4D-CRNN.

Different from 4D-CRNN trained on intrasubjects, we used all subjects' signals for training to increase the amount of dataset, which can achieve better generalization and reduce the effect of overfitting to some extent. Therefore, this experiment involved a total of 51,170 samples for emotion classification in three categories and the emotional monitoring model was the same across subjects. Furthermore, according to Hypothesis 2, we compared the effect of data from different durations (10%~100%, step: 10%) up to the end of the video as input to find the optimal classification performance.

After training and selecting, the optimal model will be used to separately monitor each trial's real-time emotions during the cognitive performance stage, which includes the look and decision phase.

2.6 Interaction Analysis From Behavior and Brain Perspectives

1) Behavioral data to measure current cognitive abilities: Following previous studies, we used reaction time and accuracy to measure subjects' current cognitive abilities. We counted the average reaction time (0-2s) and error rate (0-1) of each subject under the three emotions for subsequent statistical analysis.

2) Brain mechanisms based on activation pattern: This paper is based on the activation pattern of the brain to explore the potential explanatory brain mechanisms, which is shown in Fig. 2(c). In this study, we adopted 21 canonical electrodes, i.e., Fp1, Fpz, Fp2, F7, F3, Fz, F4, F8, T7, C3, Cz, C4, T8, P7, P3, Pz, P4, P8, O1, Oz, and O2, to reduce the effect of volume conduction.

Power spectral density (PSD) analysis can reflect activity differences among multiple brain regions. The relative power spectral density (rPSD) can reduce the differential effect between different subjects, and it has been used in emotion analysis and cognitive measures.

In this study, the Welch method was used to calculate each electrode's PSD in the theta, alpha, beta, and gamma bands. And the rPSD is defined as the ratio of the PSD in four different bands to that of the entire band, which is formulated as:

$$rPSD(i) = \frac{PSD_i}{\sum_{i \in (\theta, \alpha, \beta, \gamma)} PSD_i} \quad (3)$$

We first calculated the mean feature map under different emotions in four EEG bands to get 12 feature maps (emotions: 3×bands: 4) of each subject. To explore the underlying brain mechanisms between emotions and cognitive performance, we calculated correct and incorrect answers under different emotions so that each subject had 24 feature maps (emotions: 3×corr: 2×bands: 4) for further analysis.

3. Results

3.1 Performance of Dynamic Emotional Monitoring Model

The monitoring model was trained with a batch size of 64 and Adam with a learning rate of 0.001. The maximum number of epochs was set as 100. The model was implemented by TensorFlow and trained on an NVIDIA RTX 3060 GPU. We applied five-fold cross-validation and used the average classification accuracy (ACC) and standard deviation (STD) to represent the performance of the model, which is shown in Table II. The comparison result from different durations (10%~100%, step: 10%) up to the end of the video is shown in Fig. 4.

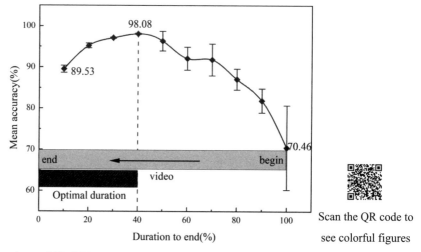

Fig. 4 Line chart of 4D-CRNN performance over time (average ACC and STD) based on five-fold cross-validation. The gray block represents the entire video and the blue block is the best performance of the last 40% of the video.

We previously hypothesized that subjects' emotional states may not be evoked at the early stage of video evocation (Hypothesis 2). Results by model performance show when training with the full amount of data, the performance is worst and unstable ($ACC=70.46\%$, $STD=10.27$). And it can be observed that performance increases as time decrease until optimal performance is reached at 40% of the data ($ACC=98.08\%$, $STD=0.16$). After 40%, the performance of the model drops once again. This is generally consistent with our Hypothesis 2. It is likely due to unevoked emotions and labeling errors in the early stage, and underfitting in the later stage owing to the reduced amount of training data, which leads to performance degradation. Therefore, the model with 40% duration was chosen as the final emotional monitoring model.

Table II Performance (average ACC and STD) of 4D-CRNN based on five-fold cross-validation during different durations

Duration(%)	ACC ± STD(%)	Duration(%)	ACC ± STD(%)
10	89.53 ± 0.85	60	92.07 ± 2.77
20	95.21 ± 0.58	70	91.79 ± 3.85
30	97.08 ± 0.14	80	87.07 ± 2.52
40	98.08 ± 0.15	90	81.88 ± 2.94
50	96.33 ± 2.40	100	70.46 ± 10.26

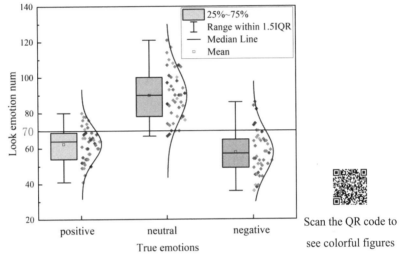

Scan the QR code to see colorful figures

Fig. 5 Comparison between true emotions and expected induced emotions during the look phase based on each subject. The expected induced emotions were 70, and different colored dots represent the number of true emotions of different subjects.

Finally, the optimal model was used to monitor separately each trial's real-time emotions during the cognitive performance stage. And we compared the number of true emotions of the subjects in the look phase with the original number of evoked video emotions, and the results are displayed in the Fig. 5. As can be seen from the figure, for the overall mean of each subject, the evoked num was 70 for each emotion, and there was a significant difference between the evoked emotions and true emotions predicted by the classifier. The increase in the number of neutral emotion and the decrease in the number of both positive and negative emotion indicates that after the end of video evocation, subjects' high arousal emotions such as positive and negative cannot be maintained for a long time and easily revert to neutral emotion during the cognitive process, which confirms

Hypothesis 1. Furthermore, the result illustrates that it is essential to conduct real-time monitoring of emotions during the emotion and cognitive abilities exploration, otherwise it is easy to cause erroneous conclusions.

3.2 Analysis of Interaction Between Emotion and Cognition

During the interaction analysis, we focused on the data from the look phase to analyze the process of problem solving based on stimulation of the brain rather than the process in which the brain focuses on testing different available solutions. Therefore, in the brain mechanisms analysis, we only analyzed the behavioral data and EEG data during the looking stage.

1) Behavior performance: The cognitive task response time and error rate can directly reflect the current state of cognitive abilities. In the analysis of behavioral data, to compare whether the reaction time and error rate under different emotions are significantly different in all subjects, we utilized one-way repeated-measurement ANOVA with three emotions as the factor. One-way repeated-measurement ANOVA can effectively control the effect of individual differences on experimental results. Post hoc multiple comparisons between each group of conditions were performed using Bonferroni correction to determine whether the effect of emotions was significant. Statistical significance was defined as P<0.05. The Bonferroni test with pairwise comparisons results is shown in Fig. 6.

Task response time: There are significant differences in the response time of the tasks ($P_{sphericity} = 0.998 > 0.05, F = 6.701, P = 0.002 < 0.05, \eta^2 = 0.138$) under the three emotions. In Fig. 6(a), the task response time is shortest in the neutral emotion, followed by the negative emotion, and the worst performance is in the positive emotion. In addition, there are significant differences between positive and neutral emotions ($P = 0.004 \leq 0.01$) and significant differences between positive and negative emotions ($P = 0.022 \leq 0.05$).

Error rate: There are significant differences in the rate of errors ($P_{sphericity} = 0.716 > 0.05, F = 6.198, P = 0.003 < 0.05, \eta^2 = 0.129$) under the three emotions. In Fig. 6(b), the performance is better in neutral emotion with the lowest error rate. Similar to the reaction time, there are significant differences between positive and neutral emotions ($P = 0.01 \leq 0.01$) and significant differences between positive and negative emotions ($P = 0.027 \leq 0.05$). In summary, neutral emotion has the best task performance both in reaction time and error rate, and this is likely because positive and negative emotions, as a category of high arousal, may occupy cognitive resources and affect the level of cognitive abilities.

Interaction Between Dynamic Affection and Arithmetic Cognitive Ability: a Practical Investigation with EEG Measurement

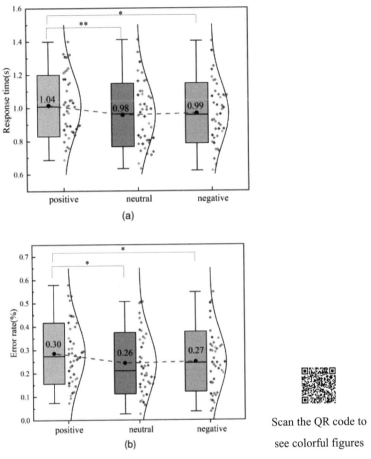

Fig. 6 Cognitive task performance based on different emotions: (a) Task response time. (b) Error rate. "*" shows a significant difference ($P \leq 0.05$); "**" shows a significant difference ($P \leq 0.01$).

Further, different question difficulties correspond to different cognitive load. To reveal this, three difficulty levels were categorized: low $(L1, L2)$, medium $(L3, L4)$, and high $(L5, L6, L7)$, according to the difference in question difficulty in Table I. Fig. 7 demonstrates the same statistical analysis steps as Fig. 6 to explore the effect of emotion on cognitive performance under different cognitive loads. In Fig. 7, the effect of emotion is slightly and not significantly different between subjects at medium difficulty. Regarding high difficulty, neutral emotions performed best. Moreover, positive emotions performed worst at all three difficulty levels in terms of reaction time and error rate, which is consistent with the findings in the overall analysis.

Meanwhile, benefiting from real-time emotion monitoring, we verified whether cognitive performance could have an impact on emotion. We counted the number of changes in emotion from positive → neutral → negative within the time of two questions for each subject when they answered the question incorrectly and calculated the percentage of total incorrect answers. Then, we also calculated the percentage of emotions that changed for the better when an incorrect answer was followed by a correct one. The result is shown in Fig. 8. Except for subjects 42, whose emotions are not affected after answers, more than 97% of the remaining subjects' emotions are affected negatively after wrong answers or positively when the answers change from the incorrect to correct to varying degrees, and the degree of the effect is related to the individual variability of the subjects. Some subjects (No. 1, 15, 7, 9) are affected by a percentage of add up to more than 80%, that is, the possibility that emotions would be affected by the answers are extremely high.

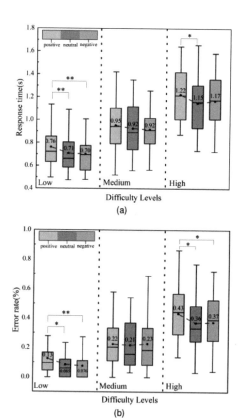

Scan the QR code to see colorful figures

Fig. 7 Cognitive task performance based on different emotions under different difficulty levels: (a) Task response time. (b) Error rate. "*" shows a significant difference ($P \leq 0.05$); "**" shows a significant difference ($P \leq 0.01$).

Interaction Between Dynamic Affection and Arithmetic Cognitive Ability: a Practical Investigation with EEG Measurement

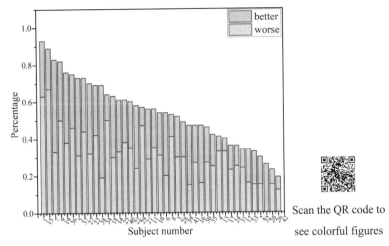

Scan the QR code to see colorful figures

Fig. 8 Percentage of emotions affected by the correct or incorrect answer. Orange indicates the percentage of emotions that changed for the worse after an incorrect answer, and green indicates the percentage of emotions that changed for the better when the answers changed from incorrect to correct. The values (orange: worse) of Subject 32 and 42 are 0, and the values (green: better) of Subject 42 is 0.

2) Brain mechanisms analysis: We first visualized the average scalp rPSD distribution under different emotions in four bands from all subjects, normalized to [0,1] according to each band, and the results are shown in Fig. 9. The energy of theta is mainly concentrated in the prefrontal lobe; the energy of gamma is mainly concentrated in the frontal and temporal lobes, and the energy of neutral and negative emotions are higher than that of positive emotions, which is consistent with the results in [46]. However, the distribution of alpha and beta bands differ from that in the above mentioned study, likely because, in our experiments, the EEG signals were analyzed not in a pure emotion-evoking experiment but in conjunction with a cognitive task. As seen from the figure, the energy distribution of the alpha band was more dispersed, located in other regions except for the prefrontal lobe, and the energy of the beta band was mainly concentrated in the left frontal lobe.

To better compare brain states under different emotions when cognitive performance is different, we calculated the differences in activation feature maps for right and wrong answers, normalized to [-1,1], as shown in Fig. 10. The beta band indicates that the brain is in high tension and attention, where the energy of beta in the high cognitive state is significantly higher in the frontal, parietal and occipital regions than in the low cognitive state under neutral emotion. Meanwhile, the brain activation maps don't differ much in positive or negative emotion, and even showed opposite states with neutral emotion. A similar situation occurs in the alpha and gamma bands, but the theta band shows the opposite result. We speculate that, under neutral emotion, it may be more favorable to focus attention when performing cognitive tasks; therefore, the brain is more active when performing well. Meanwhile, the t-test was used to analyze the differences under different emotions, and the results are shown in Table III. Statistical significance is defined as $P < 0.05$. There are significant differences between positive and neutral in all three bands except the low-frequency theta band, between positive and negative in the low-frequency theta band and the high-frequency gamma band, while between neutral and negative only in the alpha band.

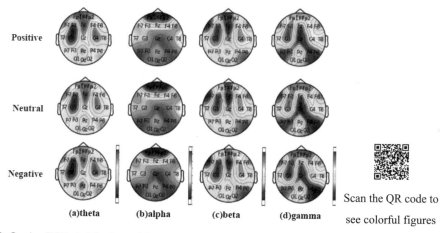

Fig. 9 Scalp rPSD distribution of three emotional states under different frequency bands.

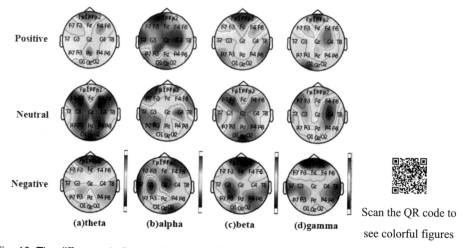

Fig. 10 The difference in Scalp rPSD distribution between correct and incorrect answers of three emotional states under different frequency bands. The red area indicates that the energy of correct answers is higher than that of wrong answers, and the opposite is shown in blue.

Table III The t-test analysis of difference Scalp rPSD distribution under different emotions.

	positive		neutral
	neutral	negative	negative
theta	\	*	\
alpha	*	\	*
beta	**	\	\
gamma	*	**	\

Note: "*" shows significance difference ($P \leq 0.05$); "**" shows significance difference ($P \leq 0.01$); "\" shows no significance difference ($P > 0.05$).

4. Discussion

With advances in neuroscience, a growing number of studies have focused on the interaction between emotion and cognitive ability. Previous studies have explored changes in behavioral data in response to emotional picture stimuli and ERP changes in response to emotional stimulation of faces. The present study aimed to investigate the effects of emotions on cognitive performance and the corresponding patterns of brain activation during a continuous cognitive task in a realistic scenario. Thus, we have reconstructed the dataset by video evocation and monitored the whole cognitive state using EEG, and the overall analysis process is shown in Fig. 2.

4.1 Analysis of Necessity of Dynamic Emotional Monitoring Model (Hypotheses 1 and 2)

Some studies proposed that emotions are not stable, and we therefore hypothesized that emotions would change during the cognitive task (Hypothesis 1). We thus constructed an emotional monitoring model to monitor emotions during the cognitive task in real time. During model construction, this study hypothesized that the subjects' emotional states might not be elicited, which may cause a decrease in accuracy during the training process due to labeling errors (Hypothesis 2). Therefore, we trained the data with different durations (10%~100%, step: 10%) separately, as shown in Fig. 4, where the performance became better as the data decreased, with the best results for the five-fold cross-validation at the last 40% of the video ($ACC = 98.08\%$, $STD = 0.16$), thus proving Hypothesis 2.

Next, we used the optimal model to monitor real-time emotions in the cognitive task. Fig. 5 compares the original labels with the predicted emotions during cognitive performance stage, which clarifies that emotions are not retained for a long time after the end of the video elicitation. Furthermore, emotions with higher arousal, such as positive and negative, are easily reverted to neutral emotions; meanwhile, combined with the behavioral data, Fig. 8 shows the percentage of the number of changes in emotion from positive → neutral → negative within the time of two questions for each subject when they answered the question incorrectly and emotion from negative → neutral → positive when the answers change from the incorrect to correct. We find that more than 97% of the subjects' emotions are affected by varying degrees in both of the above cases, and the degree of the effect is related to the individual variability of the subjects. Thus, the results confirm the conjecture in Hypothesis 1, suggesting that emotions may be difficult to maintain for long periods of time and can be affected by cognitive tasks during emotion and cognitive tasks. It assists in validating the reasonableness of the predicted emotions, and also further argues the necessity for real-time emotion monitoring to effectively explore the effect of emotion on cognitive performance during continuous cognitive activities.

4.2 Analysis of Affection-Cognition Interaction Based on Behavioral and Brain Mechanisms (Critical Question)

Based on real-time monitored emotions as well as EEG data from the look phase, we analyzed the effect of emotions on cognitive performance in terms of behavioral data and brain mechanisms, respectively. In previous studies with inconsistency of findings, reference [17] found that positive emotion can promote cognitive performance, and reference [18] found faster reaction times under negative emotion. In the present study, we first analyzed the accuracy of subjects' reaction time and error rate under different emotions and conducted a one-way repeated-measurement ANOVA both at the overall level and at different difficulty levels. The best performance is conducted for both reaction times and accuracy for neutral emotions at the overall level and at the high difficulty, which is inconsistent with the above studies but consistent with some probes in real cognitive scenarios. For example, reference [53] explored the relationship between mood and cognitive load in medical students and found that higher cognitive load was strongly associated with calmer mood. Meanwhile, reference [54] described emotions as the extrinsic cognitive load that competes for the limited resources of working memory by requiring the processing of off-task or task-irrelevant information. That is, when the difficulty of the question is higher, more cognitive load is required, and maintaining a neutral mood at this point is more conducive to cognitive performance.

Then, we used the activation pattern for brain mechanisms analysis. Fig. 9 shows the scalp rPSD distribution of three emotional states under different frequency bands, where alpha and beta waves differ from the results in the reference [46]. Alpha wave is the main expression of electrical activity when the cerebral cortex is in a waking and relaxed state, and beta wave appears after a person is stimulated, reflecting that the brain is in a state of high tension and concentration and is generally present in conscious states, such as cognitive reasoning and computation. Thus, the inconsistency of the conclusions may be because, in our experiments, the EEG was not analyzed solely under emotion evocation but together with the cognitive task. The results also suggest that cognitive activity has an impact on the activation patterns of emotions in the brain.

Finally, to better explore the brain states at different cognitive performances, we calculated the rPSD when answering the question correctly minus the question incorrectly. Fig. 10 shows the difference maps of activation patterns in the four bands under different emotions. The high cognitive state is significantly higher in the frontal, parietal and occipital regions than in the low cognitive state during neutral emotions under beta and gamma, whereas the brain activation maps don't largely differ or even shows the opposite during positive and negative emotions. Previous studies have demonstrated that the frontal and parietal lobes are potential brain regions for cognitive processes. Thus, the analysis of brain mechanisms is generally consistent with the behavioral results, in that the activation of high-frequency EEG is stronger during correct answers than during incorrect answers in neutral emotions, while positive and negative responses do not present the same findings. This suggests that concentration is more favorable in neutral emotions, and the brain is therefore more active when performing well.

In terms of different subjects, Fig. 4, Fig. 5 and Fig. 10 show that the results of the two basic hypotheses are consistent across subjects. However, the specific findings of this study are not exactly the same across subjects. In Fig. 6, the small squares of different colors represent different subjects, and it can be seen that although all of them were better in neutral emotion at the overall level, there are still differences between individuals. Moreover, Fig. 8 also shows the differences in the degree to which different subjects' moods are affected. The different specific findings across subjects may be due to the influence of intellectual differences among individuals, emotional stability and other extraneous factors. This needs to be further investigated but beyond the scope of this work.

4.3 Limitations and Future Work

In this study, there are four main limitations. Firstly, the subjects in this study were Chinese university students, and the sample size was not large enough, which may have led to the problem of sampling bias. Thus, the results may not be generalizable to other age, occupation, and cultural background groups. A more extensive and representative sample will be conducted in the future to enhance its external validity and generalizability. Secondly, this study was a simulation experiment in a laboratory scenario, which may not have fully recreated the behavior in the actual environment, and future studies will be conducted in actual scenarios. Thirdly, the techniques of deep learning are currently developing rapidly in the field of emotion recognition, more advanced methods of which, such as references [57], [58], could be adopted if preferred for achieving more accurate accuracies. Finally, only three emotions, positive, neutral and negative, were used in this study, yet emotions can vary continuously on a two-dimensional space of validity-arousal. Therefore, if the emotion monitoring model of this study is extended to a continuous space to monitor emotional states in real time, it will lay the foundation for better fitting the relationship between emotions and cognitive performance in the future.

5. Conclusion

This work has attempted to address a critical question which is to investigate the interaction between emotional state and cognitive ability by the use of EEG measurement. The current findings have illustrated that emotions are affected by cognitive activity during continuous cognitive tasks (regarding Hypothesis 1 proposed in the introduction), making real-time emotion monitoring necessary. The result further validates that, in video emotion elicitation, there is no successful elicitation in the first period and the best elicitation at the last 40% of the video, thus proving Hypothesis 2. The behavioral results have shown the best performance in terms of accuracy and response time under neutral emotions at the overall level or at a higher level of difficulty and emotions can be influenced by task performance. From the perspective of brain mechanisms analysis, the present study has found that cognitive activity influenced the activation pattern of emotions in the brain; meanwhile, under neutral emotions, the activation of high-frequency EEG is stronger in better cognitive performance than in worse cognitive performance, while positive and negative do not present the same findings. The results of this study provide a potential basis for assessing the cognitive abilities of individuals with different emotions in a variety of applications of cognitive scenarios.

Acknowledgments

This work was supported in part by the National Natural Science Foundation of China (NO. 62271455).

References

[1] BARRY R J, CLARKE A R, JOHNSTONE S J. A review of electrophysiology in attention-deficit/hyperactivity disorder: I. Qualitative and quantitative electroencephalography[J]. Clinical neurophysiology, 2003, 114(2): 171-183.

[2] CHIN-TENG L, CHUNG I F, LI-WEI K, et al. EEG-based assessment of driver cognitive

responses in a dynamic virtual-reality driving environment[J]. IEEE transactions on biomedical engineering, 2007, 54(7): 1349-1352.

[3] CLARK D M. The velten mood induction procedure and cognitive models of depression: a reply to riskind and rholes (1985)[J]. Behaviour research and therapy, 1985, 23(6): 667-669.

[4] CLOUDE E B, WORTHA F, DEVER D A, et al. Negative emotional dynamics shape cognition and performance with MetaTutor: toward building affect-aware systems[C]//2021 9th International Conference on Affective Computing and Intelligent Interaction (ACII). IEEE. 2021.10.1109/acii52823.2021.9597462.

[5] DUAN R N, ZHU J Y, LU B L. Differential entropy feature for EEG-based emotion classification[C]//2013 6th International IEEE/EMBS Conference on Neural Engineering (NER). IEEE. 2013.10.1109/ner.2013.6695876.

[6] FABRE L, LEMAIRE P. How emotions modulate arithmetic performance[J]. Experimental psychology, 2019, 66(5): 368-376.

[7] FABRE L, MELANI P, LEMAIRE P. How negative emotions affect young and older adults' numerosity estimation performance[J]. Quarterly journal of experimental psychology, 2022, 76(5): 1098-1110.

[8] FRASER K, MCLAUGHLIN K. Temporal pattern of emotions and cognitive load during simulation training and debriefing[J]. Medical teacher, 2018, 41(2): 184-189.

[9] GĄGOL A, MAGNUSKI M, KROCZEK B, et al. Delta-gamma coupling as a potential neurophysiological mechanism of fluid intelligence[J]. Intelligence, 2018, 66: 54-63.

[10] GARRISON K E, SCHMEICHEL B J. Effects of emotional content on working memory capacity[J]. Cognition and emotion, 2018, 33(2): 370-377.

[11] GONZALEZHERNANDEZ H G, PENACORTES D V, FLORESAMADO A, et al. Decreasing exam-anxiety levels with mindfulness through EEG measurements[C]//2022 IEEE Global Engineering Education Conference (EDUCON). IEEE. 2022.10.1109/educon52537.2022.9766539.

[12] GROSS J J, LEVENSON R W. Emotion elicitation using films[J]. Cognition and emotion, 1995, 9(1): 87-108.

[13] HUR J, IORDAN A D, BERENBAUM H, et al. Emotion–attention interactions in fear conditioning: Moderation by executive load, neuroticism, and awareness[J]. Biological psychology, 2016, 121: 213-220.

[14] JUNG N, WRANKE C, HAMBURGER K, et al. How emotions affect logical reasoning: evidence from experiments with mood-manipulated participants, spider phobics, and people with exam anxiety[J]. Frontiers in psychology, 2014, 5: 570.

[15] JUNG R E, HAIER R J. The parieto-frontal integration theory (P-FIT) of intelligence: Converging neuroimaging evidence[J]. Behavioral and brain sciences, 2007, 30(2): 135-154.

[16] KIRKEGAARD THOMSEN D. The association between rumination and negative affect: a review[J]. Cognition and emotion, 2006, 20(8): 1216-1235.

[17] KLADOS M A, PARASKEVOPOULOS E, PANDRIA N, et al. The impact of math anxiety on working memory: a cortical activations and cortical functional connectivity EEG study[J]. IEEE access, 2019, 7: 15027-15039.

[18] KLIMESCH W. EEG alpha and theta oscillations reflect cognitive and memory performance: A review and analysis[J]. Brain research reviews, 1999, 29(2-3): 169-195.

[19] KOELSTRA S, MUHL C, SOLEYMANI M, et al. DEAP: a database for emotion analysis; Using physiological signals[J]. IEEE transactions on affective computing, 2012, 3(1): 18-31.

[20] KUIJSTERS A, REDI J, DE RUYTER B, et al. Inducing sadness and anxiousness through visual media: measurement techniques and persistence[J]. Frontiers in psychology, 2016, 7.

[21] LALLEMENT C, LEMAIRE P. Age-related differences in how negative emotions influence arithmetic performance[J]. Cognition and emotion, 2021, 35(7): 1382-1399.

[22] LANG X, WANG Z, TIAN X, et al. The effects of extreme high indoor temperature on EEG during a low intensity activity[J]. Building and environment, 2022, 219: 109-225.

[23] LEDOUX J E. Cognitive-emotional interactions in the brain[J]. Cognition and emotion, 1989, 3(4): 267-289.

[24] LEDOUX J E, BROWN R. A higher-order theory of emotional consciousness[J]. Proceedings of the national academy of sciences, 2017, 114(10).

[25] LERNER J S, LI Y, VALDESOLO P, et al. Emotion and decision making[J]. Annual review of psychology, 2015, 66(1): 799-823.

[26] LI F, YI C, JIANG Y, et al. Different contexts in the oddball paradigm induce distinct brain networks in generating the p300[J]. Frontiers in human neuroscience, 2019, 12.

[27] LI P, LIU H, SI Y, et al. EEG based emotion recognition by combining functional connectivity network and local activations[J]. IEEE transactions on biomedical engineering, 2019, 66(10): 2869-2881.

[28] LI R, REN C, GE Y, et al. MTLFuseNet: a novel emotion recognition model based on deep latent feature fusion of EEG signals and multi-task learning[J]. Knowledge-based systems, 2023, 276: 110756.

[29] LI X, ZHANG Y, TIWARI P, et al. EEG based emotion recognition: a tutorial and review[J]. ACM computing surveys, 2022, 55(4): 1-57.

[30] LI Z, WU X, XU X, et al. The recognition of multiple anxiety levels based on electroencephalograph[J]. IEEE transactions on affective computing, 2022, 13(1): 519-529.

[31] LIU D, WANG Y, LU F, et al. Emotional valence modulates arithmetic strategy execution in priming paradigm: an event-related potential study[J]. Experimental brain research, 2021, 239(4): 1151-1163.

[32] LIU J, LI J, PENG W, et al. EEG correlates of math anxiety during arithmetic problem solving: Implication for attention deficits[J]. Neuroscience letters, 2019, 703: 191-197.

[33] LIU Y, FU Q, FU X. The interaction between cognition and emotion[J]. Chinese science bulletin, 2009, 54(22): 4102-4116.

[34] LIU YJ, YU M, ZHAO G, et al. Real-time movie-induced discrete emotion recognition from EEG signals[J]. IEEE transactions on affective computing, 2018, 9(4): 550-562.

[35] MAUSS I B, ROBINSON M D. Measures of emotion: a review[J]. Cognition and emotion, 2009, 23(2): 209-237.

[36] NIU H, ZHAI Y, HUANG Y, et al. Investigating the short-term cognitive abilities under local strong thermal radiation through EEG measurement[J]. Building and environment,

2022, 224: 109567.

[37] PEIRCE J W. PsychoPy—Psychophysics software in python[J]. Journal of neuroscience methods, 2007, 162(1-2): 8-13.

[38] PEREIRA D R, SAMPAIO A, PINHEIRO A P. Interactions of emotion and self-reference in source memory: an ERP study[J]. Cognitive, affective, & behavioral neuroscience, 2021, 21(1): 172-190.

[39] PEREIRA E T, GOMES H M, VELOSO L R, et al. Empirical evidence relating EEG signal duration to emotion classification performance[J]. IEEE transactions on affective computing, 2021, 12(1): 154-164.

[40] PESSOA L. On the relationship between emotion and cognition[J]. Nature reviews neuroscience, 2008, 9(2): 148-158.

[41] PLASS J L, KALYUGA S. Four ways of considering emotion in cognitive load theory[J]. Educational psychology review, 2019, 31(2): 339-359.

[42] PLASS J L, KAPLAN U. Emotional design in digital media for learning[C]//Emotions, Technology, Design, and Learning. Elsevier. 2016: 131-61.10.1016/b978-0-12-801856-9.00007-4.

[43] RAJENDRAN V G, JAYALALITHA S, ADALARASU K. EEG based evaluation of examination stress and test anxiety among college students[J]. IRBM, 2022, 43(5): 349-361.

[44] RUSSELL G S, JEFFREY ERIKSEN K, POOLMAN P, et al. Geodesic photogrammetry for localizing sensor positions in dense-array EEG[J]. Clinical neurophysiology, 2005, 116(5): 1130-1140.

[45] SHEN F, DAI G, LIN G, et al. EEG-based emotion recognition using 4D convolutional recurrent neural network[J]. Cognitive neurodynamics, 2020, 14(6): 815-828.

[46] SI Y, JIANG L, LI P, et al. Relationship between decision making and resting-state EEG in adolescents with different emotional stabilities[J]. IEEE transactions on cognitive and developmental systems, 2024, 16(1): 243-250.

[47] Luck S J. An introduction to the event-related potential technique[M]. Cambridge, MA, the USA: MIT press, 2014.

[48] TYNG C M, AMIN H U, SAAD M N M, et al. The influences of emotion on learning and memory[J]. Frontiers in psychology, 2017, 8: 235933.

[49] WANG H, WU X, YAO L. Identifying cortical brain directed connectivity networks from high-density EEG for emotion recognition[J]. IEEE transactions on affective computing, 2022, 13(3): 1489-1500.

[50] WANG Y, HUANG Z, MCCANE B, et al. EmotioNet: a 3-D convolutional neural network for EEG-based emotion recognition[C]//2018 International Joint Conference on Neural Networks (IJCNN). IEEE. 2018.10.1109/ijcnn.2018.8489715.

[51] WEI-LONG Z, BAO-LIANG L. Investigating critical frequency bands and channels for EEG-based emotion recognition with deep neural networks[J]. IEEE transactions on autonomous mental development, 2015, 7(3): 162-175.

[52] WELCH P. The use of fast Fourier transform for the estimation of power spectra: a method based on time averaging over short, modified periodograms[J]. IEEE transactions on audio

and electroacoustics, 1967, 15(2): 70-73.

[53] XU X, JIA T, LI Q, et al. EEG feature selection via global redundancy minimization for emotion recognition[J]. IEEE transactions on affective computing, 2023, 14(1): 421-435.

[54] YANG Y, WU Q, FU Y, et al. Continuous convolutional neural network with 3D input for EEG-based emotion recognition[C]//Neural information processing. springer international publishing. 2018: 433-43.10.1007/978-3-030-04239-4_39.

[55] YAO Y, LIAN Z, LIU W, et al. Experimental study on physiological responses and thermal comfort under various ambient temperatures[J]. Physiology and behavior, 2008, 93(1-2): 310-321.

[56] ZHANG H, SONG R, WANG L, et al. Classification of brain disorders in rs-fmri via local-to-global graph neural networks[J]. IEEE transactions on medical imaging, 2023, 42(2): 444-455.

[57] ZHU C, JIANG Y, LI P, et al. Implicit happy and fear experience contributes to computational estimation strategy execution: behavioral and Neurophysiological Evidence[J]. Neuropsychologia, 2021, 159: 107959.

[58] ZHU C, LI P, LI Y, et al. Implicit emotion regulation improves arithmetic performance: an ERP study[J]. Cognitive, affective, and behavioral neuroscience, 2022, 22(3): 574-585.

SSTM-IS: Simplified STM Method Based on Instance Selection for Real-Time EEG Emotion Recognition*

1. Introduction

Brain-Computer Interface (BCI) is a communication system that does not depend on the output path composed of peripheral nerves and muscles. The BCI allows users to control computers or other devices through brain activity and involves many research fields such as medicine, neurology, signal processing, and pattern recognition. In recent years, the affective BCI (aBCI) has attracted great interests which endows BCI system with the ability to detect, process and respond to the affective states of humans using EEG signals. The aBCI has shown great development potential in many application fields, for example, it can help the patients with psychological diseases establish effective social interaction; remind the driver to better focus on the driving to avoid traffic; make the machine analyze the emotion of human and provide emotional companionship.

Emotion is one of the most important physiological and psychological states of the human body. Modern medicine believes that emotion is directly related to human health. Prolonged negative emotion can reduce creativity and cause a loss of focus, even cause anxiety and depression. Accurate and reliable emotion monitoring can, on the one hand, help to restore emotion and enhance concentration and, on the other hand, help service providers to analyze user preferences and thus provide more personalized products and services. Currently, emotion monitoring is based on two main types of physiological signals: subjective physiological signals such as voice and expression, and objective physiological signals such as EEG and ECG. Since the EEG signal has the advantages of high temporal resolution, not easy to pretend and the popularity of non-invasive portable acquisition devices, it has been widely used for emotion detection.

As shown in Fig. 1, the workflow of EEG signal-based emotion recognition includes

---

* The paper was originally published in *Frontiers in Human Neuroscience*, 2023, 17, and has since been revised with new information. It was co-authored by Shuang Ran, Wei Zhong, Danting Duan, Long Ye, and Qin Zhang.

the acquisition of EEG signals, signal pre-processing, feature extraction and classification, and the output of recognition result. For the EEG processing, the acquired EEG signals need to be pre-processed first to improve the signal-to-noise ratio by the methods such as filtering, artifact removal and principal component analysis, because their amplitudes are very small and susceptible to be interfered by other electrophysiological signals. And then, the features in time domain, frequency domain and time-frequency domain are extracted and decoded to recognize the emotion of experimenter by using the recognition methods such as machine learning.

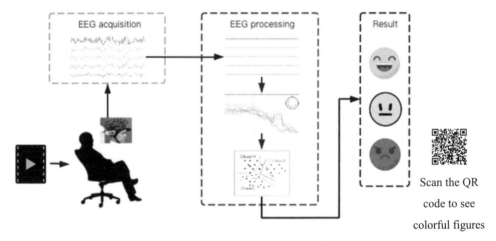

Scan the QR code to see colorful figures

Fig. 1 The workflow of EEG signal-based emotion recognition: stimulating the subject's emotions, collecting EEG signals and recognizing emotions based on the features extracted from the pre-processed EEG signals.

The existing methods of emotion recognition based on EEG signals have the following two limitations:

- Most of the methods use the models trained on existing data to test the new subject for real-time emotion recognition. However, the EEG signals are non-stationary and the training and testing samples must be collected from the same individual or even the same test environment, otherwise the recognition accuracy will drop dramatically. This means building universal models that work across subjects is challenging.
- Most of the existing cross-subject methods need to analyze the results manually or process data offline, which require all the EEG data to be saved first. While the human emotions always change in real-time, the above methods are time-consuming and user-unfriendly, difficult to support the practical applications of real-time emotion recognition.

To address both of these points, our work aims to explore an algorithm that can recognize emotion in cross-subject situation quickly and accurately. By selecting the informative instances and simplifying the update strategy of hyperparameters, we propose a simplified style transfer mapping method based on instance selection (SSTM-IS), which can use a small amount of labeled data in the target domain to make the model adapt for the new subject in a short time. To verify the effectiveness of the proposed method, we compare and analyze the performance of representative methods in terms of accuracy and computing time on SEED, SEED-IV, and collected dataset by ourselves for emotion recognition. The experimental results show that our algorithm achieves the accuracy rate of 86.78% in the computing time of 7s on the SEED dataset, 82.55% in 4s on SEED-IV and 77.68% in 10s on self-collected dataset. In addition, the work in this paper also includes the design and implementation of a real-time aBCI system to test the proposed algorithm in practical application. There are three key contributions in this work:

- By selecting the informative instances and simplifying the update strategy of hyperparameters in style transfer mapping, we propose the SSTM-IS algorithm to perform the cross-subject emotion recognition more accurately and quickly.
- We validate the proposed algorithm on both of the public and self-collected datasets. The experimental results demonstrate that the proposed algorithm can achieve higher accuracy in a shorter computing time, satisfying the needs of real-time emotion recognition applications.
- We design and implement a real-time emotion recognition system that integrates the modules of EEG signal acquisition, data processing, emotion recognition and result visualization. The applicability of the proposed algorithm is also verified online in a real case.

This paper is structured as follows: Section 2 briefly discusses the related works. The proposed emotion recognition algorithm is described in detail in Section 3. Subsequently in Section 4, the real-time emotion recognition system is developed and realized. And then Section 5 gives and analyzes both of the offline and online experimental results. Finally, some conclusions and future works are presented in Section 6.

2. Related Works

As we know, the EEG signals are non-stationary and various people to people. The emotion recognition model trained on the existing dataset is often not applicable to new subjects. While the applications of real-time emotion recognition require algorithms to accurately and quickly recognize the EEG signals of a newcomer. A common solution is to introduce transfer learning into EEG emotion recognition, so as to make the model adapt to new individuals. The core of transfer learning is to find the similarity between existing knowledge and new one, so can use the existing to learn the new. In transfer learning, the existing knowledge is denoted as the source domain, and the new knowledge to be learned is defined as the target domain. The distribution between the source domain and target domain are different but related to each other. It is necessary to reduce the distribution difference between the source and target domains for knowledge transfer.

For the EEG emotion recognition, one method is to use all unlabeled data from target domain to train the model, and enhance the model performance by reducing the differences between the source domain and target domain. Li et al. (2021) proposed the BiDANN-S framework for cross-subject emotion recognition, which reduces the differences across domains by adversarially training domain discriminators. Li et al. (2020b) used the adversarial training to adjust the marginal distribution of the shallow layers for reducing the difference between the source and target domains, while using correlation enhancement to adjust the conditional distribution of the deep layers for enhancing network generalization. Chen et al. (2021) proposed the MS-MDA network based on domain-invariant and domain-specific features, so that different domain data share the same underlying features, while preserving the domain-specific features. This kind of methods given above belong to deep learning algorithms, as we know, the deep learning algorithms use the back propagation for parameter optimization, and need a large amount of data in model calibration phase to improve the recognition performance of new subjects. As a result, when a new subject comes, the consuming time for model calibration is relatively long and thus these algorithms are not suitable for the applications of real-time EEG emotion recognition. Another method mainly obtains the domain invariant features from the differences of multiple source domains to train a general model. In the training phase, it is not necessary to obtain any data from the target domain. Ma et al. (2019) proposed a new adversarial domain generalization framework, DResNet, in which the domain information was used to learn unbiased weights across subjects and biased weights specific to subjects. However, the performance of the model obtained by this method is generally worse than that of models with target domain data participating in the training. For example, the accuracy rate of the PPDA model with all target domain data participating in training is 1.30% higher than that of the PPDA_NC without target domain data involved in training.

From the above analysis, it can be inferred that one of the ideas to obtain a real-time emotion recognition model with good performance is to integrate the two kinds of methods, that is, using a small amount of target domain data for supervised learning, and training a generalized model. Chai et al. (2017) proposed an ASFM to integrate both the marginal and conditional distributions within a unified framework, which achieves an accuracy of 83.51% on the SEED dataset. It should be noted that on the selection of source domain, the works described above use all the subjects' source domain data to train the model without selection. However, the studies by Yi and Doretto (2010) and Lin and Jung (2017) have shown that the inappropriate selection of source domain data may cause the negative transfer. In addition, the large amount of data involved in model training phase increases the computing time and is not applicable to real-time emotion recognition. Li et al. (2020a) considered each subject as a source domain for multi-source domain selection and proposed an MS-STM algorithm which reduces the domain difference when using a small amount of labeled target data. Later, Chen et al. (2022) proposed a conceptual online framework FOIT in which they selected instance data from source domains, but the resulting recognition accuracy is not satisfactory and the framework is not verified in an actual scene.

On the real-time emotion recognition system, to the best of our knowledge, there are few reports on relevant work. Jonathan et al. (2016) developed a mobile application for processing and analyzing EEG signals, which can display the EEG spectrum, classify EEG signals, visualize, and analyze the classification results. However, this system does not realize the data upload function, and only uses the offline data that has been stored on the server. Nandi et al. (2021) proposed a real-time emotion classification system based on logistic regression classifier, and carried out the experiments of simulating real-time emotion recognition on DEAP. Weiss et al. (2022) compared several classifiers and chose the logistic regression to realize a real-time EEG emotion recognition system on SEED-IV. Although the above works put forward the concept of real-time emotion recognition system, they use the models trained on existing data to test the new subject, without model calibrating. Besides they all use the offline datasets or simulated real-time emotion recognition for verification, not performing the real-time emotion recognition applications in real scene.

According to the above analysis, the human emotions always change over time, and thus it is necessary to realize a real-time emotion recognition system with good performance in practical application. The existing transfer learning methods for EEG emotion recognition either use all the subjects' source domain data to train the model without selection, or consider each subject as a source domain for multi-source domain selection, which may cause negative transfer and increase the computing time. To address this problem, this paper proposes a simplified STM algorithm by optimizing the updating strategy of hyperparameters in STM and using SVM classifier with a one-vs-one scheme for parameter optimization. On the other hand, we also refine the granularity of selecting source domain data, and obtain the most informative instances to further enhance the generalization of the model. With the above improvements, the algorithm proposed can achieve higher accuracy by using a small amount of data for model calibration in a short computing time, satisfying the need of real-time EEG emotion recognition applications.

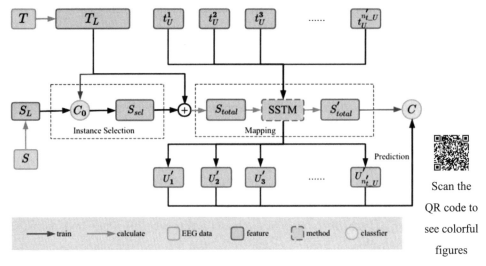

Fig. 2 Framework of the proposed SSTM-IS algorithm.

3. Methods

In this section, we present the proposed SSTM-IS method which consists of two main steps as shown in Fig. 2. Firstly, our work is to refine the granularity of selecting source domain data from different subjects to different sample instances, and select the most informative instances from the source domain through a classifier trained by the labeled data from the target domain to improve the recognition accuracy of new subject. And then by simplifying the updating strategy of hyperparameters in STM, a simplified STM (SSTM) algorithm is developed to make the distributions of source and target domains more similar.

Define the source domain as $S = \{s_i, i = 1, 2, \ldots, n_s\}$ and its corresponding label is $L = \{y_s^1, y_s^2, \ldots, y_s^{n_s}\}$. The target domain is $T = \{t_i, i = 1, 2, \ldots, n_t\}$ which is divided into the labeled and unlabeled parts, and the labeled part corresponds to the label L_T. Here, n_s and n_t are, respectively, the sample numbers of source domain and target domain data. S and T have different marginal distributions. The DE features of the source domain are extracted and denoted as $S_L = \{s_L^i \in R^m \mid i = 1, 2, \ldots, n_s'\}$, where n_s' represents the number of DE feature samples of source domain and m represents the feature dimension. Similarly, for the target domain, the DE features of the labeled and unlabeled data are $T_L = \{t_L^i \in R^m \mid i = 1, 2, \ldots, n_{t\_L}'\}$ and $T_U = \{t_U^i \in R^m \mid i = 1, 2, \ldots, n_{t\_U}'\}$, respectively, where n_{t_L}' and n_{t_U}' represent the feature numbers of the labeled and unlabeled target domain data. $U' = \{U_i' \in R^m \mid i = 1, 2, \ldots, n_{t\_U}'\}$ represents the features mapped from T_U.

3.1 Instance Selection

Using the source domain data and target domain data for transfer learning can build a robust emotion recognition model. Among the existing EEG emotion recognition algorithms, some methods use all of the subjects' data to train the model and do not perform the selection of source domain data. The others consider each subject as a source domain for multi-source domain selection, and the selected source domain data are all involved in the model training. However, the inappropriate selection of source domain data may cause the negative transfer. In addition, the large amount of data involved in model training increases the computing time and is not applicable to real-time recognition. Inspired by the idea of sample query in active learning, we propose to refine the granularity of selecting source domain data from different subjects to different sample instances. The specific strategy is training a classifier with labeled target domain data and using the classifier to select the most informative instances from the source domain to improve the emotion recognition accuracy of new subject.

INPUT: $S_L = \{s_L^i \in R^m \mid i=1,\ldots,n_s'\}$: source domain data, s_L^i is sample; $L = \{y_s^1, y_s^2, \ldots, y_s^{n_s'}\}$: the labels to S_L ;

$T = T_L \cup T_U$: target domain data, including labeled T_L (corresponding labels are $L_T = \{y_t^1, y_t^2, \ldots, y_t^{n_t'}\}$) and unlabeled T_U ; n_c : the number of emotional categories in labels;

k, β, γ : hyper parameter;

OUTPUT: $\hat{y}$: predicted labels on the unlabeled data;

1: $T_L \to C_0$

2: for each $i \in [1, n_s']$ do

3: $\quad w_i = Acc(C_0, s_L^i)$

4: end for

5: $W = \{w_1, w_2, \ldots, w_{n_s'}\}$

6: $S_L = \{S_L^i \text{ sorted by } w_i \text{ and grouped by } y_s^i\} = S_L^1 \cup S_L^2 \cup \ldots \cup S_L^{n_{cL}}$

7: $S_{sel} = S_L^1[1:k] \cup S_L^2[1:k] \cup \ldots \cup S_L^{n_c}[1:k]$

8: for each $t_L^i \in T_L$ do

9: $\quad o_i$ calculated by Equation (4)

10: end for

11: learn STM A, b with $\{t_L^i, o_i\}$

12: $S_{sel} \to C$

13: $U'_{1\sim n'_{L\_U}}$ is transformed by $A, b \to x, x \in T_U$

14: predict $\hat{y} = C\left(U'_{1\sim n'_{L\_U}}\right)$

Algorithm 1 Simplified STM based on instance selection.

For the procedure of instance selection shown in Fig. 2, we first train an SVM classifier C_0 by using the DE feature T_L of the labeled target domain data and its corresponding label Y_L. Here, C_0 is the same for all the domain subjects, and is trained by the sessions with different emotion categories from labeled target domain. The probability of each sample in source domain denoted as w_i is then predicted by C_0 and taken as the information contained in each sample instance:

$$w_i = Acc(C_0, s_L^i), i = 1, 2, \ldots, n'_s \tag{1}$$

Subsequently, the instances can be selected according to their prediction probabilities w_i. To avoid the unbalanced distribution, we sort the instances of the source domain within each emotion category according to the amount of information it contains as S_{L_sort}, and select the top k samples of instance data with the highest probabilities for each emotion category as the top k highest informative instances S_{sel}:

$$S_{L\_sort} = \{S_L^i \text{ sorted by } w_i \text{ and grouped by } y_s^i\} = S_L^1 \cup S_L^2 \cup \ldots \cup S_L^{n_c} \tag{2}$$

$$S_{sel} = S_L^1[1:k] \cup S_L^2[1:k] \cup \ldots \cup S_L^{n_c}[1:k] \tag{3}$$

where y_s^i is the emotion label of each instance in source domain, and n_c is the number of emotion categories. Thus, we can get the informative instance data S_{sel} for the subsequent transfer learning to reduce the data redundancy and improve the computing efficiency.

3.2 Simplified STM

The STM algorithm solves a style transfer mapping to project the target domain data T to another space, where the differences between the target domain T and the source domain S are reduced. In this way, the classifier C trained by the source domain data can be used for the classification of the target domain data in a specific space. As shown in Fig. 2, the category number for classifiers of C_0 and C is consistent with those of emotions recognized. Assuming the DE features of the source and target domains obey Gaussian distribution, the labeled target domain data T_L can be mapped by the Gaussian model into:

$$O = \{o^i \in R^m\}, o^i = \mu_c + \min\left\{1, \frac{\rho}{d(t^i_{L_c}, c)}\right\}, i = 1, \ldots, n'_{t\_L} \quad (4)$$

$$d(t^i_{L_c}, c) = \sqrt{(t^i_{L_c} - \mu_c)^T \Sigma_c^{-1}(t^i_{L_c} - \mu_c)} \quad (5)$$

where o^i is the mapped value of t^i_L by the Gaussian model, $d(t^i_{L_c}, c)$ is the Mahalanobis distance in emotion category of $(c = 1, 2, \ldots, n_c)$, $t^i_{L_c}$ is the target domain data corresponding to C, and μ_c is the average of the instances labeled with C in S_{sel}. Here, ρ is used to control the deviation between o^i and μ_c.

Suppose the affine transformation from o^i back to t^i_L is represented as $Ao^i + b$, the parameters $A \in R^{m \times m}$ and $b \in R^m$ can be learned by optimizing the weighted square error with regular terms:

$$\min_{A \in R^{m \times m}, b \in R^m} \sum_{i=1}^{n'_{t\_L}} \left\| Ao^i + b - t^i_L \right\|_2^2 + \beta \left\| A - I_{m \times m} \right\|_F^2 + \gamma \left\| b \right\|_2^2 \quad (6)$$

where the hyperparameters β and γ are used to control the state between non-transfer and over-transfer. The STM method updates the values of β and γ in each iteration of calculation, spending much computing time. Through the experiment, it is found that the fixed values of β and γ can be obtained to shorten the computing time without reducing the accuracy. So we select β and γ as the fixed constants, and then Equation (6) can be solved as follows:

$$A = QP^{-1}, b = \frac{1}{\hat{f}}(\hat{t} - A\hat{o}) \quad (7)$$

$$Q = \sum_{i=1}^{n'_{t\_L}} f_i t^i_L (o^i)^T - \frac{1}{\hat{f}} \hat{t}\hat{o}^T + \beta I_{m \times m} \quad (8)$$

$$P = \sum_{i=1}^{n'_{t\_L}} f_i o^i (o^i)^T - \frac{1}{\hat{f}} \hat{o}\hat{o}^T + \beta I_{m \times m} \quad (9)$$

$$\hat{o} = \sum_{i=1}^{n'_{t\_L}} f_i o^i \quad (10)$$

$$\hat{t} = \sum_{i=1}^{n'_{t\_L}} f_i t^i_L \quad (11)$$

$$\hat{f} = \sum_{i=1}^{n'_{t\_L}} f_i \gamma \quad (12)$$

where the parameter f_i is the confidence of t^i_L to o^i. Since the parameter γ is fixed, $\hat{f}$ will not change with iteration, which reduces the computing time.

The specific pseudo code of SSTM-IS algorithm is shown in Algorithm 1. According to the description given above, we first select the most informative instance samples from the source domain, and then perform the transfer learning between the selected instances and the labeled target domain data by simplifying the updating strategy of hyperparameters in STM, improving the emotion recognition accuracy and increasing the computing speed.

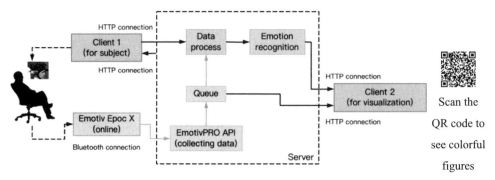

Fig. 3 The flowchart of online emotion recognition.

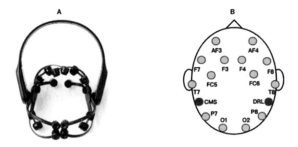

Fig. 4 The EEG acquisition device used in our experiments: (A) Emotiv EPOC X, (B) the electrode distribution.

4. Real-Time Emotion Recognition System

Most of the existing emotion recognition systems based on EEG signals are limited to manual offline for data processing and result analysis. This way of offline processing needs to load all the data first, and thus cannot perform the real-time emotion recognition obviously. The SSTM-IS algorithm proposed in this paper can establish a model suitable for new subjects from existing data in a short computing time, and can be well-applied to real-time emotion recognition. To verify the effectiveness of the SSTMIS algorithm in practical use, we simulate the real-time emotion recognition in online situation. The system framework is shown in Fig. 3.

SSTM–IS: Simplified STM Method Based on Instance Selection for Real–Time EEG Emotion Recognition

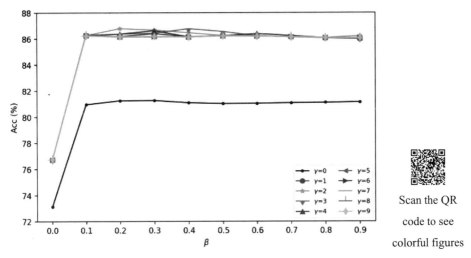

Fig. 5 The accuracy results under different values of β and γ.

Scan the QR code to see colorful figures

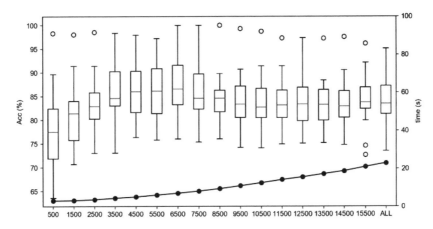

Fig. 6 The accuracy and computing time of the proposed algorithm by selecting different number of instances on SEED dataset.

For the actual online situation, we develop a real-time EEG emotion recognition system that integrates EEG signal acquisition, processing, emotion recognition and result visualization, as shown in Fig. 3. The system uses an Emotiv EPOC X with 14-channel electrodes to collect EEG signals, namely AF3, F7, F3, FC5, T7, P7, O1, O2, P8, T8, FC6, F4, F8, AF4 with additional two reference electrodes CMS and DRL, as shown in Fig. 4. The sampling frequency of Emotiv EPOC X is 256 Hz and the bandwidth is 0.20-43 Hz. The Emotiv EPOC X is connected to the server by Bluetooth, and the EEG signals of the subjects are recorded by calling the EmotivPRO API. Here, the Client 1 is the experimental interface of the subject, which is used to realize the interaction between the subject and the server. In the experiment, the necessary prompts can be given to subjects by Client 1, allowing them to control the experimental process according to their own experiences, such as the length of rest time. We collected and stored the EEG data of the subjects who have experimented on this system before as the source domain data. As shown in Fig. 3, when a new subject comes, he receives video stimulation, conducts subjective evaluation, and controls the experimental process through the interface of Client 1. The server follows the experimental paradigm to control the experimental process, stores and processes experimental data, as well as responds to the requests from pages. In the work process, the server collects the first three sessions of EEG data from the new subject as the labeled target domain data, and uses these labeled data to train a classifier for selecting the informative instance data from source domain. And then the simplified STM is called to obtain an emotion recognition model by using these labeled data and selected instance data. Subsequently, the server pre-processes the raw EEG data within every segment of 10s by using the MNE-Python library for bandpass filtering of 0.10-50 Hz and notching filter for denoising. And then the DE features are extracted to put into the trained model for real-time emotion recognition. Once the model training is completed, the system only needs several milliseconds to output an emotion prediction for a 10s data segment. Finally, the server feeds the raw EEG signals and recognition results into the Client 2 for visualization. Here, the Client 2 is used to monitor the state of the subject, present video stimuli, visualize the raw EEG signals as well as their spectral maps and topographic maps, and analyze the real-time emotion recognition results.

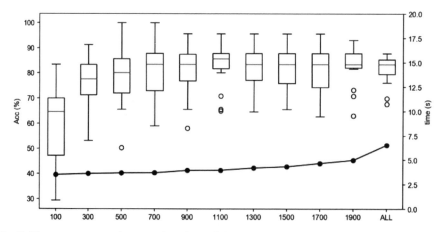

Fig. 7 The accuracy and computing time of the proposed algorithm by selecting different number of instances on SEED-IV dataset.

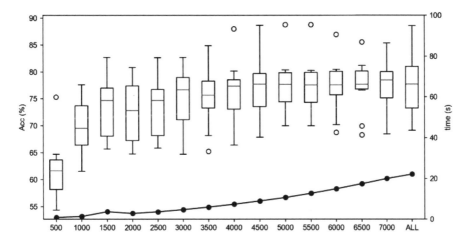

Fig. 8 The accuracy and computing time of the proposed algorithm by selecting different number of instances on self-collected dataset.

5. Experiments and Results

5.1 Datasets and Settings

In order to verify the performance of the proposed SSTM-IS algorithm, we select two public datasets SEED and SEED-IV, as well as the collected 14-channel EEG signal dataset for offline and online experiments. The SEED dataset selected 15 Chinese film clips with the types of positive, neutral, and negative as visual stimuli, and each clip is about 4 min. The 15 subjects conducted three experiments by using the same stimuli with an interval of 1 week, and 3,394 samples had been collected for each subject in one experiment. The SEED-IV dataset selected 24 movie clips with four emotions of happiness, sadness, fear, and neutral (6 clips for each emotion), and the duration of each movie clip is about 2 min. The 15 subjects conducted three groups of experiments which used completely different stimuli at different times, and 822 samples had been collected in one experiment. Both SEED and SEED-IV datasets use 62-channel acquisition equipment to record EEG signals of subjects at a sampling rate of 1,000 Hz, and pre-process the collected data as follows: downsampling to 200 Hz, using 0.3-50 Hz band-pass filter and extracting DE features.

Table I Performance comparisons with the existing typical EEG emotion recognition algorithms on SEED and SEED-IV datasets.

Dataset	Method	Acc (%)	Runtime (s)
SEED	TCA	64.24 ± 15.34	298
	CORAL	63.59 ± 7.60	19
	MS–MDA	85.04 ± 7.85	1,959
	PPDA_NC	85.40 ± 7.10	–
	PPDA	86.70 ± 7.10	–
	DResNet	85.30 ± 8.00	–
	MS–STM	83.22 ± 13.96	776
	ASFM	83.51 ± 10.18	–
	FOIT	82.05 ± 12.36	32
	SSTM	84.02 ± 5.38	22
	Instance–sel	79.66 ± 8.15	**5**
	SSTM–IS	**86.78 ± 6.65**	7
SEED_IV	TCA	29.06 ± 2.16	155
	CORAL	32.79 ± 12.25	43
	MS–STM	80.28 ± 9.93	833
	FOIT	78.15 ± 7.31	20
	SSTM	81.66 ± 6.10	7
	Instance–sel	64.22 ± 17.80	**3**
	SSTM–IS	**82.55 ± 8.48**	4
Self-collected	TCA	47.04 ± 4.91	203
	CORAL	46.26 ± 2.56	31
	MS–MDA	68.98 ± 4.91	347
	MS–STM	67.83 ± 4.75	89
	SSTM	76.96 ± 4.08	9
	Instance–sel	70.31 ± 5.62	**6**
	SSTM–IS	**77.68 ± 5.13**	10

The bold values indicate the best performance in the current dataset.

The dataset collected by ourselves recorded the EEG signals of 10 subjects at a sampling rate of 256 Hz with the Emotiv EPOC X, a 14-channel wireless portable EEG acquisition instrument. In the selection of stimuli videos, by taking into account the native language environment of the subjects, we selected 12 Chinese short videos with three emotions positive, neutral, and negative as video stimuli, after the subjective evaluation from 800 short videos. The duration of each short video is about 1-3 min. In the experiment, the emotion types of two adjacent videos are different and the videos are played pseudo-randomly, and 2,495 samples had been collected for each subject. With the 14-channel EEG signals obtained, the band-pass filter is used for 0.10-50 Hz frequency filtering, and the notch filter is employed for denoising. And then the pre-processed EEG signals are cut into 1-s segments and the DE features are also extracted.

For the experimental paradigm on the self-collected dataset, after filling in basic information, the subject needs to read the experiment description and completes a 5-min baseline recording with opening or closing eyes alternately every 15s. During a 15-s baseline recording, the subject is required to remain relaxed, blink as infrequently as possible, and look at a fixation cross "+" on the screen. Then a video stimulus is displayed and the subject is asked to stay as still as possible and blink as infrequently as possible when watching the stimulus. After that, the subject filled in the subjective evaluation scales based on his immediate true feelings. To eliminate the effect of the previous stimulus, the subject is asked to complete two simple calculation questions within ten as a distraction. Next, a more than 30-s period of rest is taken, during which a blank screen is displayed and the subject is asked to clear his brain of all thoughts, feelings and memories as much as possible. When the subject clicks the "NEXT" button on the screen, the next trial starts. The above process is repeated until 12 short videos have been played.

In the experiments, we use the leave-one-subject-out verification method on the SEED, SEED-IV, and self-collected datasets, and employ all sessions of the subjects in the source domain as the source data. For the three-category SEED and self-collected datasets, the first three sessions are taken from target domain as the labeled data. For the SEED-IV dataset with four categories, since two adjacent sessions may have the same emotion category, we use the first several sessions from target domain as the labeled data until all emotion categories have been presented. In the target domain, the data amounts used for calibration and testing are 674 and 2,720 in SEED, 499 and 323 in SEED-IV, 330 and 2,165 in self-collected dataset.

For the hyperparameters of β and γ in Equation (6), we set their values through the experiment. Take the SEED dataset as an example, Fig. 5 presents the accuracy results under different values of β and γ. It can be seen from Fig. 5 that, when $\gamma = 0$ and $\beta = 0$, the accuracy is only 73.14%; when $\gamma = 0$ and $\beta > 0$, the accuracy is about 81%. When $\gamma > 0$ and $\beta > 0$, the accuracy fluctuates in a small range around 86%, and when $\gamma = 2$ and $\beta = 0.2$, the accuracy reaches the local maximum of 86.78%. The same to SEED-IV and self-collected datasets. Therefore, we set $\gamma = 2$ and $\beta = 0.2$ in the experiments. It is found that the fixed values of β and γ can be obtained to shorten the computing time without reducing the accuracy.

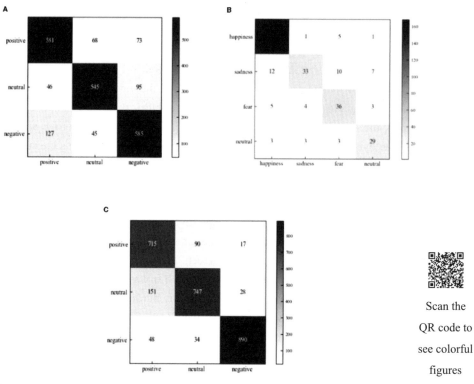

Fig. 9 Confusion matrix of the proposed method in cross-subject on (A) SEED, (B) SEED-IV, and (C) self-collected datasets.

5.2 Offline Experiments
5.2.1 Analysis on the Quantity Selection of Instances

Firstly, we test the effect of the number of instance selection on the performance of emotion recognition. Here, the incremental number of instances is selected according to the size of the dataset. Concretely, we select the number of instances from 500 to all with the incremental number being 1,000 for the SEED dataset, from 100 to all with the incremental number being 200 for the SEED-IV dataset, and from 500 to all with the incremental number being 500 for the self-collected dataset. The experimental results are shown in Figures 6-8, respectively.

In Figures 6-8, the horizontal axis corresponds to the number of selected instances k for each emotion category. The box denotes the accuracy under different values of k corresponding to the left vertical axis, and the blue line represents the computing time of model under different values of k corresponding to the right vertical axis. For the term of computing time, it is increased as the number of selected instances increases, as shown in the blue line. For the term of accuracy, it can be seen from Figures 6-8 that, it is not the more the number of instances is, the higher the recognition accuracy is. The accuracy can be rapidly improved before the number of selected instances reaches a certain value, which indicates that the selected instances during this period is informative and can be well-transferred to improve the generalization ability of the model. However, when the number of the instance exceeds this certain value, the improvement of the accuracy slows down until reaching the maximum value, tending to be flat and decreasing. This indicates that too many instances selected will also cause data redundancy and lead to negative transfer, degrading the performance of the model. It can be seen from Figures 6-8 that on the SEED, SEED-IV, and the dataset collected by ourselves, the selection number k in Equation (3) is, respectively, chosen to be 6,500, 1,100, and 5,500 for the instances with different emotion categories to train the model, which can obtain higher accuracy and lower standard deviation with 86.78±6.65%, 82.55±8.48%, and 77.68±5.13% in computing time of 7, 4, and 10s, respectively.

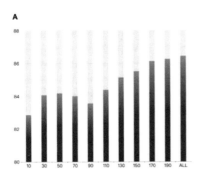 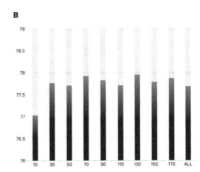

Fig. 10 The recognition accuracies corresponding to different numbers of selected instances used to calibrate the model on (A) SEED and (B) self-collected datasets.

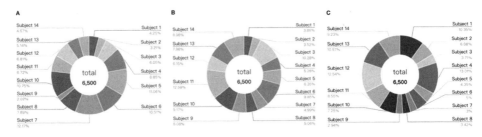

Scan the QR code to see colorful figures

Fig. 11 Distribution of the selected instances from the subjects on the SEED: (A) positive, (B) neutral, and (C) negative.

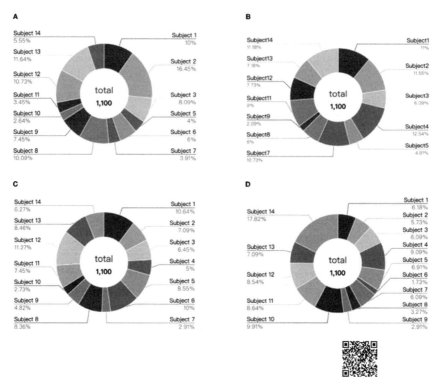

Scan the QR code to see colorful figures

Fig. 12 Distribution of the selected instances from the subjects on the SEED-IV: (A) happy, (B) sad, (C) neutral, and (D) fear.

5.2.2 Ablation Experiments

As illustrated in Fig. 2, the key components of the proposed SSTM-IS are instance selection and SSTM. To evaluate the effects of these two components on the SSTM-IS, we remove one at a time and evaluate the performance of the ablated models. The experimental results of the proposed SSTM-IS and two ablated models are summarized in Table I. Here, the model of SSTM indicates to use all the source domain data for transfer learning without the component of instance selection. And the model of instance-sel represents to use the selected instances to perform the classification directly without transfer learning. It can be seen from Table I that on the SEED, SEED-IV, and self-collected datasets, the proposed SSTM-IS achieves the best accuracies of 86.78, 82.55, and 77.68%, the model of SSTM obtains the accuracies of 84.02, 81.66, and 76.96%, and the model of instance-sel gets the accuracies of 79.66, 64.22, and 70.31%, respectively. This indicates by jointly using two components of SSTM and instance selection, the proposed SSTM-IS algorithm can achieve the best performance on the SEED, SEED-IV, and self-collected datasets. In addition, the component of SSTM has a stronger impact on the performance compared with the instance

SSTM–IS: Simplified STM Method Based on Instance Selection for Real–Time EEG Emotion Recognition

selection, verifying the effectiveness of transfer learning. Moreover, the impact of instance selection on SEED is greater than that on SEED-IV. This may be because the SEED has a large amount of data and more redundancy in source domain, and thus the instance selection is more effective than that on the SEED-IV.

5.2.3 Comparison with Other Methods

In order to prove the effectiveness of the proposed algorithm, we compare several representative algorithms with our method on the public datasets of SEED and SEED-IV as well as the dataset collected by ourselves. The experimental results are shown in Table I in terms of accuracy and runtime. Among these representative methods compared in Table I, MS-MDA, PPDA, and DResNet are incorporated as the benchmark algorithms by using deep learning. It should be noted that in Table I, the results of MS-MDA are the reproductions with the open source codes on the SEED dataset. The accuracies and runtime of TCA, CORAL, MS-STM, and FOIT on SEED and SEED-IV are referenced from the results given in). In addition, the accuracies and runtime of TCA, CORAL, MS-MDA, and MS-STM on the self-collected dataset are the reproductions with their open source codes. All the experiments are implemented by using Python and a GPU of NVIDIA GeForce GTX 1080.

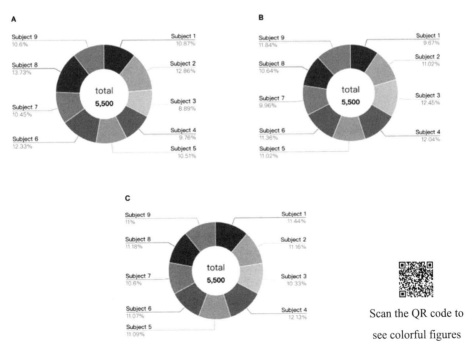

Scan the QR code to see colorful figures

Fig. 13 Distribution of the selected instances from the subjects on the self-collected dataset: (A) positive, (B) neutral, and (C) negative.

It can be seen from Table I that compared with the existing representative EEG emotion recognition algorithms, the proposed SSTM-IS algorithm can achieve an accuracy of 86.78±6.65% on the SEED dataset, which is 0.09% higher than the current best performance PPDA algorithm with the standard deviation reduced by 6.34%. On the SEED-IV dataset, the accuracy of our method reaches 82.55±8.48%, which is 2.83% higher than that of MS-STM with the standard deviation reduced by 14.60%. On the self-collected dataset, the accuracy of our method reaches 77.68±5.13%, which is 14.52% higher than that of MS-STM. On the other hand, the proposed method uses a small amount of target domain data to train the model for a new subject, and the runtime of model training is 7, 4, and 10s on the SEED, SEED-IV, and self-collected datasets, respectively. Once the model training is completed, our algorithm only needs several milliseconds to recognize the emotion state of un-labeled EEG samples. However, the existing deep learning methods use back propagation as the optimization strategy and need a long time to train the model for a new subject. Specifically, it can also be seen from Table I that the runtime of the proposed SSTM-IS algorithm is far less than that of the deep learning method of MS-MDA on the SEED dataset (7 vs. 1959s) as well as the self-collected dataset (10 vs. 347s). This indicates that our method can quickly calibrate the model suitable for new subjects, greatly shortening the runtime and more suitable for the real-time EEG emotion recognition system.

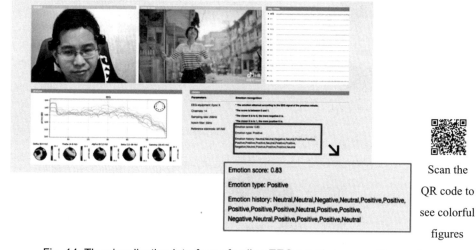

Scan the QR code to see colorful figures

Fig. 14 The visualization interface of online EEG emotion recognition system.

5.2.4 Additional Evaluations

To analyze the recognition ability of the proposed algorithm for different emotion categories, the confusion matrix of predictions made on the SEED, SEED-IV, and self-collected datasets are shown in Fig. 9. It can be seen from Fig. 9A that on the SEED dataset, with the instance number being 6,500 for different emotion categories, the projection effects from source to target show differences between the different emotion recognition. Specifically, our method achieves the recognition accuracies of 78.22, 85.76, and 95.19% for the emotion categories of positive, neutral, and negative, respectively. This difference is also shown on the SEED-IV and self-collected datasets. For the SEED-IV dataset as shown in Fig. 9B, our method can obtain the recognition accuracies of 89.36, 80.49, 66.67, and 72.50%, respectively for the emotion categories of happiness, sadness, fear and neutral. And for the self-collected dataset as shown in Fig. 9C, our method can obtain the recognition accuracies of 77.06, 82.83, and 77.69%, respectively for the emotion categories of positive, neutral, and negative. The results on the SEED, SEED-IV, and self-collected datasets indicate that the proposed algorithm has strong discriminative capability for different emotion categories.

In addition, the proposed SSTM-IS algorithm needs a small amount of the labeled target domain data for supervised learning when training the model. In the previous experiments, we used the EEG data collected under the first three stimulus videos for the supervised learning. However, in the real-time emotion recognition applications, the less supervised data needed means the better experience for user. Therefore, we explore the impact of the number of selected instances used for supervised learning on the recognition accuracy by adding the instances collected within 20s successively on the SEED and self collected datasets. The experimental results are shown in Fig. 10. It can be seen from Fig. 10 that, the recognition accuracy increases with the number of instances supervised until it tends to be flat and fluctuates in a small range. This gives us the insight, that is, a balance between the accuracy and user experience can be found in practice by selecting the number of instances where the accuracy levels off, reducing the amount of supervised data in the target domain to calibrate the model.

Considering the experimental results shown in Section 5.2.1, where the higher accuracy and lower deviation can be obtained for the selected instances of 6,500, 1,100, and 5,500 from the source domains of the SEED, SEED-IV, and self-collected datasets, respectively, here we further explore the data distributions of these selected instances in the source domain. The visualization results are presented in Figures 11-13, respectively.

We can find from Figures 11-13 that for each of the three datasets, the selected instances come from almost all the subjects of the source domain. With the number of source domain instances being the same for each emotion category, the numbers of selected instances vary from subject to subject in the source domain between the different emotion recognition. This demonstrates that the instances from subjects contribute differently to the construction of emotion recognition model on the target domain. The proposed strategy of instance selection is capable of obtaining the most informative samples from different subjects in the source domain for transfer learning on the target domain, which increases the generalizability of the model and reduces the negative transfer caused by data redundancy effectively.

5.3 Online Experiments

In order to verify the reliability of our algorithm in the actual scenes, we use the real-time EEG emotion recognition system developed in Section 4 to conduct the real-time emotion recognition experiment for practical application. For the experimental paradigm in the online experiment, the experimental process is similar to that on the self-collected dataset. It should be noted that, the first three sessions of EEG data are collected from the new subject as the labeled target domain data to calibrate the model by the server. After the model calibrating is completed, the emotion recognition results will be displayed in real time for the experimenters.

For the actual online situation as shown in Fig. 3, we preprocess the raw data within every segment of 10s and extract the DE features to put into the trained model for real-time emotion recognition. In the experiment, we divide the emotion states into three categories: positive, neutral, and negative. The visualization interface of the realistic experiment is shown in Fig. 14. Here, the three windows in the first row are used for monitoring the subject state, stimulus display and real-time raw EEG signals. The window on the left of the second row shows the spectrum diagram and topographical maps of EEG signals, which are used to visualize the changes in frequency characteristics of EEG signals. And the window on the right shows the real-time and historical emotion recognition results. In the actual experiment, when the model training is completed, the proposed system outputs an emotion prediction for every 10s EEG data segment. The real-time system developed above implements the transfer learning algorithm by using a small amount of data for model calibrating, and it has also been verified in the real-time emotion recognition applications in real scene. In addition, it should be explained that we think it is inaccurate to take the label of the whole stimulating video as the ground truth for every 10s data segment, and thus we did not conduct quantitative analysis on the online accuracy. It can be inferred that the emotion varies sometimes very fast and is effected by many factors such as the individual difference and induced effect of stimulating video, the online accuracy will be lower than the experimental results on the offline dataset.

6. Discussion and Conclusion

The above experiments indicate that the proposed SSTM-IS algorithm brings a solid improvement in the accuracy and computing time for EEG emotion recognition. The strategy of instance selection is used to obtain the informative data samples from the source domain, which shortens the computing time for training the model. At the same time, the simplified STM method can further reduce the time cost without decreasing the accuracy. The offline experiments on the public and self-collected datasets show that the proposed SSTM-IS has improved the accuracy with less time cost compared with the representative methods. Specifically, the accuracies on the SEED, SEED-IV, and self-collected datasets have been improved to 86.78±6.65%, 82.55±8.48%, and 77.68±5.13%, respectively, and the computing time has been shortened to 7, 4, and 10s. In addition, we also develop a real-time EEG emotion recognition system to carry out the actual online experiments. The online experimental results demonstrate that, the designed system with the proposed SSTM-IS can provide a practically feasible solution for the actual applications of aBCIs.

There are also some tips for future work. First, the proposed algorithm still needs to collect a small amount of data from new individuals for model calibration, which reduces the user experience to a certain extent. In the future, the real-time emotion recognition algorithms can be explored without calibration. Second, the data of new subjects can be incorporated selectively into the source domain to optimize the data composition in realistic scenarios. Besides, the developed system also reserves the acquisition interfaces of other physiological signals to facilitate the subsequent integration of multiple physiological signals to further improve the system performance.

Data availability statement

The raw data supporting the conclusions of this article will be made available by the authors, without undue reservation.

Ethics statement

The studies involving human participants were reviewed and approved by Research Project Ethical Review Application, State Key Laboratory of Media Convergence and Communication, Communication University of China. The patients/participants provided their written informed consent to participate in this study. Written informed consent was obtained from the individual(s) for the publication of any potentially identifiable images or data included in this article.

Author contributions

SR and WZ: conceptualization and validation. SR: methodology and writing-original draft preparation. SR and DD: software and formal analysis. LY: investigation and visualization. DD: data curation. WZ: writing-review and editing. QZ: funding acquisition and supervision. All authors have read and agreed to the published version of the manuscript.

Funding

This work was supported by the National Natural Science Foundation of China under Grant No. 62271455 and the Fundamental Research Funds for the Central Universities under Grant No. CUC18LG024.

Conflict of interest

The authors declare that the research was conducted in the absence of any commercial or financial relationships that could be construed as a potential conflict of interest.

Publisher's note

All claims expressed in this article are solely those of the authors and do not necessarily represent those of their affiliated organizations, or those of the publisher, the editors and the reviewers. Any product that may be evaluated in this article, or claim that may be made by its manufacturer, is not guaranteed or endorsed by the publisher.

References

[1] ACCORDINO R, COMER R, HELLER W B. Searching for music's potential: a critical examination of research on music therapy with individuals with autism [J]. Research in autism spectrum disorders, 2007, 1(1): 101-115.

[2] ALTAMIRANO ASHER WEISS K, CONCATTO F, CELESTE GHIZONI TEIVE R, et al. On-line recognition of emotions via electroencephalography [J]. IEEE latin america transactions, 2022, 20(5): 806-812.

[3] CHAI X, WANG Q, ZHAO Y, et al. A fast, efficient domain adaptation technique for cross-domain electroencephalography(EEG)-based emotion recognition [J]. Sensors, 2017, 17(5): 1014.

[4] CHEN H, HE H, CAI T, et al. Enhancing EEG-based emotion recognition with fast online instance transfer[C]//Internet of Things. Springer International Publishing. 2022: 141-60.10.1007/978-3-030-91181-2_9.

[5] CHEN H, JIN M, LI Z, et al. MS-MDA: multisource marginal distribution adaptation for cross-subject and cross-session EEG emotion recognition[J]. Frontiers in neuroscience, 2021, 15: 778488.

[6] DONG M, YUHUAN Z, HONGZHI Q, et al. Study on EEG-based mouse system by using brain-computer interface[C]//2009 IEEE International Conference on Virtual Environments, Human-Computer Interfaces and Measurements Systems. IEEE. 2009.10.1109/vecims.2009.5068900.

[7] DUAN R N, ZHU J Y, LU B L. Differential entropy feature for EEG-based emotion classification [C]//2013 6th International IEEE/EMBS Conference on Neural Engineering (NER). IEEE. 2013.10.1109/ner.2013.6695876.

[8] GAO Z, LI S, CAI Q, et al. Relative wavelet entropy complex network for improving EEG-based fatigue driving classification[J]. IEEE transactions on instrumentation and measurement, 2019, 68(7): 2491-2497.

[9] HOSSAIN I, KHOSRAVI A, HETTIARACHCHI I, et al. Multiclass informative instance transfer learning framework for motor imagery-based brain-computer interface[J]. Computational intelligence and neuroscience, 2018, 2018: 1-12.

[10] JAYARAM V, ALAMGIR M, ALTUN Y, et al. Transfer learning in brain-computer interfaces[J]. IEEE computational intelligence magazine, 2016, 11(1): 20-31.

[11] LEE J, HWANG J Y, PARK S M, et al. Differential resting-state EEG patterns associated with comorbid depression in Internet addiction[J]. Progress in neuro-psychopharmacology and biological psychiatry, 2014, 50: 21-26.

[12] LI J, QIU S, DU C, et al. Domain adaptation for EEG emotion recognition based on latent representation similarity[J]. IEEE transactions on cognitive and developmental systems, 2020, 12(2): 344-353.

[13] LI J, QIU S, SHEN Y Y, et al. Multisource transfer learning for cross-subject EEG emotion recognition[J]. IEEE transactions on cybernetics, 2019: 1-13.

[14] LI Y, ZHENG W, ZONG Y, et al. A Bi-Hemisphere domain adversarial neural network model for EEG emotion recognition[J]. IEEE transactions on affective computing, 2021, 12(2): 494-504.

[15] LI-CHEN S, YING-YING J, BAO-LIANG L. Differential entropy feature for EEG-based vigilance estimation[C]//2013 35th Annual International Conference of the IEEE Engineering in Medicine and Biology Society (EMBC). IEEE. 2013.10.1109/embc.2013.6611075.

[16] LUNESKI A, KONSTANTINIDIS E, BAMIDIS P D. Affective medicine[J]. Methods of information in medicine, 2010, 49(03): 207-218.

[17] MA B Q, LI H, ZHENG W L, et al. Reducing the subject variability of EEG signals with adversarial domain generalization[C]//Neural Information Processing. Springer International Publishing. 2019: 30-42.10.1007/978-3-030-36708-4_3.

[18] MU W, LU B L. Examining four experimental paradigms for EEG-based sleep quality evaluation with domain adaptation[C]//2020 42nd Annual International Conference of the IEEE Engineering in Medicine Biology Society (EMBC). IEEE. 2020.10.1109/embc44109.2020.9176055

[19] MURIAS M, WEBB S J, GREENSON J, et al. Resting state cortical connectivity reflected in EEG coherence in individuals with autism[J]. Biological psychiatry, 2007, 62(3): 270-273.

[20] NANDI A, XHAFA F, SUBIRATS L, et al. Real-time emotion classification using EEG data stream in E-Learning contexts[J]. Sensors, 2021, 21(5): 1589.

[21] PAN S J, TSANG I W, KWOK J T, et al. Domain adaptation via transfer component analysis[J]. IEEE transactions on neural networks, 2011, 22(2): 199-210.

[22] PARK J-S, KIM J-H, OH Y-H. Feature vector classification based speech emotion recognition for service robots[J]. IEEE transactions on consumer electronics, 2009, 55(3): 1590-1596.

[23] SCHALK G, BRUNNER P, GERHARDT L A, et al. Brain–computer interfaces (BCIs): detection instead of classification[J]. Journal of neuroscience methods, 2008, 167(1): 51-62.

[24] SUN B, SAENKO K. Deep CORAL: Correlation alignment for deep domain adaptation[C]//Lecture Notes in Computer Science. Springer International Publishing. 2016: 443-50.10.1007/978-3-319-49409-8_35.

[25] WALSH P. The clinical role of evoked potentials[J]. Journal of neurology, neurosurgery and

psychiatry, 2005, 76(suppl_2): ii16-ii22.

[26] WEILONG Z, BAOLIANG L. Investigating critical frequency bands and channels for EEG-based emotion recognition with deep neural networks[J]. IEEE transactions on autonomous mental development, 2015, 7(3): 162-175.

[27] WU D, XU Y, LU B L. Transfer learning for EEG-based brain–computer interfaces: a review of progress made since 2016[J]. IEEE transactions on cognitive and developmental systems, 2022, 14(1): 4-19.

[28] WU W, SUN W, WU Q M J, et al. Multimodal vigilance estimation using deep learning[J]. IEEE transactions on cybernetics, 2022, 52(5): 3097-3110.

[29] YAO Y, DORETTO G. Boosting for transfer learning with multiple sources[C]//2010 IEEE computer society conference on computer vision and pattern recognition. IEEE. 2010.10.1109/cvpr.2010.5539857.

[30] ZHAO L M, YAN X, LU B L. Plug-and-play domain adaptation for cross-subject EEG-based emotion recognition[J]. Proceedings of the aaai conference on artificial intelligence, 2021, 35(1): 863-870.

[31] ZHENG W L, LIU W, LU Y, et al. EmotionMeter: a multimodal framework for recognizing human emotions[J]. IEEE transactions on cybernetics, 2019, 49(3): 1110-1122.

Multi-Source Information-Shared Domain Adaptation for EEG Emotion Recognition*

1. Introduction

In recent years, EEG-based emotion recognition has aroused the interest of researchers by its advantages in real-time with a high time resolution and stable patterns over time. The field of affective Brain-Computer Interface (aBCI) refers to detect people's emotional state through EEG signals, which could contribute to detecting mental disorders and objective evaluation of depression. Due to the reason of head size, body states and experimental environment, the structural and functional variability of brain can be different across subjects, and thus there are significant individual differences between their EEG signals. The traditional machine-learning algorithms train an EEG-based recognition network with data from some subjects. However, the trained network can not generalize well to the new subject because of the domain shift problem. To resolve this problem, the conventional operation is to collect the labeled data of new subject and finetune the network parameters for calibration. But it always provides an unfriendly experience for users and is less convenient.

With the development of deep-learning technologies, the researchers attempt to use the transfer learning methods to deal with the problem caused by non-stationary nature and inter-subject structural variability of EEG signals in practical aBCI applications. The transfer learning can build the bridge between the source domain and unlabeled target domain through finding the similarity between them. The existing transfer learning methods can be mainly classified into two categories, domain adaption (DA) and domain generalization (DG). The DG methods appeal to extract the domain-invariant feature, however, without the access to target domain data, it is hard to train a network generalized well to new subject. In contrast, the DA methods use target domain information in the training phase

---

* The paper was originally published in *Lecture Notes in Computer Science*, 2022, 13535, and has since been revised with new information. It was co-authored by Ming Gong, Wei Zhong, Jiayu Hu, Long Ye, and Qin Zhang.

to solve the domain shift problem, mainly by the ways of feature transformation, instance-based and model pretraining. When applying the DA methods to the task of cross-subject EEG emotion recognition, the existing works either regard all source domains as the same or train source-specific learners directly, ignoring the relationship between individuality and commonality in different domains. Therefore, it is necessary to consider the individual differences and group commonality together among the multi-source domains, further improving the recognition performance.

In this paper, we propose a multi-source information-shared domain adaptation network for cross-subject EEG emotion recognition. In the proposed network, we firstly extract the domain-specific and domain-shared features to represent the individuality and commonality of EEG signals from different domains, respectively. And then in the training phase, besides using the maximum mean discrepancy (MMD) to make the marginal distribution between source and target domains closer, as well as the diff-loss to improve the astringency of network, another loss is also proposed to narrow the distance between target private domains, which enhances the mapping capability of private extractors by considering the information of other private domains. Furthermore, instead of artificially adjusting the weights of classifiers by experience, we integrate the outputs of classifiers by distribution distance between the source domains, realizing the adjustment of network optimization dynamically. The experimental results on SEED and SEED-IV datasets show that the performance of the proposed network outperforms the state-of-the-art domain adaptation methods. The contributions of this paper can be summarized as follows:

— We propose an efficient EEG emotional recognition network by integrating individual differences and group commonality of multi-source domains.
— The was-loss is proposed to make the marginal distribution of domain-specific features closer among target private domains. Conjuncted with the diff-loss, the domain adaptation ability of network can be enhanced.
— We propose a dynamic weighting strategy based on the MMD distribution distance between source private domains and shared domain to adjust the optimization of network.

2. Related Work

With the rapid developments of deep learning, the transfer learning has become the mainstream method in the field of EEG-based emotion recognition. There are two main branches in transfer learning to reduce inter-subject variability, DA and DG. The goal of

DG is to learn a model from several source domains that generalizes to the unseen target domains through data manipulation, representation learning and learning strategy. The scatter component analysis (SCA) extended the kernel fisher discriminant by amplifying the information from the same category and domain, while dampening from different category and domain. Ma et al. proposed a domain residual network (DResNet) by improving DG-DANN with ResNet. Since the DG methods do not utilize target domain information in training phase, they hardly achieve high recognition accuracy like DA methods.

In contrast, the DA methods use target domain information to transfer knowledge by minimizing the domain shifts between the source and target domains. Zheng et al. applied transfer component analysis (TCA) and transductive parameter transfer (TPT) to the cross-subject EEG emotion recognition on the SEED dataset. Li et al. provided another method of domain-adversarial neural networks (DANN) with adversarial training of feature extractor and domain classifier. Luo et al. proposed the Wasserstein GAN domain adaptation network (WGANDA) by using the gradient penalty to alleviate domain shift problem. Zhao et al. developed a plug-and-play Domain Adaptation (PPDA) network, which tried to disentangle the emotion information by considering the domain-specific and domain-invariant information simultaneously. Hao et al. took the source data with different marginal distributions into account and proposed a multi-source EEG-based emotion recognition network (MEERNet). Meanwhile, they used the disc-loss to improve domain adaptation ability and proposed a multi-source marginal distribution adaptation (MS-MDA) network for cross-subject and cross-session EEG Emotion recognition. It can be seen from above analysis, most of the DA methods either regard all source domains as the same or train source-specific learners directly, which may make the network learn unreasonable transfer knowledge, resulting in negative transfer. In this paper, we consider not only the domain-specific individuality and domain-shared commonality among different domains but also the relationship between them, further improving the recognition performance.

3. Methods

In this section, we propose a multi-source information-shared domain adaptation network for cross-subject EEG emotion recognition, which is defined as MISDA.

3.1 Framework

The overall framework of our proposed MISDA network is shown in Fig. 1, which consists of five components including common extractors, private extractors, shared extractor, private classifiers and shared classifier. To get a reliable network trained on the multiple source domain data to predict emotion state from newly collected data, we extract different domain-specific features and domain-invariant features for emotion recognition.

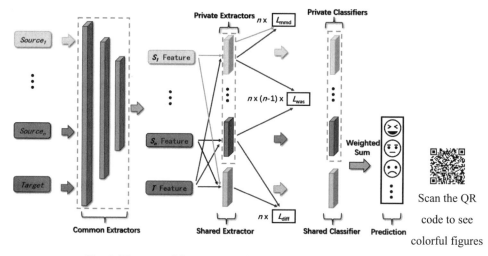

Fig. 1 The overall framework of our proposed MISDA network.

Assume that X_S is the source data matrices and Y_S is their labels, and X_T is the target unlabeled data matrices,

$$X_S = \{X_S^i\}_{i=1}^n, Y_S = \{Y_S^i\}_{i=1}^n \tag{1}$$

where n represents the numbers of subjects in source datasets. The common extractor EC maps the source data matrices X_S and target data matrices X_T to the low-level feature space,

$$X_S' = \{X_S'^i\}_{i=1}^n = \{E_C(X_S^i)\}_{i=1}^n, X_T' = E_C(X_T) \tag{2}$$

Then for low-level features of each source domain, we build a private extractor to get its source domain-specific features. And for those of target domain, the domain-specific features are extracted by n private extractors,

$$F_{SP}^i = E_P^i(X_S'^i), \quad F_{TP}^i = E_P^i(X_T'), i = 1,2,\ldots,n \tag{3}$$

In the parallel pipeline, the shared extractor extracts shared deep features from all domains,

$$F_{SS} = \{E_S(X_S'^i)\}_{i=1}^n, F_{TS} = E_S(X_T') \tag{4}$$

Subsequently, the private classifier C_P^i and shared classifier C_S take (F_{SP}^i, F_{TP}^i) and (F_{SS}, F_{TS}) as inputs and output emotional prediction ($\hat{Y}_{SP}^i$, $\hat{Y}_{TP}^i$) and ($\hat{Y}_{SS}$, $\hat{Y}_{TS}$), respectively,

$$\hat{Y}_{SP}^i = C_P^i(F_{SP}^i), \hat{Y}_{TP}^i = C_P^i(F_{TP}^i), i = 1,2,\ldots,n$$

$$\hat{Y}_{SS} = C_S(F_{SS}), \hat{Y}_{TS} = C_S(F_{TS}). \tag{5}$$

Finally, $\hat{Y}_S$ is the weighted sum by $\hat{Y}_{SP}^i$ and $\hat{Y}_{SS}$, $\hat{Y}_T$ by $\hat{Y}_{TP}^i$ and $\hat{Y}_{TS}$, respectively. We use $\hat{Y}_S$ the to calculate the classification loss, and $\hat{Y}_T$ to predict emotion category.

3.2 Modules

In this subsection, we describe the following modules included in the proposed framework in detail.

Common Extractor. Despite considering the individual differences on EEG signals, there still exist some common low-level features resulted by human brain activities and signal characteristics. And thus, we assume that the EEG signals from different subjects share the same shallow features. Similar to the pipelines of MEER-Net and MS-MDA, a common extractor is used to map all domain data into a common latent space, extracting the low-level features.

Private Extractors and Shared Extractor. Considering the difference among domains, we set up a fully connected layer for each source domain to map the data from common feature space to latent domain-specific feature space, capturing the individual domain information. Here we choose MMD measure the marginal distribution distance between the source and target domains in the reproducing kernel Hilbert space H,

$$MMD(F_{SP}, F_{TP}) = \frac{1}{n} \sum_{i=1}^{n} \left\| \left(\psi(F_{SP}^i) - \psi(F_{TP}^i) \right) \right\|_H \tag{6}$$

where ψ denotes the mapping function.

In order to make the marginal distributions of domain-specific features closer among target private domains, another metric named L_{was} is also introduced, which is inspired by the Wasserstein loss,

$$L_{was} = \sum_{i=1}^{n} \sum_{j=1, j \neq i}^{n} \left| \bar{F}_{TP}^i - \bar{F}_{TP}^j \right| \tag{7}$$

where $\bar{F}_{TP}^i$ and $\bar{F}_{TP}^j$ denote the mean vector across feature dimensions from the domain-specific features extracted by the i-th and j-th private extractors, respectively. The L_{was} aligns the target private features inside and provides a closer private domain, which contributes to extracting the shared domain.

The shared extractor maps the low-level feature space to the shared one. Inspired by the DG disentanglement, we continue to extract shared information from the low-level features. To balance the learning ability between private and shared extractors, the shared extractor is built with the same structure as the private one. Moreover, the difference loss L_{diff} is applied to encourage the shared and private extractors to extract different information of low-level features as

$$L_{diff} = \left| \bar{F}_{TP} - \bar{F}_{TS} \right|, i = 1, 2, \ldots, n. \tag{8}$$

It is worth mentioning that, since all source domains are linearly independent, the shared domain can be identified by the conjunction of L_{diff} and L_{was} in each iteration. And thus the shared extractor could extract the closer domain-invariant features from different domains, which enhances the domain adaptation ability of the network.

Private Classifiers and Shared Classifier. Following the private extractors, the private classifiers use the domain-specific features to predict emotion states. The Softmax activate function is adopted after the fully connected layer corresponding to each source domain. Similarly, the shared classifier also has the same structure as the private ones in order to balance the classification ability between them. In the training process, we choose cross-entropy to measure the classification loss of private and shared classifiers as,

$$L_{cl_p} = -Y_S^i \log \hat{Y}_{SP}^i, L_{cl_s} = -Y_S^i \log \hat{Y}_{SS}^i \tag{9}$$

where Y_S^i is the emotion label of i-th source domain

Weight Sum. For the sake of dynamically adjusting the optimization process and balancing the weight of private and shared networks, we also propose a weight sum strategy according to the similarity between the private and shared domains. In the training process, we integrate the private and shared classifiers by calculating the MMD distance between the source private and shared domains

$$\omega_S = softmax\left(\left\|\psi\left(F_{SP}^i\right) - \psi\left(F_{SS}\right)\right\|_H\right) \tag{10}$$

Then the classification loss is calculated by,

$$L_{cl} = L_{cl_p} + \omega_S * L_{cl_s} \tag{11}$$

The higher ω_S indicates that the distribution of shared domain is more similar to those of private ones, and thus more trust can be given to the shared classifier.

With L_{mmd}, L_{was}, L_{diff} and L_{cl} given above, the final loss can be represented as,

$$L = L_{cl} + \alpha L_{mmd} + \beta L_{wass} + \gamma L_{diff} \tag{12}$$

where α, β and γ are the hyper-parameters.

In the prediction phase, the proposed MISDA network predicts the final results by integrating the predictions of private and shared domains,

$$\hat{Y}_T = \frac{1}{n}\sum_{i=1}^{n}\left(C_P^i\left(F_{TP}^i\right) + \omega_T^i * C_S\left(F_{TS}\right)\right) \tag{13}$$

where ω_T^i is calculated by the MMD distance between the i-th target private and shared domains.

In summary, the work flow of the proposed MISDA framework is given as follows.

Algorithm 1: The work flow of the proposed MISDA framework

Input: Source domain dataset $X_S = \{X_S^i\}_{i=1}^n$ and the labels $Y_S = \{Y_S^i\}_{i=1}^n$, target domain dataset X_T
Output: Predicted target domain emotional state $\hat{Y}_T$

1. Random initialize $E_P^{1\sim n}, C_P^{1\sim n}, E_C, E_S$ and C_S
2. **Training phase:**
3. **for** Epochs **do**
4.   sources = 1 : n
5.   **for** sources **do**
6. $E_C(X_S) \to X'_S, E_C(X_T) \to X'_T$
7. $E_P(X'_S) \to F_{SP}, E_P(X'_T) \to F_{TP}$
8. $(F_{SP}^{1\sim n}, F_{TP}^{1\sim n}) \to (6), (F_{TP}^i, F_{TP}^j)^{1\sim n} \to (7)$
9.   **Parallel:**
10. $E_S(X'_S) \to F_{SS}, E_S(X'_T) \to F_{TS}$
11. $(F_{TP}^{1\sim n}, F_{TS}) \to (8), (F_{SP}^{1\sim n}, F_{SS}) \to (10)$
12. $C_P^{1\sim n}(F_{SP}^{1\sim n}), C_S(F_{SS}) \to (9)(11)$
13.   Update network by minimizing the total loss $\to$ (12)
14.   **end**
15. **end**
16. **Test phase:**
17. $E_P^{1\sim n}(E_C(X_T)) \to F_{TP}^{1\sim n}, E_S(E_C(X_T)) \to F_{TS}$
18. Predict target labels $(F_{TP}^{1\sim n}, F_{TS}) \to (13)$
19. **Return:** $\hat{Y}_T$

4. Experiments

4.1 Dataset and Preprocessing

We evaluate the proposed network on SEED and SEED-IV, which are commonly-used public datasets for emotion recognition based on EEG. For the datasets of SEED and SEED-IV, 15 Chinese participants were required to watch 15 film clips to elicit their emotion. The raw data are recorded at a sampling rate of 1000Hz with an international 10-20 system using the ESI NeuroScan System by a 62-channel electrode cap. The preprocessed data is downsampled 200Hz and filtered with a bandpass filter of 0-75Hz to eliminate the interference of environment. The differential entropy (DE) features are then extracted with anon-overlapping one-second Hanning window from five frequency bands, δ: 1-3Hz, θ: 4-7Hz, α: 8-13Hz, β: 14-30Hz, γ: 31-50Hz. The number of samples for each subject is 3394, and each sample has the feature dimensions of 310 (62 channels multiplied by five frequency bands). The SEED dataset includes three emotion categories, which are positive, neutral and negative. While the emotion categories in the dataset of SEED-IV are neutral, sad, fear and happy.

4.2 Implementation Details

To examine the effectiveness of handling the cross-subject emotion recognition problem and fairly compare with other research results, we evaluate the proposed network with Leave-One-Subject-Out strategy. Specifically in each epoch, we select one subject as the target one with unlabeled data and the other fourteen subjects as the source subjects.

The common extractor is a 3-layer multilayer perceptron with nodes 310-256-128-64 to extract the low-level features. We choose one linear layer for each private extractor to reduce the feature dimensions from 64 to 32 followed by a 1-D BatchNorm layer. Each private classifiers a single fully-connected network with a hidden dimension from 32 to the number of emotion categories. The shared extractor and shared classifier have the same structure as the private extractors and private classifiers, respectively. The LeakyRelu activation function is used in all hidden layers. In addition, we also normalize all the data by the electrode-wise method used in reference [20] to get better performance.

For the hyper-parameters in (12), considering the trade-off among the constituent losses, we set $\alpha = \frac{2}{1+e^{-10 \times i / \text{iteration}}} - 1$ and $\beta = \gamma = \frac{\alpha}{100}$. The different values of learning rate and batch size are compared on the SEED dataset in Fig.2. It can be seen from Fig.2 that with the learning rate being 0.01 and batch size 32, our model achieves the highest accuracy. Furthermore, the adam optimizer is applied as the optimizing function and the epoch is set to be 200. The whole framework is implemented by PyTorch.

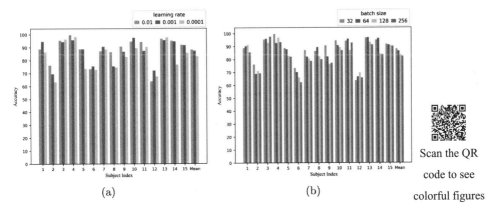

Fig. 2 The hyper-parameter selections on SEED, (a) learning rate and (b) batch size.

Table I Ablation study running on SEED and SEED-IV

Variants	SEED		SEED-IV	
	Mean	Std.	Mean	Std.
MISDA	88.1%	9.5%	73.8%	11.9%
w/o L_{mmd}	82.5%	8.4%	66.3%	12.5%
w/o L_{was}	85.3%	10.5%	70.6%	12.6%
w/o L_{diff}	84.5%	10.9%	70.8%	10.7%
w/o L_{was} and L_{diff}	83.6%	10.3%	68.6%	11.4%
w/o L_{mmd}, L_{was} and L_{diff}	76.1%	9.7%	67.5%	12.8%

w/o L* means the framework trained without the loss L*.

4.3 Ablation Study

To demonstrate the effect of each module in MISDA, we give the performance of ablated framework on the datasets of SEED and SEED-IV as shown in Table I.

Here the recognition performance is measured by the metrics of mean accuracy (Mean) and standard deviation (Std.). It can be seen from Table I that, three loss functions all contribute to improving the recognition performance, and achieve mean accuracies of 88.1% and 73.8% with standard deviations of 9.5% and 11.9% on the SEED and SEED-IV, respectively. In addition, the result without L_{mmd} has a more significant drop compared with those removing one of other two losses, showing the importance of narrowing the distance between source and target domains. Furthermore, it should be noticed that, depriving L_{was} and L_{diff} simultaneously would decrease the performance much more than removing either of them. This indicates by conjuncting L_{was} to provide the closer private domain, the L_{diff} loss could locate the shared domain more compactly, which contributes to strengthening the domain adaptation ability of network.

4.4 Comparisons with Competing Methods

In this section, we demonstrate the effectiveness of our proposed MISDA framework. Table II shows the mean accuracies and standard deviations of MISDA and compares the results with those of competing methods on the EEG cross-subject emotion recognition task of SEED and SEED-IV, respectively. It can be seen from Table II that our MISDA framework outperforms other competing methods in the metric of mean accuracy and achieves an improvement of 1% compared to the algorithms having the best performance at present on the SEED dataset. But the standard deviation of MISDA is relatively high, which may be caused by the exceptional subject with specific domain having less overlap to others. This problem could be alleviated by increasing the number of subjects in dataset, which would help to train more generalized domain-specific extractors. And for the results on SEED-IV, the proposed network also achieves the highest accuracy of 73.8%, indicating the generality of our framework for domain adaptation on EEG emotion recognition.

Table II Performance comparisons with competing methods on SEED and SEED-IV

Methods	SEED		SEED-IV	
	Mean	Std.	Mean	Std.
DResNET	85.3%	8.0%	————	————
DANN	79.2%	13.1%	————	————
DDC	————	————	54.3%	4.2%
DAN	83.8%	8.6%	69.8%	4.2%
WGANDA	87.1%	7.1%	————	————
PPDA	86.7%	7.1%	————	————
MEER-NET	87.1%	2.0%	71.0%	12.1%
MS-MDA	81.6%	9.1%	59.3%	5.5%
MISDA (Ours)	88.1%	9.5%	73.8%	11.9%

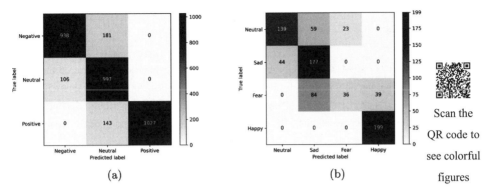

Fig. 3 The confusion matrices of predictions on (a) SEED and (b) SEED-IV, the value of square represents the number of samples.

Further in order to show the recognition ability of MISDA framework among different emotion categories, Fig.3 shows the confusion matrix of predictions made on the SEED and SEED-IV datasets. It can be seen from Fig.3(a) that on the SEED dataset, our network achieves the recognition accuracies of 83.8%, 90.4% and 87.8% respectively on the emotion categories of negative, neutral and positive, showing strong discriminative capability across emotion. The results on the SEED-IV dataset as shown in Fig.3(b) indicate that, our model can obtain decent accuracies for the emotion categories of neutral, sad and happy. While for the category of fear, the proposed MISDA confuses it with sad because these two emotions are relatively similar on EEG signals.

5. Conclusion

In this paper, we propose the MISDA network for cross-subject EEG emotion recognition, which not only divides the EEG representations into private components specific to the subject and shared features universal to all subjects, but also cares the relationship between individual difference and group commonality. In order to narrow the distance among different private domains, the was-loss is also proposed in conjunction with diff-loss, enhancing the domain adaptation ability of the network. Furthermore, we also propose the dynamic weighting strategy to balance the shared and private information, so as to adjust the optimization of network. The performance of our network has been evaluated by conducting Leave-One-Subject-Out cross-validation on the datasets of SEED and SEED-IV, which outperforms competing methods by the mean accuracies of 88.1% and 73.8%, respectively. Our future work will concentrate on reducing the variance by improving the stability of network.

References

[1] BORGWARDT K M, GRETTON A, RASCH M J, et al. Integrating structured biological data by kernel maximum mean discrepancy[J]. Bioinformatics, 2006, 22(14): e49-e57.

[2] CHEN H, JIN M, LI Z, et al. MS-MDA: multisource marginal distribution adaptation for cross-subject and cross-session EEG emotion recognition[J]. Frontiers in neuroscience, 2021, 15: 778488.

[3] CHEN H, LI Z, JIN M, et al. MEERNet: multi-source EEG-based emotion recognition network for generalization across subjects and sessions[C]// 2021 43rd Annual International Conference of the IEEE Engineering in Medicine Biology Society (EMBC). IEEE. 2021.10.1109/embc46164.2021.9630277.

[4] CHEN N, LI C. Hyperspectral image classification approach based on wasserstein generative adversarial networks[C]//2020 International Conference on Machine Learning and Cybernetics (ICMLC). IEEE. 2020.10.1109/icmlc51923.2020.9469586.

[5] GHIFARY M, BALDUZZI D, KLEIJN W B, et al. Scatter component analysis: a unified framework for domain adaptation and domain generalization[J]. IEEE transactions on pattern analysis and machine intelligence, 2017, 39(7): 1414-1430.

[6] HUANG J, SMOLA A J, GRETTON A, et al. Correcting sample selection bias by unlabeled data[C]//Advances in Neural Information Processing Systems 19. The MIT Press. 2007: 601-8.10.7551/mitpress/7503.003.0080.

[7] LI H, JIN Y M, ZHENG W L, et al. Cross-subject emotion recognition using deep adaptation networks[C]//Neural Information Processing. Springer International Publishing. 2018: 403-13.10.1007/978-3-030-04221-9_36.

[8] LUO Y, ZHANG S Y, ZHENG W L, et al. WGAN domain adaptation for EEG-based emotion recognition[C]//Neural Information Processing. Springer International Publishing. 2018: 275-86.10.1007/978-3-030-04221-9_25.

[9] MA B Q, LI H, ZHENG W L, et al. Reducing the subject variability of EEG signals with adversarial domain generalization[C]//Neural Information Processing. Springer International Publishing. 2019: 30-42.10.1007/978-3-030-36708-4_3.

[10] PAN S J, TSANG I W, KWOK J T, et al. Domain adaptation via transfer component analysis[J]. IEEE transactions on neural networks, 2011, 22(2): 199-210.

[11] PFURTSCHELLER G, MULLER-PUTZ G R, SCHERER R, et al. Rehabilitation with brain-computer interface systems[J]. Computer, 2008, 41(10): 58-65.

[12] PUTNAM K M, MCSWEENEY L B. Depressive symptoms and baseline prefrontal EEG alpha activity: a study utilizing ecological momentary assessment[J]. Biological psychology, 2008, 77(2): 237-240.

[13] SAMEK W, MEINECKE F C, MULLER K-R. Transferring subspaces between subjects in brain-computer interfacing[J]. IEEE transactions on biomedical engineering, 2013, 60(8): 2289-2298.

[14] SANEI S, CHAMBERS J A. EEG signal processing[M]. Hoboken, New Jersey, the USA:

Wiley, 2007.

[15] SUGIYAMA M, BLANKERTZ B, KRAULEDAT M, et al. Importance-weighted cross-validation for covariate shift[C]//Lecture notes in computer science. Springer Berlin Heidelberg. 2006: 354-63.10.1007/11861898_36.

[16] WANG J, CHEN Y. Deep transfer learning[C]// Introduction to transfer learning. Springer Nature Singapore. 2022: 141-62.10.1007/978-981-19-7584-4_9.

[17] WANG J, LAN C, LIU C, et al. Generalizing to unseen domains: a survey on domain generalization[C]//Proceedings of the Thirtieth International Joint Conference on Artificial Intelligence. International Joint Conferences on Artificial Intelligence Organization. 2021.10.24963/ijcai.2021/628.

[18] WANG L L, ZHENG W L, HAI W M, et al. Measuring sleep quality from EEG with machine learning approaches[C]//2016 International Joint Conference on Neural Networks (IJCNN). IEEE. 2016.10.1109/ijcnn.2016.7727295.

[19] ZHENG W L, LU B L. Investigating critical frequency bands and channels for EEG-based emotion recognition with deep neural networks[J]. IEEE transactions on autonomous mental development, 2015, 7(3): 162-175.

[20] YANG H, RONG P, SUN G. Subject-independent emotion recognition based on entropy of EEG signals[C]//2021 33rd Chinese Control and Decision Conference (CCDC). IEEE. 2021.10.1109/ccdc52312.2021.9602439.

[21] ZHAO L M, YAN X, LU B L. Plug-and-play domain adaptation for cross-subject EEG-based emotion recognition[J]. Proceedings of the aaai conference on artificial intelligence, 2021, 35(1): 863-870.

[22] ZHENG W L, ZHU J Y, LU B L. Identifying stable patterns over time for emotion recognition from EEG[J]. IEEE transactions on affective computing, 2019, 10(3): 417-429.

[23] ZHOU K, LIU Z, QIAO Y, et al. Domain generalization: a survey[J]. IEEE transactions on pattern analysis and machine intelligence, 2022: 1-20.

Emotional Quality Evaluation for Generated Music Based on Emotion Recognition Model*

1. Introduction

In the theory of emotional music evaluation, empathy is often used as a criterion to evaluate the quality of a piece of music. Particularly for evaluating the quality of emotional music generation, most of the existing methods use the subjective experiments, which require a high level of experimental environment, making it difficult to obtain reliable results. So it is critical to evaluate the emotional quality of generated music from the objective way.

In this paper, we propose to evaluate the quality of emotional music generation from the perspective of emotion recognition. In the proposed method, we firstly use the statistical chi-square test and Pearson correlation coefficient to analyze the correlation between different audio features and emotion categories. And then for feature extraction, the audio features with the most significant correlation are picked out and fused together with the Mel spectrum to send to the constructed network. Finally, we build a residual convolution network to predict the emotion category of the generated music, realizing the purpose of emotion quality evaluation. In the experiments, we evaluate the performance of the proposed emotion recognition model, and apply it to the quality evaluation of emotional music generated by the state-of-the-art methods. The experimental results show that the proposed algorithm can achieve higher recognition accuracy, and the obtained evaluation results are also in accord with the subjective method.

Our contributions are as follows.

- We analyze the correlation between the audio features and their emotional expression of music, and choose the features with the most significant correlation to represent the emotional characteristic of music.

---

* The paper was originally published in *2022 IEEE International Conference on Multimedia and Expo Workshops (ICMEW)*, 2022, and has since been revised with new information. It was co-authored by Hongfei Wang, Wei Zhong, Lin Ma, Long Ye, and Qin Zhang.

- We propose a quality evaluation algorithm from the perspective of music emotional recognition, providing an objective way for evaluating the emotional music generation task.

The remainder of the paper is organized as follows. Section 2 gives a brief review of the related works. The proposed model is formulated in Section 3. In Section 4, the experimental settings, results and discussions are presented. Finally, the conclusions are drawn in Section 5.

2. Related Works

For the task of emotional music generation, the existing methods focus on building the generation network itself, however, the quality of the generated music is mainly evaluated in the subjective way. Among the quality evaluation methods, Madhok used emotional Mean Opinion Score (MOS), which is a common measure for quality evaluation of audio-visual data. Hung rated the emotion categories on a five-point Likert scale. Mao and Ferreira conducted the subjective experiments in Amazon MTurk. These subjective evaluation methods are demanding on the experimental environment and lack of unified objective evaluation standard for quality evaluation. On the other hand, the evaluation performance is influenced by the cultural background and individual differences of the subjects, so it is difficult to obtain reliable results. Considering that the emotional expression of generated music is an important aspect of its quality evaluation, it is necessary to propose an objective evaluation method from the perspective of music emotion recognition.

For the music emotion recognition, Juan used the Mel spectrum as the feature of vocal singing music and performed transfer and multi-task learning to predict classes of music emotion. Won applied the Mel spectrum to the automatic labeling model of music emotion based on the Short Chunk ResNet network. Based on the audio data, Hung selected 20-dimensional Mel-scale Frequency Cepstral Coefficients (MFCC) as the base feature and proposed a 2D-CNN with residual convolution blocks. Zhang fused MFCC and GammaTone filtering features and combined the convolutional long- and short-time neural network and BiLSTM for the model. In addition, Zhang proposed an attention-based joint feature extraction model for high performance predictions.

It should be noticed that, the existing methods for music emotion recognition do not consider the relationship between audio features and expressed emotions. And thus they cannot extract audio features with strong representation and the accuracy of the resulting recognition model is not high enough. Therefore, it is necessary to analyze the emotion characteristic of audio features and propose a music emotion recognition model with higher accuracy, providing a reliable objective way of quality evaluation for emotional music generation tasks.

3. Method

In this paper, we propose an emotional quality evaluation algorithm for generated music based on emotion recognition model. The proposed emotion recognition model is based on residual convolutional network. As shown in Fig. 1, we choose MFCC and Mel spectrum as the most significant audio features, and then send them to our recognition network to predict the emotion of generated music. In the following part, the proposed algorithm is described in detail from the modules of emotional music feature analysis and music emotion recognition model.

3.1 Emotional Music Feature Analysis

For the quality evaluation of generated music, the existing methods do not take into account the relationship between audio features and emotion categories. Our proposed method can extract music features with the most significant correlations and better use the constructed network for learning, thus improving the accuracy of music emotion recognition. To analyze the degree of correlation between different audio features and emotional categories, we choose direct descriptive statistics and hypothesis test. More specifically, the chi-square test can discriminate the correlation between different audio features and emotion labels, and the Pearson correlation coefficient method can be used to analyze the degree of correlation between audio features and emotion categories.

For the feature selection, here we choose eight commonly-used audio features, which are MFCC, spectral centroid (Cent), spectral band-width (Spec bw), spectral contrast (Contrast), zero-crossing rate (ZCR), root mean square energy (RMSE), spectral rolloff (Rolloff) and beats per minute (Tempo). For the features of MFCC, Cent, Spec bw, Contrast, ZCR, RMSE and Rolloff, we calculate the mean, standard deviation and variance as the statistical values for feature analysis. And for the feature of Tempo, their total and average values are calculated. In the process of feature analysis, the chi-square test is firstly used to verify the existence of correlation between the above eight audio features and emotional categories. For some samples in the separated intervals (0, 1] and (1, 2] on the EMOPIA dataset, we calculate the frequencies of the mean values of MFCC as shown in Table I. Here Q1, Q2, Q3 and Q4 are four emotion categories in Russell's emotion model. If there is a correlation between the mean value of MFCC and emotion category, then the statistic value can be calculated based on the frequencies given in Table I. If the statistic value is smaller than the threshold value, then the above correlation assumption holds.

Table I Interval frequencies of mean values of MFCC on some samples of EMOPIA dataset

Emotion category	(0, 1]	(1, 2]	Total/Each
Q1	56	28	94
Q2	34	56	90
Q3	32	27	59
Q4	27	30	57

In order to obtain the correlation degree between the audio feature and emotion category, the Pearson correlation coefficient is used. Specifically, we calculate the Pearson correlation coefficient between the statistical value X_i and the emotion category Y_i for each audio feature separately as follows,

$$r = \frac{1}{n-1} \sum_{i=1}^{n} \left(X_i - \frac{\overline{X}}{\sigma_x} \right) \left(Y_i - \frac{\overline{Y}}{\sigma_y} \right) \qquad (1)$$

where r represents the correlation coefficient, $X_i - \overline{X}/\sigma_x$, $\overline{X}$ and σ_x are the standard score, mean and standard deviation of the sample X_i. Similarly, $Y_i - \overline{Y}/\sigma_y$, $\overline{Y}$ and σ_y are the standard score, mean and standard deviation of the emotion label Y_i. The value of r indicates the degree of correlation. The closer r is to $+1(-1)$, the stronger the positive (negative) linear correlation is.

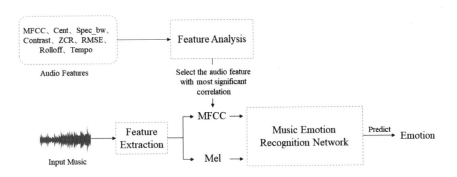

Fig. 1 Framework of the proposed emotional quality evaluation algorithm for generated music

Table II shows the top 8 audio statistical features and their statistical values. It can be seen from Table II that the correlation coefficients of MFCC in terms of standard deviation, mean and variance are all greater than 0.5, which means that the feature of MFCC has a stronger correlation with the emotion category. Therefore in the following music recognition model, we choose MFCC as the most significant audio feature to predict the emotion of generated music.

Table II Top 8 audio statistical features and their statistical values

Audio statistical features	Significant probability	Correlation coefficient	Sum
MFCC–Standard deviation	0.7307	0.6448	1.3755
MFCC–Mean	0.6251	0.6401	1.2591
ZCR–Variance	0.5708	0.4686	1.0394
Cent–Mean	0.4824	0.4604	0.9428
MFCC–Variance	0.3452	0.5905	0.9357
Spec bw–Mean	0.2712	0.6281	0.8993
RMSE–Mean	0.3222	0.5762	0.8984
Spec bw–Standard deviation	0.2597	0.4209	0.6806

3.2 Music Emotion Recognition Model

In order to keep the correlation of the temporal information of audio features between different frames, the audio feature with the most significant correlation is selected for feature fusion with Mel spectrum, and then the residual convolutional network is used for building the music emotion recognition model.

3.2.1 Feature Extraction

Considering the human auditory information transmission and the frequency of sound perception, we select the Mel spectrum as the basic audio feature of the music emotion recognition model.

For the process of feature extraction, we firstly use the Hanning window to cut each music sample into audio clips with a duration of 3s, and set an overlap interval of 1s to keep the audio clips continuous. The window length is set to 1024, the sampling rate is 22050, and the frame shift length is 275. Then, the Fourier transform with 1024 sampling points is used to perform the STFT on the audio fragment after the windowing. Finally, the number of channels of the Mel filter is set to 80, and the Mel spectrum feature vector is obtained with the dimension of [80, 721], where the dimension of 721 represents the frame length after processing. From the analysis in Section 3.1, we can see that the standard deviation of MFCC is the most relevant feature to the emotion category of music, so it is selected with Mel spectrum to fed into the subsequent network as shown in Fig. 1.

3.2.2 Music Emotion Recognition Network

In order to evaluate whether the sentiment representation of the generated music is consistent with the real sentiment, we design a music emotion recognition network as shown in Fig. 2. The proposed network uses a 1D convolution combined with three residual blocks. As shown in Fig. 2, the Mel spectrum and the most significantly correlated audio feature MFCC are selected as the input features. After the processing of three residual blocks, the features fusion is performed in different dense layers. Finally, the emotion category of generated music is predicted by the fully connected layer.

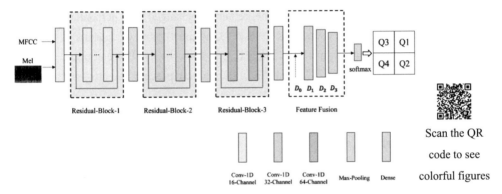

Fig. 2 The music emotion recognition model based on residual convolutional network

In the proposed network shown in Fig. 2, the colors of the 1D convolutional layers from light to dark are used to indicate that the number of channels is changing with a growing trend of base 16 and multiplicity N, where N is taken as 1, 2 and 3. The first 1D convolutional layer is used to keep the input feature vector consistent with the channels of residual blocks. And then the output features are fed sequentially into three residual convolutional blocks, and two adjacent residual blocks are connected by a maxpooling layer. The calculation of the residual convolutional blocks is given as follows,

$$x_{l+1} = h(x_l) + F(x_l + W_l) \qquad (2)$$

where x_l represents the input audio features of the current layer, $h(x_l)$ means the dimensional transformation of 1×1 convolution on x_l, and $F(x_l, W_l)$ denotes the residual convolution block. Each residual convolution block $F(x_l, W_l)$ consists of two convolution layers cascaded as

$$F(x_l, W_l) = Conv_2\left(\operatorname{Re}LU\left(BN\left(Conv_1(x_l, W_l)\right)\right)\right) \qquad (3)$$

where $Conv_{1,2}$ indicates the 1D convolutional layer, BN is the feature normalization process, and $\text{Re}\,LU$ is the activation function layer.

In order to show the effect of the feature fusion on recognition accuracy in different layers, we fuse two output features at the layer of D_0, D_1, D_2 and D_3, respectively. As shown in Fig. 2, D_0 denotes the direct fusion of the output features, D_1, D_2 and D_3 denote to perform the feature fusion in the first, second and third fully connected layers, respectively. Finally, the fused features are used to predict the emotion category of generated music by softmax.

The specific parameters are set as follows: the number of channels in the first convolutional layer is 16 and the convolutional kernel is 3. The channel number in the first residual block is 16 and the convolutional kernel is 3. The number of channels in the second residual block is 32 and the convolutional kernel remains unchanged. The channel number in the third residual block is 64 and the kernel remains unchanged. The perceptual field size of maxpooling layer is 2. The dropout is set to 0.8, and the number of neurons in the fully connected layers D_1, D_2 and D_3 are 512, 256 and 128, respectively. The emotion categories predicted by softmax are Q1, Q2, Q3 and Q4. In the training process, the cross-entropy loss function is used.

4. Experiments

In the experiments, we firstly verify the performance of the proposed music emotion recognition model on the EMOPIA dataset. And then, the proposed model is applied to evaluate the emotional quality of the generated music to show its effectiveness.

4.1 Experiment on Music Emotion Recognition

In the model training process, we use the audio data in the format of WAV from the EMOPIA emotional music dataset. The computer configuration is NVIDIA GeForce RTX208 with 10G RAM based on Tensorflow. In addition, the iteration is 50, the batch number is 32, and the learning rate is 0.001. The experiments are performed on both four-category and two-category emotion recognition tasks, where two-category classification includes the arousal and valence dimensions in Russell's emotion model.

Table III shows the experimental results of our model comparing with the existing representative algorithms Logistic regression and S-Chunk-ResNet for the four-category and two-category recognition tasks. It can be seen from Table III that by only using the Mel spectrum as the basic audio feature, the proposed model achieves the recognition accuracy of 0.843 and 0.794 respectively in the four-category and valence dimension of two-category recognition tasks. Compared to Logistic regression model, there are improvements of 61.1%

and 42.2% respectively, while compared to the S-Chunk-ResNet model, the improvements are 24.5% and 12.7%. In contrast, the Logistic regression model achieves the highest recognition accuracy of 0.919 on the arousal dimension of two-category classification task, which is 1.2% higher than the proposed model. This indicates that in the case where the same Mel spectrum feature is used, the proposed network has better generalization ability than the Logistic regression and S-Chunk-ResNet. On the other hand, compared with single Mel spectrum feature, the recognition accuracies after feature fusion have been improved to some extent. Specifically, the recognition accuracies of our network with feature fusion have been improved by 2.6%, 0.4% and 1.8% in the four-category as well as arousal and valence dimensions of two-category recognition tasks, respectively. In particular, our feature-fused network achieves a recognition accuracy of 0.912 on the arousal dimension of two-category classification task, which is comparable to the performance of Logistic regression model. This also validates that our network can further improve the recognition accuracy through feature fusion.

Table III Experimental results of our model comparing with Logistic regression and S-Chunk-ResNet

Model	4Q	A	V
Logistic regression –Mel	0.523	**0.919**	0.558
S–Chunk–ResNet –Mel	0.677	0.887	0.704
Our network–Mel	**0.843**	0.908	**0.794**
Our network–MFCC	0.743	0.612	0.786
Feature fusion– D_0	0.784	0.803	0.765
Feature fusion– D_1	**0.865**	**0.912**	0.743
Feature fusion– D_2	0.839	0.765	**0.809**
Feature fusion– D_3	0.813	0.784	0.657

4.2 Quality Evaluation Experiment of Emotional Music Generation

In order to verify the effectiveness of our model, we evaluate the emotional quality of generated music from the objective and subjective aspects.

For the objective evaluation, the generated music by the existing representative algorithms LSTM+GA, CP Transformer and GRU+SF are tested on both of four category and two-category emotion recognition tasks. During the experiment, we apply the proposed model to identify the emotion categories of generated music and calculate the accuracy as the evaluation index for its emotional quality. The experimental results are shown in Table IV. It can be seen from Table IV that compared with LSTM+GA, the emotion recognition accuracies of generated music by GRU+SF have been improved by 49.0%, 9.2% and 8.0% respectively in the four-category as well as arousal and valence dimensions of two-category recognition tasks. Compared with CP Transformer, the recognition accuracies of GRU+SF have been increased by 8.2%, 2.4% and 12.5% in the four-category as well as arousal and valence dimensions of two-category recognition tasks, respectively. The experimental results indicate that GRU+SF can better restore the audio characteristics of music and obtain higher recognition accuracy, which is consistent with the objective evaluation results as analyzed in reference [18].

To further verify the reliability of the proposed objective evaluation method, we also conduct the subjective evaluation on the generated music by LSTM+GA, CP Transformer and GRU+SF. In the subjective evaluation experiment, humanness, richness and consistency of emotional expression are selected as indicators for generated music quality evaluation. The test samples consist of 12 generated music clips, of which 4 music clips are generated by each of the above three methods. We collect the evaluation results of 10 subjects as shown in Table IV. It can be seen from Table IV that, GRU+SF generates music with the highest scores in Humanness, Richness and Consistency of emotional expression. This also verifies that by applying the proposed emotion recognition model to the objective evaluation of emotion quality for generated music, the obtained results are consistent with those of subjective evaluation. It should be noticed that, the higher the recognition accuracy of the music emotion recognition model used, the more reliable the results of its use for emotion quality evaluation of generated music. This further indicates that it is effective to design emotional quality evaluation algorithms for generated music from the perspective of music emotion recognition.

Table IV Emotion recognition accuracy of generated music by the existing representative algorithms

Model	Subjective metrics			Objective metrics		
	Humanness	Richness	Consistency	4Q	A	V
LSTM+GA	2.59	2.74	2.6	0.53	0.76	0.75
CP transformer	3.31	3.22	3.26	0.73	0.81	0.72
GRU+SF	**3.48**	**3.72**	**3.84**	**0.79**	**0.83**	**0.81**

5. Conclusion

In this paper, we propose an emotional quality evaluation method for generated music from the perspective of music emotion recognition. For the feature extraction, the correlation is analyzed between audio features and emotion category of generated music, and thus the audio features with most significant correlation are picked out. In the construction of emotional music recognition model, we employ a 1D convolution combined with three residual blocks to predict the emotion category of generated music. In the experiments, both the objective and subjective evaluation experiments are performed. The experimental results verify that by applying the proposed model to the objective evaluation of emotion quality for generated music, the obtained results are consistent with those of subjective evaluation, showing the effectiveness of evaluating emotional quality for generated music from the perspective of music emotion recognition.

References

[1] SYLVIE D V, DANILO R, BUENO JOSÉ L O, et al. Music, emotion, and time perception: the influence of subjective emotional valence and arousal[J]. Front psychol, 2013, 4:417.

[2] IRRGANG M, EGERMANN H. From motion to emotion: accelerometer data predict subjective experience of music[J]. Plos one, 2016, 11(7).

[3] ZHAO K, LI S, CAI J, et al. An emotional symbolic music generation system based on LSTM networks[C]//2019 IEEE 3rd Information Technology, Networking, Electronic and Automation Control Conference (ITNEC). IEEE, 2019: 2039-2043.

[4] MADHOK R, GOEL S, GARG S. SentiMozart: music generation based on emotions[C]// ICAART (2), 2018: 501-506.

[5] HUNG H T, CHING J, DOH S, et al. EMOPIA: a multi-modal pop piano dataset for emotion recognition and emotion-based music generation[C]//International Society for Music Information Retrieval Conference, November 7-12, Online, 2021: 318-325.

[6] MAO H H, SHIN T, COTTRELL G W. DeepJ: style-specific music generation[C]//IEEE International Conference on Semantic Computing. IEEE Computer Society, 2018: 377-382.

[7] FERREIRA L N, WHITEHEAD J. Learning to generate music with sentiment[C]// International Society for Music Information Retrieval Conference, November 4-8, Delft, The Netherlands, 2019: 384-390.

[8] GÓMEZCAÑÓN J S, CANO E, PANDREA A G, et al. Language-sensitive music emotion recognition models: are we really there yet?[C]//IEEE International Conference on Acoustics, Speech and Signal Processing (ICASSP). IEEE, 2021: 576-580.

[9] WON M, FERRARO A, BOGDANOV D, et al. Evaluation of CNN-based automatic music tagging models[J]. ArXiv, 2020, abs/2006. 00751.

[10] ZHANG C, YU J, CHEN Z. Music emotion recognition based on combination of multiple features and neural network[C]//IEEE Advanced Information Management, Communicates, Electronic and Automation Control Conference. IEEE, 2021, 4: 1461-1465.

[11] ZHANG M, ZHU Y, GE N, et al. Attention-based joint feature extraction model for static music emotion classification[C]//2021 14th International Symposium on Computational Intelligence and Design (ISCID). IEEE, 2021: 291-296.

[12] YANG X, DONG Y, LI J. Review of data features-based music emotion recognition methods[J]. Multimedia systems, 2018, 24: 365-389.

[13] PANDA R, MALHEIRO R, ROCHA B, et al. Multi-modal music emotion recognition: a new dataset, methodology and comparative analysis[C]//10th International Symposium on Computer Music Multidisciplinary Research – CMMR'2013.2013.

[14] LANCASTER H O, SENETA E. Chi-square distribution[M]//Encyclopedia of Biostatistics. Chichester, United Kingdom: John Wiley & Sons, 2005, 2: 283-335.

[15] SATORRA A, BENTLER P M. Ensuring positiveness of the scaled difference Chi-square test statistic[J]. Psychometrika, 2010, 75(2): 243-248.

[16] SAIDI R, BOUAGUEL W, ESSOUSSI N. Hybrid Feature Selection Method Based on the Genetic Algorithm and Pearson Correlation Coefficient[M]//HASSANIEN A. Machine Learning Paradigms: Theory and Application. Studies in Computational Intelligence. Berlin: Springer, Cham, 2019, vol 801: 3-24.

[17] RUSSELL J A. A circumplex model of affect[J]. Journal of personality and social psychology, 1980, 39(6): 1161.

[18] MA L, ZHONG W, MA X, et al. Learning to generate emotional music correlated with music structure features[J]. Cognitive computation and systems, 2022, 4(2): 100-107.

第二部分
媒介音频

Design of Linear-Phase Nonsubsampled Nonuniform Directional Filter Bank with Arbitrary Directional Partitioning*

1. Introduction

Directional information is important in many image processing applications. Therefore, as a powerful tool to extract directional information of images, directional filter banks (DFBs) have attracted much attention. Most methods on designing DFBs are implemented via a tree-structured construction, resulting in uniform subbands and fixed directional partitioning scheme. Due to the rich textures contained in images, the DFBs with arbitrary number of subbands and flexible direction selectivity are still highly expected.

It should be noticed that the aforementioned DFBs are mostly maximally decimated. However, the decimation is not always necessarily required in many applications, such as image denoising, enhancement, restoration, and texture detection. And thus some redundant directional transforms, such as curvelet, nonsubsampled contourlet and shearlet, have employed nonsubsampled directional filter banks (NSDFBs). Concretely, the shearlet transform can extract directional information by applying bandpass filters in the frequency domain on pseudo-polar grid. Inspired by this, the NSDFBs were proposed to decompose images in frequency domain based on discrete pseudo-polar transform (PPFT) and 1-D nonsubsampled filter banks (NSFBs), which are capable to design uniform and nonuniform NSDFBs with arbitrary number of subbands. After separating in frequency domain on pseudo-polar grid by 1-D NSFBs, each directional band is then inverted back to space domain. However, in some applications such as denoising, local variants of filters are desired in order to reduce the Gibbs type ringing present when filters of large support sizes are used. Moreover, the individual filters applied in pseudo-polar frequency domain always

---

* The paper was originally published in *Journal of Visual Communication and Image Representation*, 2018, 51, and has since been revised with new information. It was co-authored by Li Fanga, Long Yea, Yun Tieb, Wei Zhong, and Qin Zhang.

perform a phase shift on the corresponding directional bands along the angular axis instead of the Cartesian axis, introducing phase distortions to decomposed subbands and impacting on the thresholding decision. This suggests a better performance in using the space domain method for denoising.

In this paper, we propose a novel nonsubsampled NUDFB with not only nonuniform wedge-shaped subbands but also arbitrary directional partitioning. Our approach is implemented in space domain and possesses LP property, making it efficient to process images. The explicit expressions of wedge-shaped filters are derived and small support size filters are obtained by windowing in space domain. Theoretical deduction proves that no matter what window function we choose or what the window size is, perfect reconstruction (PR) is always achieved. Tow experiments show that our NUDFB can offer more flexible directional information extraction strategy and superior image denoising performance to various directional decomposition methods such as curvelet, NSCT and NSST.

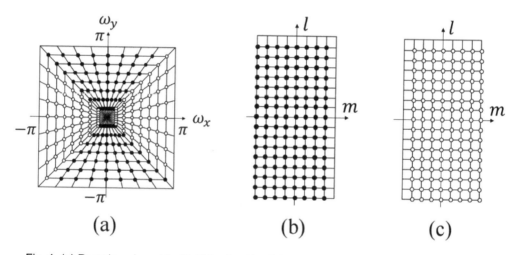

Fig. 1 (a) Pseudo-polar grid with BV subset in disks and BH subset in circles, (b) BV array, and (c) BH array.

The remainder of this paper is organized as follows. Section 2 presents the idea of the proposed NUDFB. Section 3 describes the proposed NUDFB design using window method and explores its PR conditions. Examples on image directional decomposition and denoising are given in 4 and some conclusions are drawn in Section 5.

2. Idea of the Proposed NUDFB

In this section, we first review the NUDFBs in frequency domain based on discrete

PPFT. Then we derive the idea of the proposed NUDFB in space domain.

2.1 Discrete PPFT Based NUDFBs

For a given two-dimensional signal of size $N \times N$, the discrete PPFT, proposed by Averbuch et al., evaluates the FT on the pseudo-polar grid, whose points sit at the intersection between linearly growing concentric squares and slope-equispaced rays as shown in Fig. 1. The pseudo-polar points are split into two subsets—the basically vertical (BV) and the basically horizontal (BH), given by

$$BV = \begin{cases} \omega_y = \dfrac{\pi \ell}{N}, -N \leq \ell \leq N \\ \omega_x = \omega_y \dfrac{2m}{N}, -\dfrac{N}{2} \leq m \leq \dfrac{N}{2} \end{cases}$$

$$BH = \begin{cases} \omega_x = \dfrac{\pi \ell}{N}, -N \leq \ell \leq N \\ \omega_y = \xi_y \dfrac{2m}{N}, -\dfrac{N}{2} \leq m \leq \dfrac{N}{2} \end{cases} \quad (1)$$

Fig. 1 depicts this grid, where BV points are marked with the filled disks and BH ones are marked as circles. The squares sides are of size $nk/N, k = 1, 2..., N$. The BH rays have equispaced slope: $2k/N, k = -N/2+1, -N/2+2, ..., N/2$. The BV rays are similar but with clockwise rotation of 90°. It can be seen from (1) that m represents the slope axis and ℓ the radial axis. In this way, the points in Cartesian coordinates ω_x, ω_y are converted to pseudo-polar coordinates m, ℓ. For the BV and BH arrays, the vertical axis corresponds to the index ℓ and the horizontal to m (see Fig. 1(b) and (c)). Therefore, one can apply 1-D NSFBs to the BV and BH arrays with respect to the horizontal axes, decomposing all pseudo-polar points into several subbands with nonuniform rectangle supports. Each rectangle support corresponds to a wedge-shaped region in the Cartesian coordinates. Before employing 1-D NUFBs, the two subsets should be combined and rearranged. Then subband coefficients are re-assembled and transformed back to space domain. Here the capability of multiresolution decomposition is provided by applying Laplacian pyramid at the beginning of the process, and then the aforementioned steps are performed to all subbands of Laplacian decomposition.

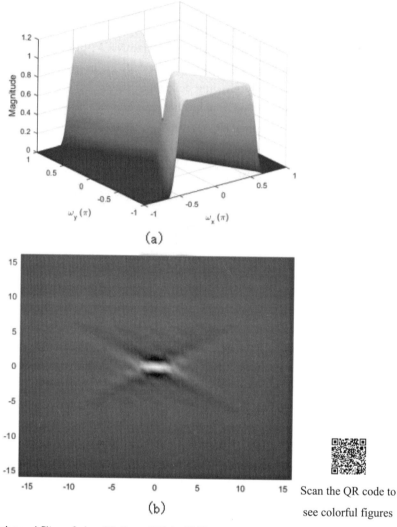

Fig. 2 Wedge-shaped filter of size 33 (from 60° to 120°, windowed by Hanning window). (a) Magnitude response. (b) Top view in space domain.

2.2 Idea of the Proposed NUDFB

The proposed NUDFB is based on the discrete time PPFT. First we modify (1) as follows,

$$BV = \begin{cases} \omega_y = \omega_\ell, -\pi \le \omega_\ell \le \pi \\ \omega_x = \omega_y \cdot \omega_m, -1 \le \omega_m \le 1 \end{cases} \quad (2)$$

$$BH = \begin{cases} \omega_x = \omega_\ell, -\pi \leq \omega_\ell \leq \pi \\ \omega_y = \omega_x \cdot \omega_m, -1 \leq \omega_m \leq 1 \end{cases} \quad (3)$$

Since BH points are similar to BV points, in our treatment the filters in the BH are obtained from converting those in the BV (see Section 3). For the rest of our description, we will refer to the BV points only. By substituting (2) into the discrete time Fourier transform for the Cartesian coordinates, we get the discrete time PPFT pair as

$$H(\omega_m, \omega_\ell) = \sum_{-\infty}^{\infty} h[x, y] e^{-j(\omega_m \omega_\ell x + \omega_\ell y)} \quad (4)$$

$$h[x, y] = \left(\frac{1}{2\pi}\right)^2 \int_{-1}^{1} \int_{-\pi}^{\pi} H(\omega_m, \omega_\ell) \times e^{j(\omega_m \omega_\ell x + \omega_\ell y)} \omega_\ell d\omega_m d\omega_\ell \quad (5)$$

Here (4) and (5) are continuous in the frequency domain, and thus the arbitrary angular partitioning can be achieved.

Next we derive the analytical expressions of wedge-shaped filters in space domain. As shown in Fig. 2(a), the wedge-shaped filter can be considered as a combination of 1-D lowpass filters along ω_x axis with 1-D highpass filters along ω_y axis. Since only the Type I FIR filters are capable to design both lowpass and highpass filters, the directional filters have to be with odd length and symmetric coefficients. Therefore, the ideal frequency response of such wedge-shaped filter is given by

$$H(\omega_m, \omega_\ell) = \begin{cases} e^{-j\frac{N}{2}(\omega_x + \omega_y)}, -1 \leq \omega_{m1} \leq \omega_m < \omega_{m2} < 1, \\ 0, \text{else} \end{cases} \quad (6)$$

where N is the order of filter, ω_{m1} and ω_{m2} are its cutoff frequencies. Such filter has the LP property with respect to Cartesian coordinates—ω_x and ω_y, instead of the pseudo-polar coordinates. But (6) is a conjugate symmetric and pure imaginary function in space domain. Then multiplying (6) by j, we can get the real-valued filter with anti-symmetry. To achieve symmetry, Hilbert transform is applied and the ideal wedge-shape response becomes

$$H(\omega_m, \omega_\ell) = \begin{cases} je^{j\left[\frac{\pi}{2} - \frac{N}{2}(\omega_x + \omega_y)\right]}, \omega_{m1} \leq \omega_m < \omega_{m2}, -\pi \leq \omega_\ell < 0 \\ -je^{j\left[\frac{\pi}{2} - \frac{N}{2}(\omega_x + \omega_y)\right]}, \omega_{m1} \leq \omega_m < \omega_{m2}, 0 \leq \omega_\ell < \pi \\ 0, \text{else} \end{cases} \quad (7)$$

Substituting (7) into (5), we get the symmetric and real-valued function as follows,

$$h[x, y] = \Psi(\omega_{m2}) - \Psi(\omega_{m1}), \quad (8)$$

where

$$\Psi(\omega_m) = \begin{cases} \dfrac{1-\cos\left[\left(x-\dfrac{N}{2}\right)\omega_m + y - \dfrac{N}{2}\right]\pi}{2\pi^2\left(x-\dfrac{N}{2}\right)\left[\left(x-\dfrac{N}{2}\right)\omega_m + y - \dfrac{N}{2}\right]}, & x \neq \dfrac{N}{2} \\[2ex] \dfrac{\cos(y-\dfrac{N}{2})\pi - 1}{2\pi^2\left(y-\dfrac{N}{2}\right)^2}\omega_m, & x = \dfrac{N}{2}, y \neq \dfrac{N}{2} \\[2ex] \dfrac{1}{4}, & x = \dfrac{N}{2}, y = \dfrac{N}{2} \end{cases} \quad (9)$$

Then by applying proper window functions to truncate (8), we can get the wedge-shaped filters with desired size (see Fig. 2). The direction and angular bandwidth of filters described by (8) are only determined by the slopes of the passband boundaries—ω_{m1}, ω_{m2}. Since it is continuous along the slope axis in the frequency domain, it is also continuous along the angle axis. Therefore, no matter how small the filter size is, the arbitrary angular partitioning can always be achieved.

2.3 Generation of Non-One Synthesis Filters

Since the wedge-shaped filter given in (8) is a rectangular window in pseudo-polar frequency domain, the corresponding synthesis filter is $G(z)=1$. In other words, the reconstruction of decomposed signal doesn't need additional filtering, making the complexity of the reconstruction minimal. However, in some applications, such as image denoising, the processing on decomposed signals may introduce non-linear distortion, which can be reduced by the non-one synthesis filters. Therefore in this section, we further derive the expression of analysis and synthesis filter pair.

In our previous work, we demonstrated the use of analysis and synthesis filters with cosine roll-off characteristics. If we replace the rectangular window in (6) with a cosine roll-off window, non-one synthesis filters will be obtained. The frequency response of such wedge-shaped filter with cosine roll-off transition band is given by

$$H(\omega_m, \omega_\ell) = \begin{cases} \sin\left[\dfrac{\omega_m - (\omega_{m1} - \omega_T/2)}{2\omega_T}\pi\right] \cdot e^{-j\frac{N}{2}(\omega_x + \omega_y)}, \\ \quad \omega_{m1} - \omega_T/2 \leq \omega_m < \omega_{m1} + \omega_T/2, \\[1ex] e^{-j\frac{N}{2}(\omega_x + \omega_y)}, \\ \quad \omega_{m1} + \omega_T/2 \leq \omega_m < \omega_{m2} - \omega_T/2, \\[1ex] \cos\left[\dfrac{\omega_m - (\omega_{m2} - \omega_T/2)}{2\omega_T}\pi\right] \cdot e^{-j\frac{N}{2}(\omega_x + \omega_y)}, \\ \quad \omega_{m2} - \omega_T/2 \leq \omega_m < \omega_{m2} + \omega_T/2 \\[1ex] 0, \text{ else} \end{cases} \quad (10)$$

where ω_T is the bandwidth of transition band. Substituting (10) into (5), unfortunately the result is a nonelementary integral. Thus, to compute (10) in the discrete domain, it suffices to compute the inverse PPFT or directly re-assemble the Cartesian sampled values and apply the inverse two-dimensional FFT. For odd size, the resulting space domain filters are also real-valued and symmetric, i.e., possessing LP property, as discussed in Section 2.2. Fig. 3 shows an example of such designed filters. The analysis and synthesis filters hold the time-reverse relation, i.e., $g(x,y) = h(N-x, N-y)$. Considering the symmetry of $h(x,y)$, we get $g(x,y) = h(x,y)$.

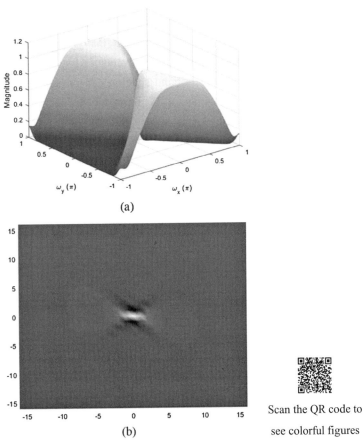

Scan the QR code to see colorful figures

Fig. 3 Cosine roll-off wedge-shaped filter of size 33 (from 60° to 120°, windowed by Hanning window). (a) Magnitude response. (b) Top view in space domain.

3. Design of the Proposed NUDFB

In this section, we focus on the design of the proposed nonsubsampled NUDFB using the wedge-shaped filters designed in Sections 2.2 and 2.3. Fig. 4(a) depicts its construction scheme. Since the wedge-shaped filter given in (8) is a rectangular window in pseudo-polar frequency domain, the corresponding synthesis filter is $G(z)=1$. Therefore, the analysis and synthesis filters can be combined into one filter, as shown in Fig. 4(b).

With the simplified structure, the reconstruction of this directional decomposition only requires a summation of the subbands rather than inverting a DFB. This results in an implementation that is most efficient computationally. In addition, this efficient inversion and the flexible decomposition may have advantages for applications such as compression routines. In the encoder, the subband spacing can be adapted to the image frequency distribution, achieving best sparsity. While in the decoder, no matter how complex the NUDFB is, the reconstruction only need to sum all subbands up, making the complexity of the decompression algorithm minimal. While in some other applications, such as image denoising, the non-one synthesis filters can suppress the nonlinear distortion caused by thresholding, as mentioned in Section 2.3. In these cases, traditional structure [see Fig. 4(a)] with cosine rolloff filters will perform better.

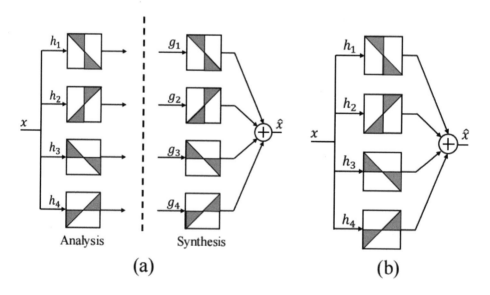

Fig. 4 (a) Construction scheme of NSDFBs, (b) simplified structure of the proposed NUDFB.

As discussed in reference [17], the cosine roll-off analysis and synthesis filters can combined into rectangular windows in frequency domain. Thus we only need to consider the PR condition for the filter banks designed by windowing (8). To simplify the analysis, the reconstruction process is split into two steps. First, we only consider the PR condition for the *BV* band. We assume the *BV* band consists of *M* wedge-shaped subbands $h_i[x,y]$ bounded by $\omega_{mi}, \omega_{m(i+1)}, i = 0,\ldots,M-1$, where $\omega_{m0} = -1, \omega_{mM} = 1$. By using 2-D windows w_i and summing up together, we get

$$\hat{h}[x,y] = \sum_{i=0}^{M-1} h_i[x,y] \cdot w_i = \sum_{i=0}^{M-1} \left[\Psi(\omega_{m(i+1)}) - \Psi(\omega_{mi}) \right] \cdot w_i$$

$$= \Psi(\omega_{mM}) \cdot w_{M-1} - \Psi(\omega_{m0}) \cdot w_0 + \sum_{i=1}^{M-1} \Psi(\omega_{mi})[w_{i-1} - w_i] \quad (11)$$

Since $h_{BV}[x,y] = [\Psi(\omega_{mM}) - \Psi(\omega_{m0})]w$, to have $h_{BV}[x,y] = \hat{h}_{BV}[x,y]$, the same 2-D window should be applied to all subbands. And thus is perfectly reconstructed. Similarly, the *BH* band $h_{BV}[x,y]$ can also be perfectly reconstructed as well.

The next step is to reconstruct the whole input signal. To achieve PR, the overall transfer function should be uniform over the whole frequency plane. From (8) we get

$$h_{BV}[x,y] = \begin{cases} \dfrac{1}{2\pi^2\left(x-\dfrac{N}{2}\right)} \left[\dfrac{1-\cos\left[\left(x-\dfrac{N}{2}\right)+\left(y-\dfrac{N}{2}\right)\right]\pi}{\left(x-\dfrac{N}{2}\right)+\left(y-\dfrac{N}{2}\right)} + \dfrac{1-\cos\left[\left(x-\dfrac{N}{2}\right)-\left(y-\dfrac{N}{2}\right)\right]\pi}{\left(x-\dfrac{N}{2}\right)-\left(y-\dfrac{N}{2}\right)} \right], x \neq \dfrac{N}{2} \\[2ex] \dfrac{\cos\left(y-\dfrac{N}{2}\right)\pi - 1}{\pi^2\left(y-\dfrac{N}{2}\right)^2}, x = \dfrac{N}{2}, y \neq \dfrac{N}{2} \\[2ex] \dfrac{1}{2}, x = \dfrac{N}{2}, y = \dfrac{N}{2} \end{cases} \quad (12)$$

By exchanging x, y in (12), we get

$$h_{BH}[x,y] = \begin{cases} \dfrac{1}{2\pi^2\left(y-\dfrac{N}{2}\right)}\left\{\dfrac{1-\cos\left[\left(x-\dfrac{N}{2}\right)+\left(y-\dfrac{N}{2}\right)\right]\pi}{\left(x-\dfrac{N}{2}\right)+\left(y-\dfrac{N}{2}\right)} + \dfrac{1-\cos\left[\left(x-\dfrac{N}{2}\right)-\left(y-\dfrac{N}{2}\right)\right]\pi}{\left(x-\dfrac{N}{2}\right)-\left(y-\dfrac{N}{2}\right)}\right\}, x \neq \dfrac{N}{2} \\[2ex] \dfrac{\cos\left(y-\dfrac{N}{2}\right)\pi - 1}{\pi^2\left(y-\dfrac{N}{2}\right)^2}, x \neq \dfrac{N}{2}, y = \dfrac{N}{2} \\[2ex] \dfrac{1}{2}, x = \dfrac{N}{2}, y = \dfrac{N}{2} \end{cases} \quad (13)$$

Comparing (12) with (13), we can obtain

$$\hat{h}[x,y] = h_{BV}[x,y] \cdot w + h_{BH}[x,y] \cdot w = \sin c\left(x - \dfrac{N}{2}\right)\pi \cdot \sin c\left(y - \dfrac{N}{2}\right)\pi \cdot w \quad (14)$$

Since all filters have the same order N being even, as mentioned in Section 2.2, (14) becomes

$$\hat{h}[x,y] = \delta\left[x - \dfrac{N}{2}, y - \dfrac{N}{2}\right] \cdot w \quad (15)$$

To make $\hat{h}[x,y]$ being a Dirac function in space domain, $w\left[x - \dfrac{N}{2}, y - \dfrac{N}{2}\right]$ should be equal to one. Meanwhile it is more convenient to convert slopes to angles. For filters that astride both *BV* and *BH* bands, it can be split into two, designed separately and add up together. Here we summarize the PR conditions as follows:

Let h be a filter designed by windowing (8) and located at the wedge-shaped region from θ_1 to θ_2, where $\theta_2 - \theta_1 \leq \dfrac{\pi}{2}$. For the *i*-th sub-band, the filter h_i is converted from h as

$$h_i = \begin{cases} h\big|_{\omega_{m1}=\tan\theta_1, \omega_{m2}=\tan\theta_2}, \\ -\dfrac{\pi}{4} \leq \theta_1 < \theta_2 < \dfrac{\pi}{4} \text{ or } \dfrac{3\pi}{4} \leq \theta_1 < \theta_2 < \dfrac{5\pi}{4} \\ h^T\big|_{\omega_{m1}=\cot\theta_1, \omega_{m2}=\cot\theta_2}, \\ \dfrac{\pi}{4} \leq \theta_1 < \theta_2 < \dfrac{3\pi}{4} \text{ or } \dfrac{5\pi}{4} \leq \theta_1 < \theta_2 < \dfrac{7\pi}{4} \end{cases} \quad (16)$$

Then if all the subband filters have the same odd size and share the same 2-D window, and the entry in the middle of the window equals one, the PR property of the proposed NUDFB can be achieved.

4. Applications

In this section, we give two examples to show the performance of the proposed NUDFB. The first example illustrates a two-level non-uniform directional decomposition of a test image using simplified structure [see Fig. 4(b)]. And the second one shows the potential of the proposed NUDFB in the application of image denoising using traditional structure [see Fig. 4(a)]. In the experiments, the multiscale property of the proposed NUDFB is obtained by cascading the Laplacian pyramid structure, as shown in Fig. 5.

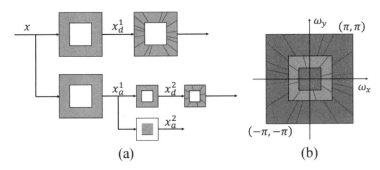

Fig. 5 Combination of Laplacian pyramid with the proposed NUDFB. (a) Implementation structure. (b) Idealized frequency partitioning obtained with the proposed structure.

For the first example, we take the "Foreman" image as an example to demonstrate the capability of extracting directional information of the proposed NUDFB. Since no processing is applied after decomposition, simplified structure [see Fig. 4(a)] with filters designed in Section 2.2 are employed. In the experiment, we first employ the Laplacian pyramid structure to perform two-level multiscale decomposition, obtaining the highpass component x_d^1, bandpass x_d^2 and lowpass x_a^2. Since the highpass component x_d^1 has the similar directional distribution as bandpass x_d^2, we design an 8-subband NUDFB to extract directional information for both of them, with the angular partitioning {9°, 60°, 75°, 85°, 95°, 125°, 144°, 171°, 189°} as shown in Fig. 6(d) and (f). Each filter is truncated by the Hanning window of length 33. Fig. 6(e) and (g) show the directional decomposition results of x_d^1 and x_d^2, respectively. Finally, by summing up all the subband components, the input image can be perfectly reconstructed with PSNR being infinity, as shown in Fig. 6(c).

It can be seen from Fig. 6(d) and (f) that, the proposed NUDFB can exactly match the spectrum directional distribution of x_d^1 and x_d^2. Concretely, the strong directional components in subbands 1, 3, 5, 7 can be exactly extracted, and the corresponding subbands have significant directional information as shown in Fig. 6(e) and (g). This cannot be done by using the traditional tree-structured DFBs, which have the fixed uniform directional partitioning scheme. Especially for the vertical subband 3 and horizontal subband 7 in Fig. 6(d) and (f), if we use those tree-structured methods, one subband with the same directional information would be split into two different directional sub-bands, while our NUDFB can realize arbitrary directional partitioning according to the directional distribution of images. The LP and PR properties are also achieved by the proposed space-domain implementation. In addition, we can process the images by applying the filters to each distinct block with an overcomplete decomposition to accelerate, while those frequency domain methods cannot.

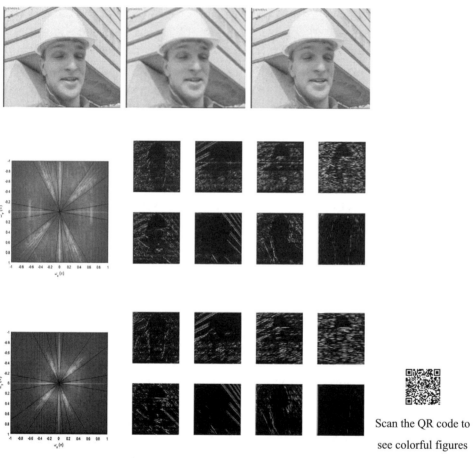

Scan the QR code to see colorful figures

Fig. 6 Two-level image directional decompositions. (a) Original image; (b) lowpass component x_a^2; (c) reconstructed image; (d) spectrum of the highpass x_d^1 and its directional partitioning; (e) the eight directional subbands at the first level; (f) spectrum of the bandpass x_d^2 and its directional partitioning; and (g) the eight directional subbands at the second level.

For the second example, we compare the proposed NUDFB with other directional decomposition methods in the application of image denoising. Here we use traditional structure (see Fig. 4(a)) with non-one synthesis filters to reduce nonlinear distortion introduced by thresholding, as discussed in Section 2.3. To compare the denoising performance of our NUDFB with other directional decomposition methods, we perform hard threshold on the subband coefficients. Taking note of the good performance of nonsubsampled contourlet transform (NSCT) for image denoising, we replicate their experiments and choose the threshold as

$$T_{i,j} = K\sigma_{Nij} \quad (17)$$

for each subband, where σ_{Nij} is the standard deviation of noise at scale j and direction i, which is inferred using a Monte Carlo technique where the variances are computed for a few normalized noisy images and then averaged to stabilize the results. This has been termed K-sigma thresholding in [11]. In the experiment, we set $K = 0$ for the lowpass coefficients, $K = 3$ and $K = 4$ for the bandpass and highpass coefficients respectively. The comparison results of our NUDFB, curvelet, NSCT and nonsubsampled shearlet transform (NSST) are shown in Table I for the test images "Lena" and "Barbara". Here for each method we use five-scale decomposition and 4, 8, 8, 16, 16 directions in the scales from coarser to finer, respectively. For the proposed NUDFB, the filter size is 33 and Kaiser window is used.

Table I Comparisons of our NUDFB with various directional decomposition methods in denoising (the best results are highlighted in bold).

Lena			PSNR (dB)		
σ	Noisy	Curvelet	NSCT	NSST	Ours
10	28.15	34.15	34.67	35.08	**35.11**
20	22.14	31.52	32.03	32.35	**32.42**
30	18.71	29.95	30.31	30.59	**30.68**
40	16.35	28.64	28.75	29.26	**29.38**
50	14.60	27.38	27.70	28.02	**28.14**
Lena			SSIM		
σ	Noisy	Curvelet	NSCT	NSST	Ours
10	0.6100	0.8929	0.8994	0.9042	**0.9044**
20	0.3408	0.8526	0.8614	0.8663	**0.8673**
30	0.2197	0.8228	0.8300	0.8357	**0.8368**
40	0.1553	0.7947	0.7972	0.8094	**0.8104**
50	0.1162	0.7648	0.7727	0.7815	**0.7818**

Table I Continued

Barbara	PSNR (dB)				
σ	Noisy	Curvelet	NSCT	NSST	Ours
10	28.14	32.28	33.01	33.18	**33.29**
20	22.19	28.83	29.35	29.62	**29.76**
30	18.80	26.90	27.10	27.48	**27.63**
40	16.47	25.31	25.49	26.00	**26.07**
50	14.76	23.92	24.26	24.68	**24.77**
Barbara	SSIM				
σ	Noisy	Curvelet	NSCT	NSST	Ours
10	0.7140	0.9088	0.9177	0.9196	**0.9209**
20	0.4781	0.8408	0.8538	0.8593	**0.8632**
30	0.3449	0.7838	0.7904	0.8026	**0.8086**
40	0.2610	0.7267	0.7336	0.7515	**0.7557**
50	0.2046	0.6692	0.6839	0.7014	**0.7063**

Table I shows the PSNR and SSIM results for various noise intensities. The results show the proposed NUDFB is consistently superior to curvelet, NSCT and NSST, in both PSNR and SSIM measures. Since the proposed NUDFB has flexible partitioning scheme, the subband spacing can be adapted to the image frequency distribution. Preliminary experiments show that the partitioning scheme can improve the denoising performance. As future work we consider to design an algorithm that can adaptively decompose images for best performance.

5. Conclusion

In this paper, we have proposed an efficient method for the design of 2-D nonsubsampled NUDFBs with arbitrary number of subbands and flexible direction selectivity. The design methodology proposed is based on window method and allows for individual LP filtering and PR property with small support filters. Application of our proposed NUDFB in image directional information extraction and noise removal were studied. In image directional information extraction, our NUDFB can realize arbitrary directional partitioning according to the directional distribution of images. In denoising, experimental results indicate that the proposed NUDFB outperforms various directional decomposition methods such as curvelet, NSCT and NSST in PSNR and SSIM.

References

[1] BAMBERGER R H, SMITH M J T. A filter bank for the directional decomposition of images: theory and design[J]. IEEE transactions on signal processing, 1992, 40(4): 882-893.

[2] PARK S, SMITH M J T, MERSEREAU R M. A new directional filter bank for image analysis and classification[C]//1999 IEEE International Conference on Acoustics, Speech, and Signal Processing. Proceedings. (Cat. No. 99CH36258). IEEE, 1999, 3: 1417-1420.

[3] PARK S I, SMITH M J T, MERSEREAU R M. Improved structures of maximally decimated directional filter banks for spatial image analysis[J]. IEEE transactions on image processing, 2004, 13(11): 1424-1431.

[4] DO M N, VETTERLI M. The contourlet transform: an efficient directional multiresolution image representation[J]. IEEE transactions on image processing, 2005, 14(12): 2091-2106.

[5] NGUYEN T T, ORAINTARA S. Multiresolution direction filterbanks: theory, design, and applications[J]. IEEE transactions on signal processing, 2005, 53(10): 3895-3905.

[6] NGUYEN T T, ORAINTARA S. A class of multiresolution directional filter banks[J]. IEEE transactions on signal processing, 2007, 55(3): 949-961.

[7] ESLAMI R, RADHA H. A new family of nonredundant transforms using hybrid wavelets and directional filter banks[J]. IEEE transactions on image processing, 2007, 16(4): 1152-1167.

[8] TANAKA Y, IKEHARA M, NGUYEN T Q. Multiresolution image representation using combined 2-D and 1-D directional filter banks[J]. IEEE transactions on image processing, 2009, 18(2): 269-280.

[9] MA N, XIONG H, SONG L. 2-D dual multiresolution decomposition through NUDFB and its application[C]//2008 IEEE 10th Workshop on Multimedia Signal Processing. IEEE, 2008: 509-514.

[10] DA CUNHA A L, ZHOU J, DO M N. The nonsubsampled contourlet transform: theory, design, and applications[J]. IEEE transactions on image processing, 2006, 15(10): 3089-3101.

[11] STARCK J L, CANDÈS E J, DONOHO D L. The curvelet transform for image denoising[J]. IEEE transactions on image processing, 2002, 11(6): 670-684.

[12] GUO K, LABATE D. Optimally sparse multidimensional representation using shearlets[J]. SIAM journal on mathematical analysis, 2007, 39(1): 298-318.

[13] EASLEY G, LABATE D, LIM W Q. Sparse directional image representations using the discrete shearlet transform[J]. Applied and computational harmonic analysis, 2008, 25(1): 25-46.

[14] AVERBUCH A, COIFMAN R R, DONOHO D L, et al. Fast and accurate polar Fourier transform[J]. Applied and computational harmonic analysis, 2006, 21(2): 145-167.

[15] SHI G, LIANG L, XIE X. Design of directional filter banks with arbitrary number of subbands[J]. IEEE transactions on signal processing, 2009, 57(12): 4936-4941.

[16] LIANG L, SHI G, XIE X. Nonuniform directional filter banks with arbitrary frequency partitioning[J]. IEEE transactions on image processing, 2010, 20(1): 283-288.

[17] FANG L, ZHONG W, ZHANG Q. Design of M-channel linear-phase non-uniform filter banks with arbitrary rational sampling factors[J]. IET signal processing, 2016, 10(2): 106-114.

Multi-Source Separation Using over Iterative Empirical Mode Decomposition*

1. Introduction

Multi-source separation of single-channel audio signal has been one of the most challenging problems in audio signal processing and addressed wide range interesting of researchers. Estimation of multiple fundamental frequencies plays a key role in polyphonic signal separation problem, but automatically extracting the fundamental frequencies, commonly perceived as pitch, from recorded audio signals is a difficult task.

A lot of multiple fundamental frequency estimation methods use iterative-based approach, in which the fundamental frequency is estimated as the predominant signal and removed before the next iteration until all harmonic sources are extracted. These methods are easy to implement because they allow using single pitch estimation algorithms in each iteration. The joint estimation methods can yield better results with higher complexity. Other kinds of methods using correlogram, probabilistic framework with harmonic model and non-negative matrix factorization (NMF) also have their advantage respectively.

Huang et al. raised a novel signal processing method named Hilbert-Huang Transform (HHT) based on EMD (Empirical Mode Decomposition), which is suitable for analyzing nonlinear and non-stationary signal. Different from the traditional signal processing methods, the HHT is an adaptive decomposition method and can yield more physical results. EMD is a complete, approximately orthogonal and self-adaptive method which has the ability to decompose signal by time scale. Some numerical experiments show that EMD behaves as a dyadic filter bank.

The method proposed in this paper is based on the over iterative EMD which uses large iteration times to yield unity envelop IMF. Although over iterative will decrease the physical meaning of IMFs, it can concentrate energy into predominant frequency component which contains pure frequency modulation signal of the fundamental frequency.

---

* The paper was originally published in *ICEC*, 2013, and has since been revised with new information. It was co-authored by Yutian Wang, Hui Wang, and Qin Zhang.

After this processing, the harmonic sources are allotted into individual IMF components. In each particular IMF, there always have one predominant harmonic source. By means of a salience estimation method, the corresponding fundamental frequencies can be extracted.

In this paper, the EMD method and its filtering property are reviewed firstly. Then the proposed algorithm which is the combination of over iterative EMD and salience estimation is described. Finally, a two instruments mixture signal is used to demonstrate the performance of our method. The experimental results agree that our algorithms can extract fundamental frequency efficiently.

2. Filter Bank Property of EMD

To analyze the nonlinear and non-stationary signal, the definition of instantaneous frequency and energy is needed, which must be functions of time. Empirical Mode decomposition (EMD) was proposed to decompose a complicated signal into a group of IMFs. Then the instantaneous frequency can be obtained by Hilbert Transform. The detail algorithm of EMD was summarized in reference [7].

The EMD result is highly determined by the sifting stop criterion. Different stop criterions make results vary. Although there are several kinds of stop criterion, Wu et al. suggest fixed sifting time criterion. In separate research, Flandrin and Wu point out that EMD is in fact a dyadic filter bank. In the research by Wu et al., the dyadic property is available only shift times are about 10. Too many or too few iteration numbers would decrease the dyadic property. When the EMD shifting iteration number becomes higher, the filter banks gradually change narrower and their central frequency gets closer, which enables the EMD to split the predominant frequency components. The filter-bank property with different iteration sifting number shows in Fig. 1. From the Fig. 1 we can reach the conclusion that with the sifting number increasing from 10 to 100, the filter bands decrease proportionally.

Unfortunately, with increasing number of iterations, the amplitude envelops trend to a straight line which will decrease the physical mean of IMFs. Furthermore, the Fourier transform can be recognized as special case of EMD when the sifting number goes to infinity. G. Rilling analyzes the EMD separation ability of two sinusoidal functions with different amplitude and frequency ratio and finds out the interval of parameters which can yield correct results.

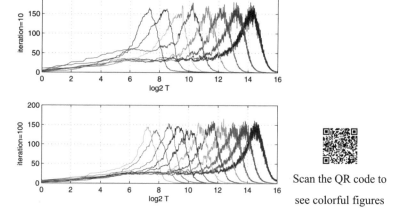

Scan the QR code to see colorful figures

Fig. 1 Filter bank property with different iteration number.

3. Fundamental Frequency Estimate Algorithm

For every harmonic source in the polyphonic signal, separating their primal component is the first and the most important step. To detect the pitches in a mixture signal, we use over iterative EMD method with sifting number 1000. From the knowledge introduced previously, we can reach the conclusion that large iteration number will reduce the physical mean of IMF but provide extremely sharp filter ability of the filter-bank. Beside it, due to the ability of decomposition signal by time scale, EMD makes the predominant signal components raise in particular IMFs. It should be noticed that with the increasing of iteration number, the centroid frequencies of the filter-bank are getting closer but the leakage also becoming serious. To overcome this problem, we should pick up available IMFs before estimating the fundamental frequencies. Our method is set a threshold of signal power ratio between the IMF and the whole signal.

After getting the available IMFs, the fundamental frequencies of each IMF can be obtained by:

1. Find the highest spectral peak f_{max} from the FFT spectrum of IMF.
2. Calculate $f_{ca} = f_{max} / n$, in which the initial value of n is 10. If the amplitude of f_{ca} is higher than a given threshold μ, the f_{ca} is considered as a fundamental frequencies. Then terminate the process, otherwise go to step 3.
3. Set $n = n - 1$, go to step 2.

After this process, each IMF will produce a individual fundamental frequency. To find out the essential harmonic source in this candidate value set, a judgment method is designed as two steps. Firstly, if some of the value in the candidate set have the relationship of approximate multiple times, we can assert that these values belong to one harmonic source. If candidate values $f_j > f_i$, and

$$f_j = mf_i \pm \sigma, m = 2, 3, \ldots, M, \qquad (1)$$

these two value are categorized into one harmonic source set H_i. σ denotes the frequency tolerance. Then the whole set is divided into several subsets $H_1, H_2, \ldots, H_n$. To determine which value is the real fundamental frequency in each subset H_i, candidates are ordered decreasingly by the sum of their harmonic amplitudes. It should be noticed that the candidates value must be in the range $[f_{min}, f_{max}]$, which is used to avoid the high and low order harmonics influence the result. Then all the possible candidate combinations are evaluated and the combination with the best salience is chosen. The salience is a function of the harmonic amplitudes in each IMF function.

$$S(i) = \sum_{j=1}^{C} A_{i,j} \qquad (2)$$

where C is the number of IMFs, $S(i)$ is the salience of ith harmonic source H_i and $A_{i,j}$ denotes the corresponding harmonic peak amplitude of each harmonic source in every IMF.

To verify our algorithm, an mixture signal of two violin samples with pitches of A5 and D4 was evaluated. The samples are from the MUMS of McGill University. The mixture signal spectrum is shown in Fig. 2 in which the amplitude is drawn in dB scale. In the diagram the harmonics distribute in a complex way which is hard to identify and extract the fundamental frequency.

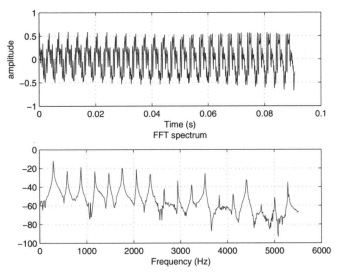

Fig. 2 The FFT spectrum of mixture signal of A5 and D4 violin.

Then the over iterative EMD with sifting number 1000 is used to yield a set of IMF function, of which the power lower than 5% of total power are abandoned. After that, they are transformed into FFT spectrum. The final results are shown in Fig. 3.

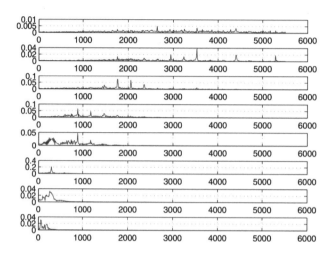

Fig. 3 The FFT spectrum of the available IMFs.

From the Fig. 3 we can clear see that all the harmonic peaks are separated into different IMF and distribute in sparse and regular way. Consequently, to estimate the fundamental frequency from these spectrums become easier than directly from the original signal spectrum. By using the algorithm introduced before, we can obtain the fundamental frequency at 882Hz and 297Hz.

4. Conclusion

In this paper, we propose a single channel polyphonic music signal multiple fundamental frequency estimation method based on over iterative EMD. By using over iterative sifting number, the filter-bank property change sharply which can yield narrower band of energy. The combination of over iterative EMD and salience estimation method can obtain good results especially for multiple times fundamental frequencies mixture signal. The experiment agrees that the proposed algorithm is efficient in both same and different instruments mixture.

Acknowledgments

This work is supported by the National Natural Science Foundation of China (Grant No. 61071149), the State Administration of Radio Film and Television of China (Grant No. 2011-11) and Program for New Century Excellent Talents of Ministry of Education of the

People's Republic of China (Grant No. NCET-10-0749).

References

[1] DE CHEVEIGNÉ A. Separation of concurrent harmonic sounds: fundamental frequency estimation and a time-domain cancellation model of auditory processing[J]. The journal of the acoustical society of America, 1993, 93(6): 3271-3290.

[2] MAHER R C, BEAUCHAMP J W. Fundamental frequency estimation of musical signals using a two-way mismatch procedure[J]. The journal of the acoustical society of america, 1994, 95(4): 2254-2263.

[3] WU M, WANG D L, BROWN G J. A multipitch tracking algorithm for noisy speech[J]. IEEE transactions on speech and audio processing, 2003, 11(3): 229-241.

[4] CHRISTENSEN M G, STOICA P, JAKOBSSON A, et al. The multi-pitch estimation problem: Some new solutions[C]//2007 IEEE International Conference on Acoustics, Speech and Signal Processing. IEEE, 2007, 3: III-1221-III-1224.

[5] SMARAGDIS P, BROWN J C. Non-negative matrix factorization for polyphonic music transcription[C]//2003 IEEE Workshop on Applications of Signal Processing to Audio and Acoustics (IEEE Cat. No. 03TH8684). IEEE, 2003: 177-180.

[6] HUANG N E, SHEN Z, LONG S R, et al. The empirical mode decomposition and the Hilbert spectrum for nonlinear and non-stationary time series analysis[J]. Proceedings of the royal society of london. series a: mathematical, physical and engineering sciences, 1998, 454(1971): 903-995.

[7] WU Z, HUANG N E. A study of the characteristics of white noise using the empirical mode decomposition method[J]. Proceedings of the royal society of london. series a: mathematical, physical and engineering sciences, 2004, 460(2046): 1597-1611.

[8] FLANDRIN P, RILLING G, GONCALVES P. Empirical mode decomposition as a filter bank[J]. IEEE signal processing letters, 2004, 11(2): 112-114.

[9] WU Z, HUANG N E. On the filtering properties of the empirical mode decomposition[J]. Advances in adaptive data analysis, 2010, 2(04): 397-414.

[10] RILLING G, FLANDRIN P. One or two frequencies? The empirical mode decomposition answers[J]. IEEE transactions on signal processing, 2008, 56(1): 85-95.

A Two-Stage Complex Network Using Cycle-Consistent Generative Adversarial Networks for Speech Enhancement[*]

1. Introduction

Speech enhancement (SE) can be described as the technique to separate the speech components from the background noise interference. It intends to improve speech quality and intelligibility in many communication applications such as front-ends for automatic speech recognition (ASR) systems and hearing aids. In recent years, due to the tremendous ability of deep neural networks (DNNs) to deal with non-stationary noise in low signal-to-noise ratio (SNR) conditions, many DNN-based approaches have demonstrated superior performance in single-channel SE. These DNN-based methods can be divided into two categories, namely masking-based approaches and mapping-based approaches. The masking-based approaches estimate a time–frequency (T-F) mask that is applied to a noisy speech signal for enhancement [e.g., ideal binary mask (IBM), ideal ratio mask (IRM)]. The mapping-based approaches are proposed to train a mapping network that directly transforms noisy-speech features (e.g., magnitude spectra, log-power spectra) to clean ones.

More recently, generative adversarial networks (GANs) have demonstrated comparable performance for SE. The same as most other DNN-based approaches, they also require a large number of paired training data, which may be difficult for practical applications. To solve the difficulty of obtaining the parallel recordings of speech and noise from the real scenarios, it is suggested using cycle-consistent GAN (CycleGAN) for SE. Moreover, CycleGAN-based approaches have also demonstrated their promising performance with parallel data, for their capability of preserving the speech structure and reducing speech distortion.

---

[*] The paper was originally published in *Speech Communication*, 2021, 134, and has since been revised with new information. It was co-authored by Guochen Yu, Yutian Wang, Hui Wang, Qin Zhang, and Chengshi Zheng.

Nevertheless, conventional CycleGAN-based approaches have two intractable limitations for SE tasks. Firstly, to ensure cycle-consistency of the original noisy speech domain and target clean speech domain, the enhanced signal always contains the original noise information. In other words, the cycle-consistency-based methods remain audible residual noise, which is challenging to eliminate. This essentially implies that background noise reduction is challenging in these algorithms. Secondly, previous CycleGAN-based SE systems only estimated the magnitude spectrum, log-power spectrum or Mel-cepstral coefficients features while combining the non-oracle (i.e., noisy) phase to reconstruct the time-domain waveform. This leads to severe phase distortion under low SNR scenarios, resulting in serious performance degradation in speech quality and intelligibility.

Multi-stage learning approaches can decompose the original difficult task into multiple more manageable sub-tasks and have demonstrated better performance than single-stage methods in many areas, such as image inpainting and image deraining. In this paper, we incorporate a CycleGAN-based magnitude spectrogram mapping network (dubbed CycleGAN-MM) and a deep complex-valued denoising network (dubbed DCD-Net) as a two-stage approach for SE. In the first stage, we utilize CycleGAN-MM to only estimate the magnitude of clean spectra, where we introduce a relativistic average loss in discriminators to stabilize the training. Motivated by recent studies in SE area for better sequence modeling, we employ temporal-frequency attention (T-FA) in generators to capture global dependency along temporal and frequency dimensions, respectively. More recently, it has been demonstrated that the magnitude and phase are difficult to be optimized simultaneously, especially in extremely low SNR conditions. In our preliminarily investigated one-stage CycleGAN-based complex mapping SE system (dubbed CycleGAN-CM), optimizing both real and imaginary (RI) components may cause the unstable training of the generators and discriminators, and consequently its performance gets even worse than only estimating the magnitude of clean speech spectrum. That is because when optimizing phase information by estimating the complex spectrum, magnitude estimation may deviate its optimal convergence path by degrees. Therefore, we only optimize the magnitude of the clean spectra in the first stage, which is then coupled with its corresponding noisy phase to obtain a coarsely enhanced complex spectrum. Subsequently, in the second stage, we introduce a deep complex denoising net to further suppress the intractable remaining noise in the previous stage, while simultaneously reconstructing the clean speech phase by estimating both real and imaginary components of the clean spectra. Instead of using real-valued neural networks, DCD-Net is designed by complex-valued convolutional networks and complex T-FA blocks to refine the coarsely enhanced complex spectrum. To the best of our knowledge, this is the first attempt to handle the complex spectral mapping in CycleGAN-based SE systems. To validate the superiority of the proposed scheme, we compare our model with recent state-of-the-art (SOTA) GAN-based and Non-GAN based

SE systems on two public datasets. Experimental results demonstrate the proposed two-stage approach outperforms the one-stage CycleGAN-based systems by a significant margin and achieves SOTA performance.

The remainder of this paper is organized as follows. Section 2 introduces the related works, including CycleGAN for SE, deep complex-valued SE systems, and multi-stage SE approaches. In Section 3, the proposed framework is described in detail. The experimental setup is presented in Section 4, and the experimental results and analysis are provided in Section 5. Finally, some conclusions are drawn in Section 6.

2. Related Works

2.1 CycleGAN-Based Methods for SE

A recent breakthrough in the SE area comes from the application of GANs as feature mapping networks. GAN consists of a generator network (G) and a discriminator network (D) that play a min-max game between each other. By using adversarial training, the objective of G is to synthesize the fake samples which are indistinguishable from the target data distribution, whilst D attempts to discriminate between the real and fake samples. SEGAN is the pioneering work employing GAN for SE task, which directly maps raw waveform of the clean speech from the mixed raw waveform in time domain. More recently, Other GAN-based SE algorithms in the time domain have been proposed to leverage different loss functions or generator structures. Another mainstream of GAN-based SE algorithms operates on the time-frequency (T-F) domain, where G is used to estimate a T-F mask.

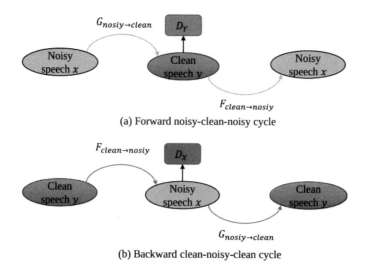

Fig. 1 (a) The forward noisy-clean-noisy cycle. (b) The backward clean-noisy-clean cycle.

However, conventional GAN-based methods may map noisy features to any random permutation of the clean features in the target domain with only adversarial losses, thus there is no guarantee that the individual enhanced feature is exactly paired with the target clean one. In other words, these methods cannot restrict that the contextual information of noisy features and enhanced features are always cycle-consistent. As a variant of GAN, CycleGAN is widely used for the unpaired image-to-image translation task, while in speech area it also demonstrates excellent performance on voice conversion, music style transfer, and SE. Incorporating CycleGANs for SE, these methods demonstrate their effectiveness in improving SE performance especially in maintaining speech integrity and reducing speech distortion. CycleGAN-based approaches for SE contain a noisy-to-clean generator G and an inverse clean-to-noisy generator F, which transforms the noisy features into the enhanced ones for the former, and vice versa for the latter. As illustrated in Fig. 1, a forward noisy-clean-noisy cycle and a backward clean-noisy-clean cycle jointly constrain G and F to be cycle-consistent, which are optimized with the adversarial loss, a cycle-consistency loss, and an identity-mapping loss, respectively. Discriminators D_X and D_Y are trained to classify the target speech features as real and the generated speech features as fake.

Nevertheless, in the standard CycleGAN-based approaches, the enhanced signal always contains the original noise information and remains audible residual background noise due to the constraint of cycle-consistency. Moreover, to the best of our knowledge, phase recovery has not been well investigated in previous CycleGAN-based SE approaches.

2.2 Two-Stage Approaches for Noise Reduction

Recently, some studies focus on conducting the original SE task by multi-stage networks, which can significantly improve the estimated speech quality. In Hao et al. (2020), Hao et al. proposed a masking and inpainting SE approach for low SNR and non-stationary noise. This two-stage approach consists of binary masking and spectrogram inpainting. In the first stage, a binary masking model is trained to remove the T-F points dominated by severe noise in low SNR conditions and obtain the spectra with the T-F points dominated by clean speech, while an inpainting model is used to recover the missing T-F points in the second stage. To solve the difficulty of estimating phase spectrum, Du et al. introduced a joint framework composed of a Mel-domain denoising autoencoder and a deep generative vocoder for monaural speech enhancement, in which the clean speech waveform is reconstructed without using the phase. The first stage enhances the Mel-power spectrum of noisy speech by denoising autoencoder, while a deep generative vocoder focuses on synthesizing the speech waveform in the second stage. More recently, Li et al. proposed a two-stage approach named CTS-Net for SE in complex domain, in which a coarse magnitude estimation network (CME-Net) and a complex spectrum refined network (CSR-Net) jointly optimize the noisy spectra. In the first stage, the target spectral magnitude is coarsely estimated, which is then coupled with the noisy phase to obtain a coarsely estimated complex spectrum. In the second stage, CSR-Net is trained to estimate both RI components of the clean complex spectrum, thus further reducing the residual noise, restoring the clean speech phase, and inpainting missing details of the estimated spectrum.

2.3 Deep Complex Neural Network Based for Phase-Aware SE

Conventional SE methods only estimate the magnitude of the speech, and the time-domain speech waveform is reconstructed by using the noisy phase and the estimated magnitude. The reason for not enhancing the phase is that it was believed that phase is not so important for SE, as well as it is intractable to directly estimate the clean speech phase because it is unstructured. However, recent studies Paliwal et al. (2011) show the importance of the accurate phase as it can significantly improve perceptual speech quality, especially in low SNR conditions. Subsequently, some SE algorithms are developed to solve this problem, and consistently show objective speech quality improvements when the phase is enhanced.

More recently, the deep learning-based phase-aware SE algorithms can be divided into two categories. The first one operates in the time domain by estimating speech signals from raw-waveform noisy signals without using any explicit T-F representation, thereby avoiding the problem of phase estimation. Recently, it is reported that the fine-detailed structures of noise and speech components are more separable with T-F representation. Hence, another main-stream phase-aware SE approaches work on optimizing both RI components of the complex spectrum by using complex-valued ratio mask (CRM) or a direct complex-mapping network. Thereby, these methods estimate both magnitude and phase information in the frequency domain. Although the above approaches have been proposed to address this issue, they are limited as the neural network still conducts real-valued operations. To this end, deep complex u-net (named DCUNET) and Deep Complex Convolution Recurrent Network (named DCCRN) are proposed to conduct phase-aware SE via complex-valued neural network. DCUNET incorporates a deep complex network and u-net structure to enhance the complex spectra of noisy speech. Note that DCUNET is trained to estimate bounded CRM and optimizes the weighted source-to-distortion ratio (wSDR) loss after reconstructing the enhanced time-domain waveform by inverse STFT. DCCRN effectively combines both the advantages of DCUNET and CRN to estimate the complex spectra of clean speech, in which LSTM is utilized to model temporal context with significantly reduced trainable parameters. However, it may cause the memory bottleneck issue when using LSTM to model temporal dependencies, which may result in reducing training efficiency.

Motivated by these studies, we propose a two-stage deep complex approach, which incorporates a complex-valued denoising network (named DCD-Net) with a CycleGAN-based magnitude mapping network (named CycleGAN-MM). Specifically, CycleGAN-MM is adopted to coarsely estimate the clean spectral magnitude in the first step, while DCD-Net aims to further suppress the intractable residual background noise and simultaneously recover the clean phase information implicitly by estimating both RI components of the clean spectrum.

3. Method

3.1 Training Target

The overall network architecture is presented in Fig. 2, which is comprised of two sub-networks, namely CycleGAN-MM and DCD-Net. In our SE task, the mixture signal in the time domain is formulated as $x(n) = s(n) + z(n)$, where x, S and Z denote noisy speech, clean speech and noise, respectively. With the STFT, the noisy speech in the time–frequency domain can be modeled as,

$$X_{t,f} = S_{t,f} + Z_{t,f} \tag{1}$$

where $X_{t,f} = |X_{t,f}|e^{j\theta_{X_{t,f}}} \in \mathbb{C}$, $S_{t,f} = |S_{t,f}|e^{j\theta_{S_{t,f}}} \in \mathbb{C}$ and $Z_{t,f} = |Z_{t,f}|e^{j\theta_{Z_{t,f}}} \in \mathbb{C}$ denote the time-frequency (t, f) representations of noisy speech, clean speech and noise, respectively. The input to CycleGAN-MM is the magnitude of the noisy spectrum $|\tilde{X}_{t,f}| = G(|X_{t,f}|)$. After the first-stage, the estimated magnitude $|\tilde{X}_{t,f}|$ is then coupled with the original noisy phase $e^{j\theta_{X_{t,f}}}$ to obtain a coarsely enhanced complex spectrum $\tilde{X}_{t,f}$. In the second stage, DCD-Net receives the enhanced complex spectrum to estimate the CRM, which can be defined as:

$$CRM = \frac{X_r S_r + X_i S_i}{X_r^2 + X_i^2} + j\frac{X_r S_i - X_i S_r}{X_r^2 + X_i^2} = \tilde{M}_r + j\tilde{M}_i \tag{2}$$

where X_r and X_i denote the RI components of the noisy speech spectrum, respectively. Similarly, S_r and S_i indicate the RI components of the clean speech spectrum. The real and imaginary parts of the CRM are represented by $\tilde{M}_r$ and $\tilde{M}_i$. Alternatively, the polar coordinate representation of $\tilde{M}$ can be presented as,

$$\tilde{M} = \tilde{M}_{mag} \cdot e^{j\theta_{\tilde{M} phase}} \tag{3}$$

where $\tilde{M}_{mag}$ and $\theta_{\tilde{M}_{phases}}$ denote the magnitude and phase of the complex-valued mask, respectively. To make the mask bounded in a unit-circle at the complex space, we use the tanh activation function to limit the magnitude mask $\tilde{M}_{mag}$ ranging from 0 to 1 like in Choi et al. (2019). Hence, the final estimated clean complex spectrum $\tilde{S}$ of DCD-Net in polar coordinates can be calculated by,

$$\tilde{S} = |\tilde{X}_{t,f}| \cdot \tilde{M}_{mag} \cdot e^{j\left(\theta_{\tilde{X}_{t,f}} + \theta_{\tilde{M}\mu phase}\right)} \tag{4}$$

Note that DCD-Net is optimized by signal approximation (SA), which directly minimizes the difference between the complex spectrum of the clean speech and that of the noisy speech applied with bounded CRM.

3.2 CycleGAN-Based Magnitude Mapping Network

As shown in Fig. 2, two generators, dubbed $G_{X \to Y}$ and $F_{Y \to X}$ and two discriminators, dubbed D_X and D_Y, are employed in CycleGAN-MM simultaneously. The generator is composed of three components, namely three downsampling layers, six dilated residual attention (DRA) blocks and three homologous upsampling layers, where the detailed structure of this generator is illustrated in Fig. 3. Each downsampling/upsampling layer block is composed of a 2D convolution/deconvolution layer, followed by instance normalization (IN), parametric Relu activation function (PRelu) and gated liner units (GLUs). GLUs can control the information flows throughout the network, showing the effectiveness of modeling speech sequential structure. In generators, we introduce Temporal-Frequency self-attention (T-F SA) and Temporal-Frequency attention gates (dubbed T-F AG) in DRA block and upsampling layers, respectively. T-F SA and T-F AG are utilized herein to capture relative contextual dependencies along the time and frequency dimensions and directly pass the salient information of source speech features. The discriminator is composed of six 2D convolutions, each of them followed by spectral normalization (SN) and PRelu, compressing the feature maps into a high-level representation. SN is proposed to stabilize the training process of the discriminator and demonstrates the effectiveness to avoid vanishing or exploding gradients. The kernel size for each 2D convolutions layer is (3, 5) except (1, 1) for the last layer in the temporal and frequency axis, respectively. The stride is set to (1, 2) along the temporal and frequency axis for the first five downsampling layers, while it is set to (1, 1) for the last layer. The number of channels throughout the 2D convolutions is (32, 32, 64, 64, 128, 1).

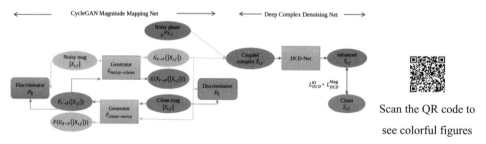

Scan the QR code to see colorful figures

Fig. 2 The diagram of the proposed CycleGAN-DCD. The left part denotes the architecture of CycleGAN-MM, while the right part denotes the architecture of DCD-Net.

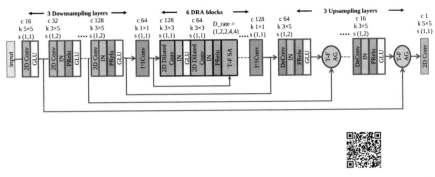

Scan the QR code to see colorful figures

Fig. 3 The generator of CycleGAN-MM. c, S and k represent output channel numbers, kernel size, and stride of 2D convolution, respectively. IN, PRelu, GLU indicate instance normalization, gated linear unit, and parametric Relu activation, respectively.

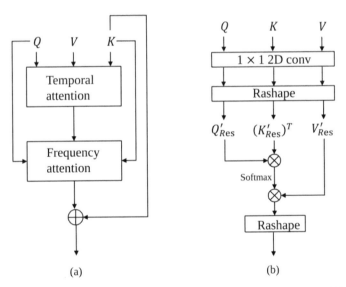

Fig. 4 (a) The diagram of T-FA. (b) The detailed diagram of TA and FA.

3.2.1 Temporal-Frequency Attention

Attention mechanism Vaswani et al. (2017) has been widely used in speech processing tasks, as it can leverage the contextual information in the time–frequency dimension and further enhance the salient speech information importing in the feature learning procedure. Moreover, using attention in SE task can simulate the human auditory perception, so that more attention is put to speech while less is put to the surrounding background noise. Following the terminology in Tang et al. (2020), we compute the attention function on a set of queries, keys, and values simultaneously, and pack them together into feature maps $Q \in \mathbb{R}^{B \times T \times F' \times C}$, $K \in \mathbb{R}^{B \times T \times F' \times C}$ and $V \in \mathbb{R}^{B \times T \times F' \times C}$, respectively. Here, B denotes the batch size of input features, T denotes the number of frames, F' denotes the number of frequency bins and C denotes the number of channels in each feature map. Inside the attention module, the feature maps Q and K are first projected into two feature spaces Q' and K' by 1×1 convolutions to calculate attention maps β. However, the size of the attention weight matrix β in the original attention mechanism is $(T \times F') \times (T \times F')$, which would be extremely large and cost heavy computational complexity. To address this problem, we introduce temporal-frequency attention (T-FA) to capture the global dependencies along temporal and frequency dimensions, respectively. As discussed in Tang et al. (2020), Zheng et al. (2020), by factorizing the original attention into temporal attention (TA) and frequency attention (FA), the large attention weight matrix can be subdivided into two much smaller ones, i.e., $(T \times T)$ and $(F' \times F')$. As shown in Fig. 4, the output y^t of $TA(Q, K, V)$ can be computed as,

$$
\begin{aligned}
& Q^t = W_Q^t * Q, K^t = W_K^t * K, V^t = W_V^t * V \\
& Q_{Res}^t, K_{Res}^t, V_{Res}^t = \mathrm{Re}\,shape^t \left(Q^t, K^t, V^t\right), \\
& \beta^t = soft\max\left(\left(Q_{Res}^t\right) \cdot \left(K_{Res}^t\right)^T\right) \\
& O^t = \mathrm{Re}\,shape^{t'}\left(\beta^t \cdot V_{Res}^t\right) \\
& y^t = \lambda O^t + K
\end{aligned}
\tag{5}
$$

where $Q_{Res}^t, K_{Res}^t, V_{Res}^t \in R^{(B \times F') \times T \times C^t}$, $C^t = \left\{\dfrac{C}{8}, \dfrac{C}{8}, C\right\}$, and $O^t \in \mathbb{R}^{B \times T \times F' \times C}$, respectively. Here, λ is a learnable scalar coefficient and initialized as 0. Similarly, $FA(Q, K, V)$ is employed after TA block, which can be expressed as,

$$
\begin{aligned}
& Q^f = W_Q^f * Q, K^f = W_K^f * K, V^f = W_V^f * V, \\
& Q_{Res}^f, K_{Res}^f, V_{Res}^f = \mathrm{Re}\,shape^f \left(Q^f, K^f, V^f\right), \\
& \beta^f = soft\max\left(\left(Q_{Res}^f\right) \cdot \left(K_{Res}^f\right)^T\right), \\
& O^f = \mathrm{Re}\,shape^{f'}\left(\beta^f \cdot V_{Res}^f\right), \\
& y^f = \lambda O^f + K,
\end{aligned}
\tag{6}
$$

where $Q_{Res}^f, K_{Res}^f, V_{Res}^f \in R^{(B \times T) \times F'^* \times C^f}$, $C^f = \left\{\dfrac{C}{8}, \dfrac{C}{8}, C\right\}$, and $O^f \in \mathbb{R}^{B \times T \times F' \times C}$, respectively. In the T-F AGs, the memory keys K come from the output of the previous layer or the final DRA block, while the queries Q and values V come from the output of the homologous downsampling layers. For the T-F SA in DRA blocks, all the Q, K, V come from the same output of the previous dilated residual layers.

A Two-Stage Complex Network Using Cycle-Consistent Generative Adversarial Networks for Speech Enhancement

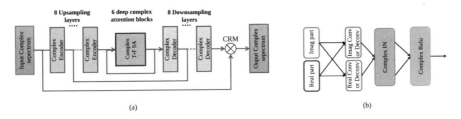

Fig. 5 (a) The topology of DCD-Net. (b) The diagram of complex-valued encoder/decoder in DCD-Net.

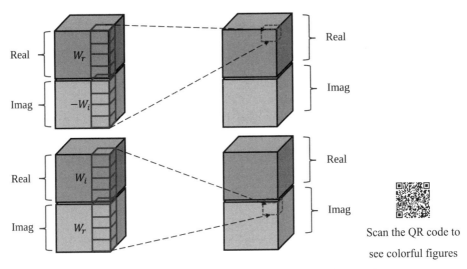

Fig. 6 The diagram of complex-valued convolutional network.

Scan the QR code to see colorful figures

3.2.2 CycleGAN-MM Loss Function

CycleGAN-MM uses the following three losses to jointly optimize the magnitude estimation process, namely relativistic adversarial losses, cycle-consistency losses, and an identity mapping loss.

Relativistic adversarial loss: For the noisy-to-clean mapping, the relativistic average least-square (RaLS) adversarial loss (Jolicoeur-Martineau, 2018) is used to make the enhanced magnitude spectra indistinguishable from the clean ones $|S_{t,f}|$, which can be expressed as below.

$$L_{Radv}(D_Y) = \\ E_{y \sim P_Y(y)}\left[\left(D_Y(y) - E_{x \sim P_X(x)} D_Y(G_{X \to Y}(x)) - 1\right)^2\right] + \\ E_{x \sim P_X(x)}\left[\left(D_Y(G_{X \to Y}(x)) - E_{y \sim P_Y(y)}(D_Y(y)) + 1\right)^2\right] \tag{7}$$

$$L_{Radv}(D_{X \to Y}) =$$
$$E_{x \sim P_X(x)}\left[\left(D_Y(G_{X \to Y}(x)) - E_{y \sim P_Y(y)} D_Y(y) - 1\right)^2\right] + \qquad (8)$$
$$E_{y \sim P_Y(y)}\left[\left(D_Y(y) - E_{x \sim P_X(x)}(D_Y(G_{X \to Y}(x))) + 1\right)^2\right]$$

where x and y is the magnitude spectrum of noisy speech and that of clean speech (e.g. $|X_{t,f}|$, $|S_{t,f}|$), respectively. Here, $L_{Radv}(D_Y)$ indicates the adversarial loss of discriminator D_Y, and $L_{Radv}(D_{X \to Y})$ indicates the adversarial loss of noisy-to-clean generator $D_{X \to Y}$. In the above equations, the generator $D_{X \to Y}$ tries to generate the enhanced magnitude spectra that can deceive the discriminator D_Y, and D_Y attempts to find the best decision boundary between the clean magnitude spectra $|S_{t,f}|$ and enhanced ones $G_{X \to Y}(|X_{i,f}|)$. Similarly, we impose two relativistic average adversarial losses $L_{Radv}(D_X)$ and $L_{Radv}(D_{Y \to X})$ for the inverse noisy-to-clean mapping.

Cycle-consistency loss: Due to high randomness, G can map noisy feature space to any random permutation of the clean feature space with only adversarial loss, so any learned mapping functions can produce an output distribution that matches the target distribution. Hence, we apply the cycle-consistency loss to limit space of possible mapping functions and preserve speech context integrity, which can be defined as follows:

$$L_{cycle}(G_{X \to Y}, F_{Y \to X}) = E_{x \sim P_X(x)}\left[\|F_{Y \to X}(G_{X \to Y}(x)) - x\|_1\right] \qquad (9)$$
$$+ E_{y \sim P_Y(y)}\left[\|G_{X \to Y}(F_{Y \to X}(y)) - y\|_1\right]$$

where $\|\cdot\|_1$ indicates the L_1-norm reconstruction error.

Identity-mapping loss: We regularize generators G and F to be close to identity mappings by minimizing identity-mapping loss as in Zhu et al. (2017), which can be given by:

$$L_{id}(G_{X \to Y}, F_{Y \to X}) = E_{x \sim P_X(x)}\left[\|F_{Y \to X}(x) - x\|_1\right] + \qquad (10)$$
$$E_{y \sim P_Y(y)}\left[\|G_{X \to Y}(y) - y\|_1\right],$$

where magnitude spectrum y and x of the target domain (i.e., $|S_{t,f}|$) and ($|X_{t,f}|$) are provided as the input to the generators (i.e., $G_{X \to Y}$ and $F_{Y \to X}$) respectively. It helps to preserve the compositions (i.e., linguistic information) of the source domain and the target domain, enforcing the generators to better map the target distribution simultaneously.

Finally, the total loss function of CycleGAN-MM can be summarized as follows:

$$L_{CycleGAN\text{-}MM} = L_{Radv}(G_{X \to Y}, D_Y) + L_{Radv}(F_{Y \to X}, D_X)$$
$$+ \lambda_{cycle} L_{cycle}(G_{X \to Y}, F_{Y \to X}) + \lambda_{id} L_{id}(G_{X \to Y}, F_{Y \to X}) \qquad (11)$$

where λ_{cycle} and λ_{id} are tunable hyper-parameters, which are initialized as 5 and 10, respectively.

3.3 Deep Complex Denoising Net (DCD-Net)

After the previous stage of enhancement, we compose the estimated magnitude with the noisy phase to get the coarsely enhanced complex spectrogram. In the second stage, the DCD-Net is proposed to further suppress the background noise and simultaneously recover the clean phase spectrum. As shown in Fig. 5, DCD-Net consists of eight complex encoder/decoder layers, and six complex Temporal-Frequency self-attention (CT-F SA) blocks.

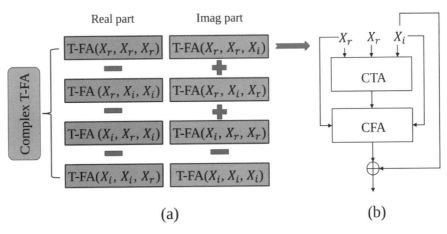

Fig. 7 (a) The calculation of Complex T-F attention. (b) The manner of T-FA (X_r, X_r, X_i).

3.3.1 Complex-Valued Encoder-Decoder

The complex encoder/decoder layers are composed of complex-valued 2D convolutions/deconvolution blocks, followed by complex instance normalization (IN) and complex PReLU (CPRelu). The complex IN and CPRelu operate instance normalization and Parametric Relu activation respectively on both real and imaginary values. We design the complex 2D convolution block according to that in DCUNET and DCCRN. The number of the channels for the encoder layers is (32, 32, 64, 64, 128, 128, 256, 256), while the kernel size and strides are set to (3, 5) and (1, 2) along the time and frequency axis, respectively. In practice, complex 2D convolution can be implemented as four traditional real-valued convolutional operations, which can be presented in Fig. 6. For complex-valued convolution operation, we define the input complex vector $X = X_r + jX_i$. Meanwhile, the complex-valued convolutional filter W is defined as $W = W_r + jW_i$, where W_r and W_i represent the real and imaginary parts of a complex convolution kernel, respectively. The complex output receives from the complex 2D convolution operation $W \otimes X$, which can be expressed as,

$$Y_{out} = (W_r * X_r - W_i * X_i) + j(W_r * X_i + W_i * X_r) \tag{12}$$

where $Y_{out} \in \mathbb{C}$ is the output of the complex-valued 2D convolution.

3.3.2 Deep Complex T–F Attention

To make DCD-Net capable of capturing long temporal dependency with complex-valued features, we introduce complex-valued attention blocks as proposed in Yang et al. (2020). Following the temporal-frequency attention as mentioned in Section 3.2.1, we propose complex temporal-frequency self-attention (CT-F SA) blocks between the complex encoder–decoder in the DCD-Net, which is calculated by using T-FA in the complex-valued manner. As illustrated in Fig. 7, given the complex-valued input $X = X_r + jX_i$, we first project them to the query matrix Q_c, the key matrix K_c and the value matrix V_c, and then calculate the complex attention output, which is defined as:

$$\begin{aligned}
Q_c &= W_Q * X, K_c = W_K * X, V_c = W_V * X, \\
\beta_c &= \text{soft}\max\left(Q_c \cdot (K_c)^T\right) = \beta_r + j\beta_i \\
&= \text{soft}\max\left(W_Q * X_r + jW_Q * X_i\right) \cdot \left(W_K * X_r + jW_K * X_i^T\right), \\
O_c &= \beta_c \cdot V_c = (\beta_r + j\beta_i) \cdot (W_V \cdot X_r + jW_V * X_i), \\
y_c &= \lambda O_c + K_c = y_r + jy_i,
\end{aligned} \tag{13}$$

where W_Q, W_K, and W_V denote the 1×1 real-valued convolutional filter. y_r and y_i denote the real and the imaginary part of the complex attention result, respectively. Hence, the complex temporal attention (CTA) can be expressed as:

$$\begin{aligned}
CTA = &(TA(X_r, X_r, X_r) - TA(X_r, X_i, X_i) - TA(X_i, X_r, X_i) \\
&- TA(X_i, X_i, X_r)) + i(TA(X_r, X_r, X_i) + TA(X_r, X_i, X_r) \\
&+ TA(X_i, X_r, X_r) - TA(X_i, X_i, X_i))
\end{aligned} \tag{14}$$

where TA denotes the temporal attention as mentioned above. Similarly, the complex frequency attention (CFA) is calculated by using FA in the complex-valued attention manner.

3.4 Loss Function

The loss function of the proposed two-stage model is defined as below. In the first step, we pretrain the CycleGAN-MM alone with $L_{CycleGAN-MM}$ until convergence, which is calculated as Eq. Then, the CycleGAN-MM and DCD-Net are jointly trained, where the parameters of the first sub-network are initialized with the pretrained CycleGAN-MM, and the overall loss function is expressed as:

$$\begin{aligned}
L_{DCD}^{Mag} &= \left\| \sqrt{|\tilde{S}_r|^2 + |\tilde{S}_i|^2} - \sqrt{|S_r|^2 + |S_i|^2} \right\|_2, \\
L_{DCD}^{RI} &= \left\| \tilde{S}_r - S_r \right\|_2 + \left\| \tilde{S}_i - S_i \right\|_2, \\
L_{Full} &= L_{DCD}^{RI} + L_{DCD}^{Mag} + \gamma L_{CycleGAN-MM},
\end{aligned} \tag{15}$$

where L_{DCD}^{Mag} and L_{DCD}^{RI} denote the loss function on the spectral magnitude and RI components, respectively. Here, $\tilde{S}_r$ and $\tilde{S}_i$ represent the RI components of the estimated clean speech spectrum, while S_r and S_i represent the target RI components of the clean speech spectrum.

4. Experiments

4.1 Datasets

In our experiments, we choose two public datasets for comparison. We first evaluated the proposed models as well as several SOTA baselines on a widely used dataset simulated on VoiceBank+DEMAND, and further evaluated our model on the WSJ0-SI84 dataset+DNS challenge.

VoiceBank + DEMAND: This dataset is widely used for evaluation as proposed in Valentini-Botinhao et al. (2016), which is a selection of the Voice Bank corpus with 30 speakers. The training dataset includes 28 speakers' 11572 utterances in the same accent region (England), while the test set contains two speakers' (one male and one female) 824 utterances. The total duration of the training set is around 10 h and the duration of the test set is around 30 mins. For both the training and testing sets, the average speech signal length was three seconds. For the training set, audio samples are added with one of the 10 noise types (2 artificial and 8 from the DEMAND database at four SNRs of 0, 5, 10 and 15 dB). The noise is from different environments including offices, public spaces, transportation stations, and streets. The test set is created with 5 test-noise types (all from the DEMAND database, but totally unseen in the training set) at SNRs of 2.5, 7.5, 12.5 and 17.5 dB. The five types of chosen noise are living room, office, bus, and street noise.

WSJ0-SI84 + DNS challenge: For further evaluation, we use the WSJ0-SI84 dataset (Paul and Baker, 1992), which includes 7138 utterances by 83 speakers (42 males and 41 females). In our experiments, we split 5428 and 957 utterances by 77 speakers for the training set and validation set, respectively. For the test set, we use two types with 150 utterances of each (seen and unseen speakers). For the first type, the speakers are totally unseen in the training set, while the speakers are within the training dataset for the second type. For the mixed set, we randomly select 20000 noises from the DNS-Challenge[①] to obtain a 55 h noise set for training. During each mixed process, a random cut vector of noise is mixed with randomly selected clean utterances. Hence, we established a 15000, 1500 noisy-clean pairs at the SNR range of [5 dB, −4 dB, −3 dB, −2 dB, −1 dB, 0dB] for training and validation, respectively. The total duration for the training set is about 30 h. For testing, we select two noises (i.e., babble and factory1) from NOISEX92 to obtain totally

---

① https://github.com/microsoft/DNS-Challenge.

900 utterances (450 for seen speakers, 450 for unseen speakers) at three SNRs of -5 dB, 0 dB and 5 dB.

4.2 Implementation Setup

The original raw waveforms were downsampled from 48 kHz to 16 kHz. The Hanning window of length 20 ms is utilized to produce a set of time frames, with 50% overlap between adjacent frames and the STFT length is 320. When training on VoiceBank + DEMAND dataset, we randomly crop a fixed-length segment (128 frames) with batch size set to 8. As for the WSJ0-SI84 dataset, the maximum utterance length is chunked to 8 s and the batch size is set to 16. We adopt the Adam optimizer with the momentum term $\beta_1 = 0.9$, $\beta_2 = 0.999$. In the first stage (20 epochs), we only train the CycleGAN-MM with an initial learning rate (LR) of 0.0002 for discriminators and 0.0005 for generators, respectively. We use L_{id} only for the first 20 epochs to guide the composition reservation. In the second stage, DCD-Net model is jointly trained with pre-trained CycleGAN-MM, while the learning rates are set to 0.001 and 0.0001 for DCD-Net and CycleGAN-MM, respectively. The same learning rates are maintained for the first 50 epochs, while they linearly decay in the remaining iterations. We train the proposed model for 80 epochs on WSJ0-SI84 + DNS dataset and 100 epochs on VoiceBank + DEMAND dataset, respectively.

4.3 Evaluation Metrics

We use the following metrics to evaluate the objective and subjective quality of the enhanced speech. The objective metrics measure the similarity between the enhanced signal and the clean reference of the test set files. The subjective quality is evaluated by DNSMOS, which is a robust nonintrusive perceptual speech quality metric designed to stack rank noise suppressors with great accuracy. Higher values of all metrics indicate better performance.

(1) PESQ: Perceptual evaluation of speech quality (PESQ) score is the most commonly used metric to evaluate speech quality, especially using the wide-band version recommended in ITU-T P.862.2 (from -0.5 to 4.5).

(2) STOI: Short-Time Objective Intelligibility (STOI) is used as a robust measurement index for nonlinear processing of noisy speech, e.g., noise reduction on speech intelligibility. The value of STOI ranges from 0 to 1.

(3) SSNR: Segmental signal-to-noise ratio, ranging from 0 to ∞.

(4) CSIG: The mean opinion score (MOS) predicts the speech signal distortion, ranging from 1 to 5.

(5) CBAK: The MOS predicts the intrusiveness of background noise, ranging from 1 to 5.

(6) COVL: The MOS predicts the overall effect, ranging from 1 to 5.

(7) DNSMOS: The speech quality ratings of the processed clips varied from very poor

(MOS=1) to excellent (MOS=5).

4.4 Baselines

For the comparison on VoiceBank + DEMAND dataset, we adopt both GAN-based and Non-GAN based methods, which are summarized as follows:

(1) **SEGAN** is the first SE approach based on the adversarial framework and works end-to-end with the raw audio. It applies to skip connection to generators, connecting each encoding layer to its homologous decoding layer.

(2) **MMSE-GAN** introduces a time–frequency masking-based SE approach based on a modified GAN and learns the mask implicitly while predicting the clean T–F representation.

(3) **RSGAN** and **RSaGAN** introduce relativistic GANs with a relativistic cost function at its discriminators and use gradient penalty to improve speech enhancement performance in the time domain. Note that RSGAN-GP employs relativistic binary cross-entropy loss while RaSGAN-GP employs relativistic average binary cross-entropy loss.

(4) **CP-GAN** is a novel GAN-based SE system for coarse-to-fine noise suppression, which contains a densely-connected feature pyramid generator and a dynamic context granularity discriminator.

(5) **MetricGAN** aims to optimize the generator with respect to one or multiple evaluation metrics such as PESQ and STOI, thus guiding the generators in GANs to generate data with improved metric scores.

(6) **Wave-U-Net** uses the U-Net architecture for speech enhancement, which performs end-to-end audio source separation directly in the time domain.

(7) **LSTM** and **BiLSTM** are two RNN-based speech enhancement approaches. Both of them have two layers of RNN cells and the third layer of fully connected NNs.

(8) **DFL-SE** proposes a fully-convolutional context aggregation network using a deep feature loss at the raw waveform level, which is based on comparing the internal feature activations in a different network.

(9) **CRN-MSE** is a typical convolutional recurrent network with encoder–decoder architecture, and MSE loss is used on the estimated and clean log-magnitude spectrogram. Note that we directly use the reported scores of RSGAN, RaSGAN, LSTM, BiLSTM and CRN-MSE from Zhang et al. (2020).

(10) **TFSNN** proposes a time–frequency smoothing neural network for SE, which effectively models the correlation in the time and frequency dimensions by using LSTM and CNN, respectively.

To further evaluate the proposed method at different SNRs On WSJ0-SI84 + DNS, we re-implement three state-of-the-art baselines, namely **Noncausal-GCRN, DCUNET-20** and **Noncausal-DCCRN**. Noncausal-GCRN is a complex spectral mapping network based on CRN, where both the real and imaginary components are estimated. Notably, to make a fair comparison, we reimplement GCRN as a non-causal configuration, where the Bi-directional LSTM is utilized for time sequencing. DCUNET-20 adopts the complex-valued building blocks and bounded CRM to deal with the complex-valued spectrum. We use the structure of DCUNET-20 with the stride along with the time dimension set to 1. The channels in encoders are set to [8, 16, 32, 32, 64, 64, 128, 128, 256, 256]. Noncausal-DCCRN introduces a deep complex convolution recurrent network for speech enhancement, where both CNN and RNN structures handle complex-valued operations. We reimplement DCCRN and employ the Bi-directional LSTM to train the model with a non-causal configuration. Note that we also implement tanh bounded CRM and optimize the system with SI-SNR loss.

5. Results and Analysis

5.1 Ablation Study

We first investigate the effectiveness of different attention mechanisms and loss functions based on VoiceBank + DEMAND dataset. As shown in Table I, we take the CycleGAN with self-attention (SA) and least-square (LS) GAN loss as the baseline. Besides, we implement a CycleGAN-based one-stage network for complex spectral mapping (dubbed CycleGAN-CM), which set the same configuration as CycleGAN-MM (III) except two decoders to separately decode real and imaginary components. From the results, we can have the following observations. When only the first stage is trained, CycleGAN-MM using RaLS loss achieves 0.07 PESQ, 0.49 dB SSNR, 0.08 CSIG, 0.06 CBAK and 0.09 COVL improvements over using conventional LS loss, while using T-FA results in better performance than using SA. CycleGAN-MM (III) integrating RaLS loss and T-FA obtains 0.12 PESQ, 1.03 dB SSNR, 0.5% STOI, 0.13 CSIG, 0.05 CBAK and 0.17 COVL improvements than CycleGAN-MM (I). Additionally, when handling one-stage complex spectral mapping, CycleGAN-CM achieves worse performance than CycleGAN-MM. This reveals the difficulty for conventional GAN-based systems to deal with directly mapping complex spectrum, which is likely caused by the unstable training of the generators and discriminators. In other words, optimizing both RI components is an intractable challenge for the GAN-based SE approaches, and thus achieving worse performance than only estimating the magnitude (see Table I).

Table I Ablation study on both CycleGAN-MM and CycleGAN-DCD. SA and w/o denote self-attention and without any attention mechanism, respectively. LS and RaLS in stage one denote least-square loss and relativistic average least-square loss, while RI and Mag in stage two denote the MSE loss of RI components and spectral magnitude, respectively.

Models	Attention	Loss	PESQ	SSNR	STOI (%)	CSIG	CBAK	COVL
Unprocessed	N/A	N/A	1.97	1.68	92.1	3.35	2.44	2.63
One-stage system								
CycleGAN-MM(I)	SA	LS	2.57	5.69	92.7	3.85	3.11	3.10
CycleGAN-MM(II)	SA	RaLS	2.64	6.18	92.8	3.93	3.17	3.19
CycleGAN-MM(III)	T-FA	RaLS	2.69	6.72	93.2	3.98	3.16	3.27
CycleGAN-CM	T-FA	RaLS	2.24	3.37	92.0	3.54	2.68	2.71
Two-stage system								
CycleGAN-DCD(I)	w/o	RI	2.72	8.21	92.9	3.91	3.41	3.29
CycleGAN-DCD(II)	w/o	RI+Mag	2.82	8.92	94.1	4.14	3.48	3.39
CycleGAN-DCD(III)	CT-F SA	RI+Mag	**2.90**	**9.33**	**94.3**	**4.24**	**3.57**	**3.49**

Table II Comparison with different complex-spectrum networks and two-stage combinations.

Methods	PESQ	STOI (%)	CSIG	CBAK	COVL
Noisy	1.97	92.1	3.35	2.44	2.63
One-stage complex systems					
Noncausal-GCRN	2.51	93.7	3.71	3.21	3.09
Noncausal-DCCRN	2.72	93.9	3.90	3.20	3.29
DCD-Net	2.70	93.8	3.92	3.24	3.21
Two-stage complex systems					
CycleGAN-GCRN	2.75	93.9	3.93	3.24	3.18
CycleGAN-DCCRN	2.84	94.1	**4.26**	3.45	3.41
CycleGAN-DCD	**2.90**	**94.3**	4.24	**3.57**	**3.49**

Subsequently, in the two-stage systems, DCD-Net is jointly trained with CycleGAN-MM (III) as the first-stage system. When only using RI components loss in DCD-Net, we obtain a marginal improvement over CycleGAN-MM on reducing background noise (e.g., 1.49 dB SSNR and 0.25 CBAK improvements), while achieving similar PESQ and STOI. This reveals the necessity and significance of the second stage in residual noise suppression. After adding the magnitude MSE loss of the estimated and clean spectrum, CycleGAN-DCD (II) improves PESQ by 0.10, STOI by 1.2% and CSIG by 0.23, respectively. This indicates that using both RI and Mag loss in DCD-Net achieves a better performance on speech quality (i.e., PESQ), speech intelligibility (i.e., STOI) and speech distortion (i.e., CSIG). Note that both CycleGAN-DCD (I) and CycleGAN-DCD (II) are trained without any attention block in DCD-Net. Compared with CycleGAN-DCD (II), CycleGAN-DCD (III) trained with CT-F SA blocks provides 0.08 gain on PESQ, 0.41 dB gain on SSNR, 0.10 gain on CSIG, 0.09 gain on CBAK and 0.10 gain on COVL, respectively. These results verify the effectiveness of the proposed attention mechanism and loss function for improving speech quality in terms of all objective metrics.

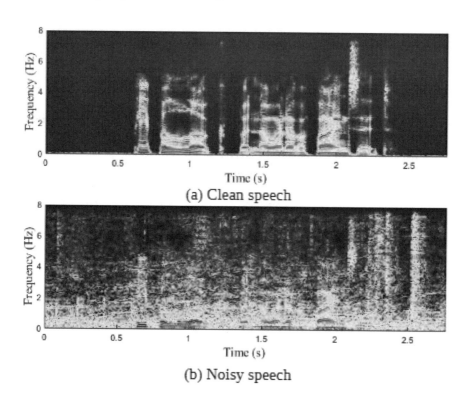

(a) Clean speech

(b) Noisy speech

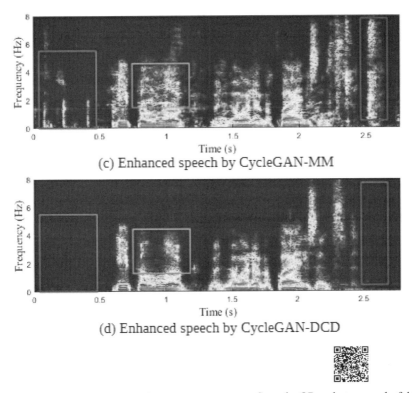

Scan the QR code to see colorful figures

Fig. 8 Illustration of the enhanced results using our proposed models. (a) The speech spectrogram of clean speech. (b) The speech spectrogram of noisy utterance. (c) The speech spectrogram of the enhanced utterance by CycleGAN-MM. (d) The speech spectrogram of the enhanced utterance by CycleGAN-DCD.

Fig. 8 shows the spectrograms of the clean utterance, the noisy utterance and the utterance enhanced by CycleGAN-MM and CycleGAN-DCD. From the figure, we observe that CycleGAN-DCD can effectively suppress the noise components, which are intractable to be eliminated in CycleGAN-MM. For example, as shown in the red sign area of Fig. 8(c) and (d), CycleGAN-DCD achieved better performance under the pure background noise condition. Besides, the green sign area shows CycleGAN-DCD can also effectively suppress the unnatural residual noise while well preserving the speech components in the case of background noise and speech are heavily mixed.

Table III Experimental results among different models including GAN-based systems and Non-GAN based systems on VoiceBank + DEMAND dataset. We directly use previously reported results. N/A denotes the result is not provided in the original paper.

Methods	Causality	PESQ	STOI (%)	CSIG	CBAK	COVL
Noisy	–	1.97	92.1	3.35	2.44	2.63
GAN-based systems						
SEGAN	×	2.16	92.5	3.48	2.94	2.80
MMSEGAN	×	2.53	93.0	3.80	3.12	3.14
RSGAN	×	2.51	93.7	3.78	3.23	3.16
RaSGAN	×	2.57	93.7	3.83	3.28	3.20
CP-GAN	×	2.64	94.0	3.93	3.29	3.28
MetricGAN	×	2.86	N/A	3.99	3.18	3.42
Non-GAN based systems						
Wave-U-Net	×	2.64	N/A	3.56	3.08	3.09
LSTM	√	2.56	91.4	3.87	2.87	3.20
BiLSTM	×	2.70	92.5	3.99	2.95	3.34
DFL-SE	√	N/A	N/A	3.86	3.33	3.22
CRN-MSE	√	2.74	93.4	3.86	3.14	3.30
TFSNN	√	2.79	N/A	4.17	3.27	**3.49**
Proposed CycleGAN-based approaches						
CycleGAN-MM	×	2.69	93.2	3.98	3.16	3.27
CycleGAN-DCD	×	**2.90**	**94.3**	**4.24**	**3.57**	**3.49**

5.2 Comparison with Different Two-Stage Structures

To validate the efficacy of DCD-Net and the proposed two-stage complex structure, we also conduct experiments on VoiceBank + DEMAND dataset. Specifically, we first compare DCD-Net with other existing complex spectrum enhancing networks (i.e., Noncausal-GCRN and Noncausal-DCCRN) and then design their corresponding two-stage structures (i.e., CycleGAN-GCRN and CycleGAN-DCCRN) for comparison. Note that the trainable parameter of Noncausal-GCRN, that of Noncausal-DCCRN and that of DCD-Net are 9.8

million, 3.7 million and 3.5 million, respectively. As shown in Table II, one can observe following phenomena. Firstly, when compared with real-valued complex-mapping network, DCD-Net outperforms GCRN in all metrics by a large margin. For example, DCD-Net provides average 0.19 PESQ, 0.21 CSIG, 0.03 CBAK and 0.12 COVL improvements than GCRN with relatively lower model complexity, which indicates the merit of complex-valued networks. Secondly, DCD-Net surpasses another complex-valued network (i.e., DCCRN) in terms of COVL and CBAK scores, while providing similar PESQ and STOI scores. This indicates DCD-Net can effectively suppress the background noise and reduce the speech distortion simultaneously. Finally, to demonstrate the merit of the proposed two-stage combination, we combine CycleGAN-MM with GCRN, DCCRN and DCD-Net as different two-stage structures (i.e., CycleGAN-GCRN, CycleGAN-DCCRN and the proposed CycleGAN-DCD) for comparison. Note that we also use bounded CRM in CycleGAN-GCRN for better performance. Compared with different two-stage methods, CycleGAN-DCD achieves consistently better performances than real-valued CycleGAN-GCRN by a significant margin, while CycleGAN-DCD outperforms complex-valued CycleGAN-DCCRN in terms of PESQ, STOI, CBAK and COVL. This validates that the proposed CycleGAN-DCD surpasses other two-stage structures including real-valued and complex-valued methods.

5.3 Comparison with the State-of-the-Art Baselines

Table III shows the comparisons with mentioned baselines on VoiceBank + DEMAND dataset. Note that CycleGAN-MM and CycleGAN-DCD employ the best configurations from the ablation study. First, we observe that CycleGAN-DCD achieves a notable improvement over most existed GAN-based methods. For example, CycleGAN-DCD exceeds SEGAN by a large margin in PESQ, STOI, CSIG, CBAK and COVL, which are 0.73, 1.8%, 0.76, 0.63 and 0.69, respectively. Compared with more recently proposed CP-GAN and Metric-GAN, our model still achieves better performance in speech quality and speech intelligibility. Note that the consistent improvements in CSIG, CBAK and COVL also indicate that CycleGAN-DCD performs better in preserving speech integrity while removing the background noise. Then, when it comes to recently proposed Non-GAN based methods, the proposed model also achieves better performance across most metrics. For example, CycleGAN-DCD provides 0.38 CSIG, 0.24 CBAK and 0.27 COVL improvements than DFL-SE, while CycleGAN-MM gets lower CBAK and similar COVL over DFL-SE. This indicates the two-stage denoising system demonstrates consistently superior performance on background noise suppression, while further improving the speech distortion and overall effect.

Table IV shows the comparisons with DCUNET, GCRN and DCCRN on WSJ0-SI84 + DNS dataset. Firstly, we observe that DCUNET, Noncausal-DCCRN and Noncausal-GCRN obtain better performance than CycleGAN-MM under different noise conditions. This is because CycleGAN-MM only estimates the magnitude spectrum and reuses the noisy phase to reconstruct waveform, which causes severe phase distortion under low SNRs. For example, Noncausal-GCRN provides average 0.28 and 0.27 PESQ improvements than CycleGAN-MM on Babble and Factory1 noises, while 3.71% and 1.86% improvements in terms of STOI. Secondly, when adding the denoising net to refine the coarsely enhanced complex spectrum, CycleGAN-DCD outperforms the one-stage model by a large margin in all metrics. For example, CycleGAN-DCD provides average 0.32 and 0.30 PESQ improvements than CycleGAN-MM on Babble and Factory1 noises, while providing 0.99 dB and 1.03 dB gain SSNR. This indicates the necessity and significance of the proposed DCD-Net in improving the speech quality and intelligibility, while further suppressing the residual noise. It can also be observed that the proposed two-stage model consistently outperforms the baselines in terms of all metrics. For example, compared with the best baseline Noncausal-GCRN, we notice that CycleGAN-DCD obtains average 0.06, 0.33% and 0.21 dB improvements in terms of PESQ, STOI and SSNR, respectively.

Table IV Objective result comparisons among different models in terms of PESQ, STOI and SSNR for both seen and unseen speaker cases. BOLD indicates the best score in each case.

Noise	Metrics	Causality	PESQ Seen -5	0	5	Avg.	PESQ Unseen -5	0	5	Avg.	STOI (%) Seen -5	0	5	Avg.	STOI Unseen -5	0	5	Avg.	SSNR (dB) Seen -5	0	5	Avg.	SSNR Unseen -5	0	5	Avg.
Babble	Noisy	N/A	1.41	1.56	1.80	1.60	1.38	1.52	1.78	1.56	59.10	74.18	86.38	73.49	60.25	73.65	85.30	73.07	-6.70	-4.09	-0.97	-3.92	-6.77	-4.21	-1.12	-4.03
	DCUNET	×	1.75	2.34	2.88	2.32	1.71	2.23	2.82	2.25	79.28	91.82	95.74	88.95	78.53	91.26	95.28	88.36	1.98	4.12	5.89	3.95	2.01	4.06	5.79	4.01
	Noncausal-DCCRN	×	1.77	2.39	2.91	2.36	1.73	2.26	2.83	2.27	82.12	91.19	95.97	89.76	81.72	91.13	95.93	89.59	1.19	2.47	4.40	2.67	1.32	2.38	4.33	2.69
	Noncausal-GCRN	×	1.83	2.29	2.88	2.34	1.78	2.29	2.87	2.31	83.09	92.55	96.32	90.65	82.61	92.51	96.22	90.44	2.15	4.47	6.95	4.53	2.16	4.84	6.97	4.66
	CycleGAN-MM(Pro.)	×	1.55	2.04	2.60	2.06	1.52	2.02	2.58	2.04	76.02	89.54	94.98	86.85	75.70	89.72	95.03	86.82	1.12	3.82	6.22	3.72	1.02	3.84	6.34	3.74
	CycleGAN-DCD(Pro.)	×	**1.85**	**2.37**	**2.92**	**2.38**	**1.81**	**2.36**	**2.90**	**2.35**	**83.43**	**92.78**	**96.34**	**90.86**	**82.95**	**92.73**	**96.25**	**90.63**	2.30	4.91	**7.18**	**4.80**	2.35	**5.03**	**7.23**	**4.87**
Factory1	Noisy	N/A	1.35	1.56	1.75	1.59	1.32	1.47	1.73	1.51	59.93	74.22	87.14	73.52	59.97	74.22	86.37	73.52	-6.70	-0.05	-1.08	0.02	-6.82	-4.27	5.03	-1.20
	DCUNET	×	1.88	2.36	2.90	2.37	1.82	2.28	2.85	2.31	81.33	91.60	95.12	89.25	80.24	91.24	95.58	89.02	2.12	4.60	6.01	4.23	2.16	4.67	6.12	4.32
	Noncausal-DCCRN	×	1.89	2.39	2.90	2.40	1.81	2.30	2.88	2.33	82.21	91.20	95.84	89.75	81.61	90.88	96.01	89.50	1.42	2.50	4.64	2.86	1.45	2.72	4.70	2.96
	Noncausal-GCRN	×	1.90	2.36	2.91	2.39	**1.92**	2.41	2.93	2.42	82.48	91.74	95.92	90.04	82.44	92.04	96.07	90.18	2.41	4.78	6.70	4.63	2.94	4.85	6.90	4.90
	CycleGAN-MM(Pro.)	×	1.69	2.16	2.68	2.18	1.66	2.15	2.65	2.15	79.26	90.37	96.10	88.58	78.23	90.34	95.20	87.92	1.79	4.27	6.38	4.14	1.80	4.23	6.45	4.16
	CycleGAN-DCD(Pro.)	×	**1.93**	**2.56**	**2.96**	**2.49**	1.91	**2.44**	**2.94**	**2.44**	**83.99**	**92.38**	**96.25**	**90.88**	**82.53**	**92.09**	**96.10**	**90.24**	2.51	**4.97**	**6.98**	**4.82**	3.05	**5.04**	**7.18**	**5.09**

A Two-Stage Complex Network Using Cycle-Consistent Generative Adversarial Networks for Speech Enhancement

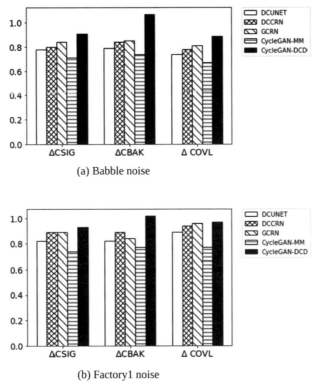

Fig. 9 Average CSIG, CBAK and COVL improvements over the unprocessed mixtures for different noise types at SNRs of −5, 0 and 5 dB

Besides, CSIG, CBAK and COVL improvements (i.e., $\triangle CSIG$, $\triangle CBAK$ and $\triangle COVL$) over the unprocessed mixtures are shown in Fig. 9. We can observe that our proposed approach produces considerable improvements than all the baselines in all metrics, especially in CBAK. This reveals the superior capability of CycleGAN-DCD on reducing residual noise and speech distortion, while consistently improving the speech overall quality.

The evaluation of the subjective speech quality on WSJ0-SI84 + DNS dataset is presented in Table V. We can observe that our method yields the best performance on both seen and unseen speakers with Babble and Factory1 noise types. For example, CycleGAN-DCD yields average 0.27 and 0.30 DNSMOS scores over one-stage CycleGAN-MM for Factory1 and Babble noise types, respectively. It indicates that our two-stage system can dramatically improve the speech perception of enhanced speech over CycleGAN-MM and other baseline systems under various noisy conditions.

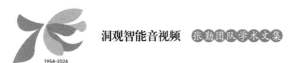

Table V The average DNSMOS scores of CycleGAN-DCD and the baseline approaches on the test set at SNRs of −5, 0 and 5 dB.

Noise type	Factory1		Babble	
Speakers	Seen	Unseen	Seen	Unseen
Mixture	2.30	2.31	2.62	2.65
DCUNET	3.12	3.13	3.16	3.16
Noncausal-DCCRN	3.15	3.15	3.18	3.18
Noncausal-GCRN	**3.30**	3.31	3.36	3.38
CycleGAN-MM(Pro.)	3.03	3.06	3.11	3.15
CycleGAN-DCD(Pro.)	**3.30**	**3.33**	**3.42**	**3.43**

6. Conclusion and Future Work

In this work, a deep complex-valued denoising sub-net is integrated into a CycleGAN-based magnitude mapping sub-net as a two-stage SE approach, which aims at estimating both the magnitude and phase of the clean speech spectrum. In the first stage, a CycleGAN-based network is first trained to estimate the spectral magnitude with relativistic average least-square losses, cycle-consistency losses and an identity mapping loss. Then, the coarsely estimated magnitude is coupled with the original noisy phase as the input to a complex denoising net, which aims to suppress the residual noise and recover the clean phase. Notably, the denoising net directly estimates both RI components of the clean spectrum applied with a complex ratio mask. Additionally, the temporal-frequency attention mechanism is employed in both two stages for modeling the global dependencies along temporal and frequency dimensions, respectively. To the best of our knowledge, this is the first CycleGAN-based approach to estimate both the clean magnitude and phase information for single-channel SE. Experiments results on VoiceBank and WSJ0-SI84 datasets verify that the proposed method outperforms the conventional one-stage CycleGAN-based SE model and other state-of-the-art GAN-based as well as Non-GAN based baselines by a considerable margin.

In future work, we will investigate the proposed CycleGAN-DCD as a complex spectral mapping network for multi-microphone speech enhancement, in which accurate phase estimation is likely more essential. Considering the promising performance of power compression and phase estimation on speech dereverberation task as discussed in Li et al. (2021b), we will investigate to use the compressed spectral magnitude as the input feature to the first stage. Besides, we will attempt to decompose the two-stage SE task into two much easier sub-tasks. In the first task, we plan to employ a CycleGAN-based network to transform the non-stationary noise type to stationary noise type like noise-whitening, while we plan to utilize a denoising net to suppress the stationary noise in the second sub-task.

Credit authorship contribution statement

Guochen Yu: Software, Draft. Yutian Wang: Data, Investigation. **Hui Wang:** Supervision, Methodology, Editing. Qin Zhang: Grammar correction, Conceptualization. Chengshi Zheng: Supervision, Methodology, Writing reviewing.

Declaration of competing interest

The authors declare that they have no known competing financial interests or personal relationships that could have appeared to influence the work reported in this paper.

Acknowledgments

This work was supported in part by the National Natural Science Foundation of China under Grant 61631016 and Grant 61501410, and in part by the Fundamental Research Funds for the Central Universities, China under Grant 3132018XNG1805. This work was also supported by the Open Research Project of the State Key Laboratory of Media Convergence and Communication, Communication University of China, China (No. SKLMCC2020KF005).

References

[1] ANDERSON S, WHITE-SCHWOCH T, PARBERY-CLARK A, et al. A dynamic auditory-cognitive system supports speech-in-noise perception in older adults[J]. Hearing research, 2013, 300: 18-32.

[2] BABY D, VERHULST S. Sergan: speech enhancement using relativistic generative adversarial networks with gradient penalty[C]//ICASSP 2019-2019 IEEE International Conference on Acoustics, Speech and Signal Processing. IEEE, 2019: 106-110.

[3] BRUNNER G, WANG Y, WATTENHOFER R, et al. Symbolic music genre transfer with cyclegan[C]//2018 IEEE 30th International Conference on Tools with Artificial Intelligence . IEEE, 2018: 786-793.

[4] CHOI H S, KIM J H, HUH J, et al. Phase-aware speech enhancement with deep complex U-Net[C]//7th International Conference on Learning Representations, May 6-9, 2019: 1-20.

[5] DAUPHIN Y N, FAN A, AULI M, et al. Language modeling with gated convolutional networks[C]//International Conference on Machine Learning. PMLR, 2017: 933-941.

[6] DU Z, ZHANG X, HAN J. A joint framework of denoising autoencoder and generative vocoder for monaural speech enhancement[J]. IEEE/ACM transactions on audio, speech, and language processing, 2020, 28: 1493-1505.

[7] ERDOGAN H, HERSHEY J R, WATANABE S, et al. Phase-sensitive and recognition-boosted speech separation using deep recurrent neural networks[C]//2015 IEEE International Conference on Acoustics, Speech and Signal Processing. IEEE, 2015: 708-712.

[8] FU S W, LIAO C F, TSAO Y, et al. Metricgan: Generative adversarial networks based black-box metric scores optimization for speech enhancement[C]//International Conference on Machine Learning. PMLR, 2019: 2031-2041.

[9] GERMAIN F, CHEN Q, KOLTUN V. Speech denoising with deep feature losses[C]//Proceedings of the Annual Conference of the International Speech Communication Association. INTERSPEECH, 2019.

[10] GOODFELLOW I, POUGET-ABADIE J, MIRZA M, et al. Generative Adversarial Nets[C]//Proceedings of the 27th International Conference on Neural Information Processing Systems. MIT Press, Cambridge, MA, USA, 2014, 2: 2672-2680.

[11] HAO X, SU X, WEN S, et al. Masking and inpainting: a two-stage speech enhancement approach for low SNR and non-stationary noise[C]//ICASSP 2020-2020 IEEE International Conference on Acoustics, Speech and Signal Processing. IEEE, 2020: 6959-6963.

[12] HEDJAZI M A, GENC Y. Texture-aware multi-resolution image inpainting[J]. arXiv preprint arXiv: 2020, 2009.14721.

[13] HU Y, LIU Y, LV S, et al. DCCRN: Deep complex convolution recurrent network for phase-aware speech enhancement[J]. arXiv preprint arXiv: 2020, 2008.00264.

[14] HU Y, LOIZOU P C. Evaluation of objective quality measures for speech enhancement[J]. IEEE transactions on audio, speech, and language processing, 2007, 16(1): 229-238.

[15] HUMMERSONE C, STOKES T, BROOKES T. On the ideal ratio mask as the goal of computational auditory scene analysis[M]. Blind source separation: advances in theory, algorithms and applications. Berlin, Heidelberg: Springer, 2014: 349-368.

[16] JOLICOEUR-MARTINEAU A. The relativistic discriminator: a key element missing from standard GAN[J]. arXiv preprint arXiv: 2018, 1807.00734.

[17] KANEKO T, KAMEOKA H. Cyclegan-vc: Non-parallel voice conversion using cycle-consistent adversarial networks[C]//2018 26th European Signal Processing Conference. IEEE, 2018: 2100-2104.

[18] KINGMA D P, BA J. Adam: A method for stochastic optimization[J]. arXiv preprint arXiv: 2014, 1412.6980.

[19] KULMER J, MOWLAEE P. Phase estimation in single channel speech enhancement using phase decomposition[J]. IEEE signal processing letters, 2014, 22(5): 598-602.

[20] LI A, LIU W, LUO X, et al. ICASSP 2021 Deep noise suppression challenge: decoupling magnitude and phase optimization with a two-stage deep network[J]. arXiv preprint arXiv: 2021, 2102.04198.

[21] LI X, WU J, LIN Z, et al. Recurrent squeeze-and-excitation context aggregation net for single image deraining[C]//Proceedings of the European conference on computer vision. 2018: 254-269.

[22] LI A, ZHENG C, PENG R, et al. On the importance of power compression and phase estimation in monaural speech dereverberation[J]. JASA express letters, 2021, 1(1): 014802.

[23] LIAO C F, TSAO Y, LEE H Y, et al. Noise adaptive speech enhancement using domain adversarial training[J]. arXiv preprint arXiv: 2018, 1807.07501.

[24] LIU G, GONG K, LIANG X, et al. Cp-gan: context pyramid generative adversarial network

[25] LOIZOU P C. Speech enhancement: theory and practice[M]. Boca Raton: CRC Press, 2013.

[26] LU X, TSAO Y, MATSUDA S, et al. Speech enhancement based on deep denoising autoencoder[C]//Interspeech. 2013, 2013: 436-440.

[27] MENG Z, LI J, GONG Y, et al. Adversarial Feature-Mapping for Speech Enhancement[C]//Interspeech (2018), 2018: 3259-3263.

[28] MENG Z, LI J, GONG Y, et al. Cycle-consistent speech enhancement[J]. ArXiv: 2018, abs/1809. 02253.

[29] MIYATO T, KATAOKA T, KOYAMA M, et al. Spectral normalization for generative adversarial networks[J]. ArXiv: 2018, 1802.05957: 1-26.

[30] SAEIDI R, MOWLAEE P, MARTIN R. Phase estimation for signal reconstruction in single-channel source separation[C]//13th Annual Conference of the International Speech Communication Association, INTERSPEECH, September 9-13, 2012: 1548-1551.

[31] PALIWAL K, WÓJCICKI K, SHANNON B. The importance of phase in speech enhancement[J]. Speech Communication, 2011, 53(4): 465-494.

[32] PANDEY A, WANG D L. Densely connected neural network with dilated convolutions for real-time speech enhancement in the time domain[C]//ICASSP 2020-2020 IEEE International Conference on Acoustics, Speech and Signal Processing. IEEE, 2020: 6629-6633.

[33] PASCUAL S, BONAFONTE A, SERRÀ J SEGAN: Speech Enhancement Generative Adversarial Network[C]//Interspeech (2017), 2017: 3642-3646.

[34] PASCUAL S, SERRA J, BONAFONTE A. Time-domain speech enhancement using generative adversarial networks[J]. Speech communication, 2019, 114: 10-21.

[35] PAUL D B, BAKER J M. The design for the wall street journal-based CSR corpus[C]//Proceedings of the Workshop on Speech and Natural Language. 1992: 357-362.

[36] REDDY C K A, GOPAL V, CUTLER R. DNSMOS: A non-intrusive perceptual objective speech quality metric to evaluate noise suppressors[J]. arXiv preprint arXiv: 2020, 2010.15258.

[37] RIX A W, BEERENDS J G, HOLLIER M P, et al. Perceptual evaluation of speech quality (PESQ)-a new method for speech quality assessment of telephone networks and codecs[C]//2001 IEEE international conference on acoustics, speech, and signal processing. Proceedings (Cat. No. 01CH37221). IEEE, 2001, 2: 749-752.

[38] SONI M H, SHAH N, PATIL H A. Time-frequency masking-based speech enhancement using generative adversarial network[C]//2018 IEEE international conference on acoustics, speech and signal processing (ICASSP). IEEE, 2018: 5039-5043.

[39] STOLLER D, EWERT S, DIXON S. Wave-u-net: a multi-scale neural network for end-to-end audio source separation[J]. arXiv preprint arXiv: 2018, 1806.03185.

[40] TAAL C H, HENDRIKS R C, HEUSDENS R, et al. A short-time objective intelligibility measure for time-frequency weighted noisy speech[C]//2010 IEEE international conference on acoustics, speech and signal processing. IEEE, 2010: 4214-4217.

[41] TAN K, WANG D L. A convolutional recurrent network for real-time speech

enhancement[C]//Interspeech 2018, 2018: 3229-3233.

[42] TAN K, WANG D L. Complex spectral map with a convolutional recurrent network for monaural speech enhancement[C]//IEEE International Conference on Acoustics, Speech and Signal Processing. IEEE, 2019: 6865-6869.

[43] TAN K, WANG D L. Learning complex spectral map with gated convolutional recurrent networks for monaural speech enhancement[J]. IEEE/ACM transactions on audio, speech, and language processing, 2019, 28: 380-390.

[44] TANG C, LUO C, ZHAO Z, et al. Joint time-frequency and time domain learning for speech enhancement[C]//Proceedings of the twenty-ninth international conference on international joint conferences on artificial intelligence. 2021: 3816-3822.

[45] THIEMANN J, ITO N, VINCENT E. The diverse environments multi-channel acoustic noise database: A database of multichannel environmental noise recordings[J]. Acoustical society of america journal, 2013, 133(5): 3591.

[46] TRABELSI C, BILANIUK O, ZHANG Y, et al. Deep complex networks[J]. arXiv preprint arXiv: 2017, 1705.09792.

[47] VALENTINI-BOTINHAO C, WANG X, TAKAKI S, et al. Investigating RNN-based speech enhancement methods for noise-robust Text-to-Speech[C]//SSW. 2016: 146-152.

[48] VARGA A, STEENEKEN H J M. Assessment for automatic speech recognition: II. NOISEX-92: a database and an experiment to study the effect of additive noise on speech recognition systems[J]. Speech communication, 1993, 12(3): 247-251.

[49] VASWANI A, SHAZEER N, PARMAR N, et al. Attention is All You Need[C]//Proceedings of the 31st International Conference on Neural Information Processing Systems. Curran Associates Inc., Red Hook, NY, USA, 2017: 5998-6008.

[50] VEAUX C, YAMAGISHI J, KING S. The voice bank corpus: design, collection and data analysis of a large regional accent speech database[C]//2013 international conference oriental COCOSDA held jointly with 2013 conference on Asian spoken language research and evaluation. IEEE, 2013: 1-4.

[51] VENKATARAMANI S, CASEBEER J, SMARAGDIS P. Adaptive front-ends for end-to-end source separation[C]//Proc. NIPS. 2017.

[52] WANG D L, CHEN J. Supervised speech separation based on deep learning: an overview[J]. IEEE/ACM transactions on audio, speech, and language processing, 2018, 26(10): 1702-1726.

[53] WANG D, LIM J. The unimportance of phase in speech enhancement[J]. IEEE transactions on acoustics, speech, and signal processing, 1982, 30(4): 679-681.

[54] WANG Y, NARAYANAN A, WANG D L. On training targets for supervised speech separation[J]. IEEE/ACM transactions on audio, speech, and language processing, 2014, 22(12): 1849-1858.

[55] WANG Z Q, WANG P, WANG D L. Complex spectral mapping for single-and multi-channel speech enhancement and robust ASR[J]. IEEE/ACM transactions on audio, speech, and language processing, 2020, 28: 1778-1787.

[56] WANG Y, YU G, WANG J, et al. Improved relativistic cycle-consistent gan with dilated residual network and multi-attention for speech enhancement[J]. IEEE access, 2020, 8:

[57] WENINGER F, HERSHEY J R, LE ROUX J, et al. Discriminatively trained recurrent neural networks for single-channel speech separation[C]//2014 IEEE global conference on signal and information processing (GlobalSIP). IEEE, 2014: 577-581.

[58] WILLIAMSON D S, WANG Y, WANG D L. Complex ratio masking for monaural speech separation[J]. IEEE/ACM transactions on audio, speech, and language processing, 2015, 24(3): 483-492.

[59] XIANG Y, BAO C. A parallel-data-free speech enhancement method using multi-objective learning cycle-consistent generative adversarial network[J]. IEEE/ACM transactions on audio, speech, and language processing, 2020, 28: 1826-1838.

[60] XU Y, DU J, DAI L R, et al. A regression approach to speech enhancement based on deep neural networks[J]. IEEE/ACM transactions on audio, speech, and language processing, 2014, 23(1): 7-19.

[61] YANG M, MA M Q, LI D, et al. Complex transformer: a framework for modeling complex-valued sequence[C]//ICASSP 2020-2020 IEEE International Conference on Acoustics, Speech and Signal Processing. IEEE, 2020: 4232-4236.

[62] YUAN W. A time-frequency smoothing neural network for speech enhancement[J]. Speech communication, 2020, 124: 75-84.

[63] ZHANG Z, DENG C, SHEN Y, et al. On loss functions and recurrency training for GAN-based speech enhancement systems[J]. arXiv preprint arXiv: 2020, 2007.14974.

[64] ZHENG C, PENG X, ZHANG Y, et al. Interactive speech and noise modeling for speech enhancement[J]. arXiv preprint arXiv: 2020, 2012.09408.

[65] ZHU J Y, PARK T, ISOLA P, et al. Unpaired image-to-image translation using cycle-consistent adversarial networks[C]//Proceedings of the IEEE international conference on computer vision. 2017: 2223-2232.

Analysis of Music Rhythm Based on Bayesian Theory*

1. Introduction

Rhythm characteristic is the skeleton of music, which plays an important role in music analysis and can be widely applied in automatic music transcription, music information retrieval systems and music emotion analysis. There are numerous computational models of rhythm, which may be classified into two categories with regard to the type of input: the first operates symbolic signs and the second operates directly from audio input.

Cemgil et al. presented several models, there were linear dynamic system, graphical probabilistic model, and Bayesian Model. Raphael presented a similar method as graphical probabilistic model and Kasteren made some improvement based on linear dynamic system. Temperley proposed a series of music feature models, including the rhythmic model which is in term of metric structure and the statistic characteristics of the western folk music. Takeda et al. proposed a probabilistic top-down approach on the joint estimation of rhythm and tempo from the performed onset events in MIDI data.

It is clearly that audio music takes the majority in the music world. However, extracting music features from audio music signal has been proved a difficult work. Klapuri et al. extracted onsets based on a bank of band-pass comb filter resonators, and then they estimated the periods and phases of three metrical pulses trained by HMM. Antonopoulos et al. proposed a method of locating rhythmic patterns, which were also trained by HMM, which operated upon the assumption that the music meter and a rough estimate of the tempo are known.

In this paper, a rhythm extraction model based on Bayesian theory is proposed. Rhythm consists of tempo and meter. In the above approaches, only ref. [7] extracted the meter finally, while most of them only provided the tempo data. But the input of the model mentioned in ref. [7] must be a list of notes in the form of pitch, ontime and offtime. Thus

---

* The paper was originally published in *2009 International Forum on Computer Science-Technology and Applications*, 2009, and has since been revised with new information. It was co–authored by Xiaolan Lin, Chuanzhen Li, Hui Wang, and Qin Zhang.

it required to pick up features before finding meter. In our work, we extract both tempo and meter to represent rhythm. First, we adopt the onset detection method to obtain onset sequence. Then the sequence is input to a Dynamic Kalman system in order to track tempo. Finally a novel meter inference scheme is proposed by combining onset data with tempo. As far as we know, it is the first time to infer the meter without complex computation. The structure of this paper is as follows: the rhythm model and its algorithm are presented in section 2. In the section 3, the proposed system is evaluated, and at last the results and prospect the future research direction are summarized in section 4.

2. Proposed Method

Given an onset sequence, it will correspond to a variety of metric structure.

Metric structure is a property of music, which represents the meter of music. Meter and tempo are two characteristics to describe rhythm. The process of rhythm extraction is depicted as figure 1.

The steps of our system are as follows: first we detect onsets of an audio musical signal. Then a tracking method is used to estimate the tempo. After that we propose a meter inference scheme. Finally, the rhythm characteristic is described as a metric graphic.

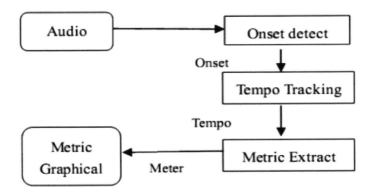

Fig. 1 The process of rhythm extraction

2.1 Onset Detection

The major step of onset detection includes signals preprocessing, detective function generation and the peak-picking.

The first step of the detection is schemed to detect the absolute value of signal. This gives the signal a set of non-zero absolute and form a detectable envelope.

Next, this signal is convolved with an edge detection filter using fast-convolution techniques. The filter is the first derivative of a Gaussian pulse. A Gaussian Function is defined as:

$$h(n) = \frac{1}{\sqrt{2\pi}\sigma} e^{-\frac{\left(n-\frac{\mu}{2}\right)^2}{2\sigma^2}} \tag{1}$$

The detection function is then formed by the following equation:

$$h'(n) = -\frac{\left(n-\frac{\mu}{2}\right)}{\sqrt{2\pi}\sigma^3} e^{-\frac{\left(n-\frac{\mu}{2}\right)^2}{2\sigma^2}} \tag{2}$$

Here, we suppose that μ is equal to the length of the data of filter so that the filter covers all samples.

The last step is to search for peaks after setting a threshold. Basically speaking, every relative minimum that occurs above a certain threshold value is counted as a note depression.

2.2 Tempo Tracking

In this subsection we employ the approach in the ref. [6] to obtain tempo. In ref. [6], the model for tempo tracking uses a linear dynamical system. The process of temp tracking is given as below:

First we define the state vector, which completely describes the system at a certain moment, and should include the onset time and the tempo. The state vector will be written as follows:

$$x_k = \begin{pmatrix} \tau_k \\ \Delta_k \end{pmatrix} \tag{3}$$

Where τ_k is the onset time of a note at time k, $k \in \{1, 2, ..., n\}$. Δ_k is the period which denotes the tempo, which can represent the duration through two consecutive beats. The score difference is the difference between two consecutive notes, referred to as γ_k. Combining the period Δ with the score difference γ, the consecutive onset time can be formulated as follow:

$$\tau_k = \tau_{k-1} + \gamma_k \Delta_{k-1} \tag{4}$$

The next step is to define the value of the transition matrix A. Assuming a constant tempo, we can define the transition matrix A as:

$$A = \begin{pmatrix} 1 & \gamma \\ 0 & 1 \end{pmatrix} \tag{5}$$

As music is full of intentional and unintentional deviations from the actual score, these deviations erode the exact data, thus they can be viewed as noise. Here, we suppose the noise as Gaussian White noise with a distribution $\omega \sim N(0, Q)$. Then the final model for generating onset time can be expressed as:

$$x_k = Ax_{k-1} + \omega_k \qquad (6)$$

The last step is to describe the observation value which can be written in a mathematical equation:

$$y_k = Cx_k + v_k \qquad (7)$$

Where C is the observation matrix, y_k is the observation value at time k, and $\mathcal{V}$ is Gaussian White noise with a distribution $v \sim N(0, R)$.

2.3 Meter Inference

Rhythm is the pattern produced by emphasis and duration of notes in music. It is comprised of rhythmic unit, a durational pattern which occupies a period of time equivalent to a pulse or pulses on an underlying metric level. Meter describes the whole concept of measuring rhythmic units. It is the abstraction of the rhythm surface of music as it unfolds in time. Meter is also described as an underlying division of time characteristic, which provides duple, triple, quadruple beats long, and each beat may be normally divided into 2 or 3 basic subdivisions named simple or compound beats. The concept can be illustrated by metrical grids, as Fig. 2. There are four common time signatures in figure 2, include the simple duple, simple triple, compound duple, compound triple (2/4, 3/4,6/8, 9/8). Take the simple duple as example. Level 2, the beat level, is the metric level where pulses are viewed as the basic time unit of the piece. Every beat on level 3 occurs behind a bar line, and every second level 1 beat is a level 2 beat, and every second level 2 beat is a level 3 beat. The principle of the metric levels of other signatures is the same as the simple duple.

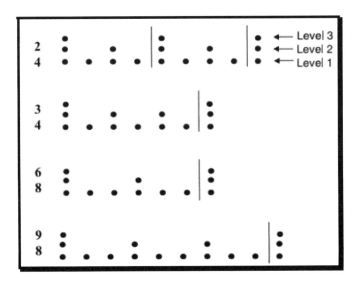

Fig. 2 Metrical grids for four common time signatures

Generally speaking, onsets may only occur at beat points, with the greatest probability to occur at level 3 beat, followed by level 2 beat, finally level 1 beat. We present the meter inference scheme in terms of the metrical structure. (Considering the practicality and the computational complexity, we only focus on four common time signatures including 3/4, 4/4, 6/8, 8/9 in this paper.)

(1) Calculate the total play time t and the tempo v of an audio music.

(2) According to t and v, calculate the value of every level beat $N_i\,(i=1,2,3)$.

(3) According to v and onset sequences detected in the 2.1 subsection, calculate the number of current onset on every level n_i.

(4) Every pair of $\dfrac{n}{N}$ indicates one meter, take the maximum value as the current beat.

3. Simulation Results

3.1 Dataset Description

Our experiments are based on two datasets. The first one consists of 46 excerpts in MIDI format, which is provided by Temperley and has been used in reference [12]. The excerpts are entirely taken from the common-practice Western repertoire. The second dataset containing three types of musical instruments (piano, violin, piccolo) is generated by the software Nuendo3 according to scores of the first dataset. Its format is wave. The reasons of generating second datasets can list as follow:

Our system aims at extracting metric structure of audio music. We need to get the comparable evaluation dataset, whose meter has been annotated. However, the involved evaluation dataset is unavailable on the website. Fortunately, we can gain the annotated information from midi excerpts. By generating the WAV clips from midi experts, the music clips will share the same meter as the midi ones.

3.2 Evaluation Method and Results

To evaluate our meter inference scheme, we extract all the features mentioned in Section 2. Fig. 4 shows that an excerpt of "The Cherry" by piano is input to our rhythm extract system. Then the metric structure is extracted as is shown in Fig. 3(c). Fig. 3 (a) is the waveform of the audio music signal. Fig. 3 (b) shows onsets of the music, which are signed by symbol '×' when onsets occur. Figure3(c) shows that there are three levels labeled by symbol '×'. The three-level structure in Fig. 3(c) corresponds with the one shown in figure 2. In figure3(c), every 2nd '×' on level 1 meets one '×' on level 2, and every '×' on 4th level 2 meets one '×' on level 3, so its meter is 4/4, which matches with the original score.

Analysis of Music Rhythm Based on Bayesian Theory

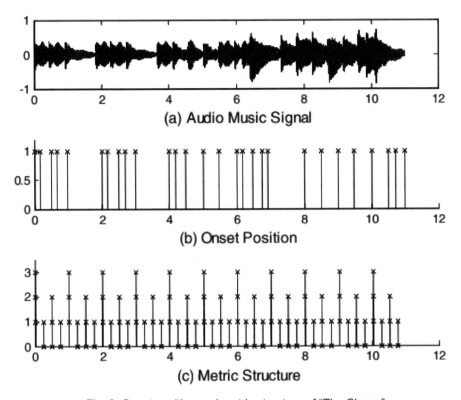

Fig. 3 Onset position and metric structure of "The Cherry"

Fig. 4 Score of "The Cherry"

In order to evaluate the performance of the presented method, we re-implemented the method proposed by Temperley in [7], which is suitable for MIDI input. So we use the MIDI Toolbox as preprocess for the extraction of pitch and onset and the result is shown in Table I. In order to evaluate the audio input, we employ an onset detection method which we add in to Temperley's source code1 as well as ours to fulfill pre-processing for comparison. Temperley's method requires accurate onsets and offsets, but the current onset detection method can not perform excellent in every kind of music. As can be seen from Table I, it performs worse in Violin because of the inaccurate of onsets. From the table we can learn that our scheme performs with higher accuracy.

Table I The accuracy of meter extraction algorithm

Timbre	Temperley's program	My program
MIDI	45.7%	69.6%
Piano	58.7%	68.2%
Violin	9.1%	56.8%
Piccolo	13.6%	43.2%

4. Conclusion

In this paper we present a new method for meter inference from audio music. We firstly detect onset time sequence, secondly track tempo, finally infer meter. Our experiments show that the proposed scheme obtains higher accuracy for piano music, but lower for violin and piccolo. Note the fact that there is no exact algorithm of onset detection, especially in non-percussive music, whose deviation of onset detection is wider. To the tempo tracing algorithm, our methods meet the half/double/ quadruple tempo problems occasionally. The inaccuracy of onset detection and tempo extraction affect the inference of the meter. Considering these factors, we will focus on improving onset detection algorithm and solving the half/multiple tempo problems in the future.

Acknowledgment

This research is supported by National Science Foundation of China (60572041), the 211 project of CUC and the Asia Media project (AM06082)

References

[1] CEMGIL A T, KAPPEN H J, DESAIN P W M, et al. On tempo tracking: tempogram representation and Kalman filtering[J]. Journal of New Music Research, 2000, 29(4): 259-273.

[2] CEMGIL A T, DESAIN P, KAPPEN B. Rhythm quantization for transcription[J]. Computer music journal, 2000, 24(2): 60-76.

[3] WHITELEY N, CEMGIL A T, GODSILL S J. Bayesian modelling of temporal structure in musical audio[C]//ISMIR. 2006: 29-34.

[4] WHITELEY N, CEMGIL A T, GODSILL S. Sequential inference of rhythmic structure in musical audio[C]//2007 IEEE International Conference on Acoustics, Speech and Signal Processing. IEEE, 2007, 4: IV-1321-IV-1324.

[5] RAPHAEL C. A hybrid graphical model for rhythmic parsing[J]. Artificial intelligence, 2002, 137(1-2): 217-238.

[6] VAN KASTEREN T. Realtime tempo tracking using kalman filtering[D]. Amsterdam: University of Amsterdam, 2006.

[7] TEMPERLEY D. Music and probability[M]. Cambridge, MA, the USA: Mit Press, 2007.

[8] TAKEDA H, NISHIMOTO T, SAGAYAMA S. Rhythm and tempo analysis toward automatic music transcription[C]//2007 IEEE International Conference on Acoustics, Speech and Signal Processing. IEEE, 2007, 4: IV-1317-IV-1320.

[9] KLAPURI A P, ERONEN A J, ASTOLA J T. Analysis of the meter of acoustic musical signals[J]. IEEE transactions on audio, speech, and language processing, 2005, 14(1): 342-355.

[10] ANTONOPOULOS I, PIKRAKIS A, THEODORIDIS S. Locating rhythmic patterns in music recordings using hidden Markov models[C]//2007 IEEE International Conference on Acoustics, Speech and Signal Processing. IEEE, 2007, 1: I-221-I-224.

[11] TEMPERLEY D. An evaluation system for metrical models[J]. Computer music journal, 2004, 28(3): 28-44.

[12] EEROLA T, TOIVIAINEN P. MIDI toolbox: MATLAB tools for music research[M]. Jyväskylä: Department of Music, University of Jyväskylä, 2004.

Learning to Generate Emotional Music Correlated with Music Structure Features*

1. Introduction

Emotion is the soul of music. The change of intensity, speed and even the length of the notes may make the emotional expression of music change dramatically, thus making music structures and emotion inseparable. Recently, the research studies on emotional music have attracted increasing attention in machine composition and its applications, such as affective music generation and music emotion analysis. However, for the application of music generation, most of the existing studies ignore the relationship between the music structure features and its emotional expression.

Inspired by the viewpoint that the internal structure of music is the main factor of inducing emotion, we propose a music generation model correlated to a given emotion based on its structure features. In the proposed model, the 'skeleton' of music-melody is regarded as the music structure feature. Firstly, we extract the melody track from the emotional MIDI dataset, and encode the emotional labels and melody track as the conditional input. Then, the GRU network takes the conditional embedding as input and outputs the MIDI event notes. Furthermore, we leverage a pre-trained emotional classification network as a loss network to promote the emotional expression of the generated MIDI pieces closer to the real music through optimising a perception loss. Finally, we conduct the subjective and objective experiments to test the emotion expression of the generated music by our model.

Our contributions are summarised as follows.

- We propose to generate emotional music by considering both the emotion and structure characteristics.
- We leverage a pre-trained emotional classification network and apply a perceptual loss to refine our model during training.

---

* The paper was originally published in *Cognitive Computation and Systems*, 2022, and has since been revised with new information. It was co-authored by Lin Ma, Wei Zhong, Xin Ma, Long Ye, and Qin Zhang.

- We collect a fine-grained emotional MIDI dataset with four emotion labels.

The remainder of the paper is organised as follows. Section 2 gives a brief review of the related works in recent years. The proposed music generation model is formulated in Section 3. In Section 4, the experimental settings, results and discussions are presented. Finally, the conclusions and future works are drawn in Section 5.

2. Related Works

From the late 19th century, the music psychologists have stated the influence of internal structure of music on its emotion. Through the combination of music structures, the quality of affective music can be further improved. Consequently, a key issue is to create a piece of music with perfect 'skeleton' in a specified emotion. In the research field of music, there is a significant distinction between the induced emotion and perceived emotion. From the perspective of affective algorithmic composition (AAC), listening to music is a way of inducing emotion. The emotional judgement of music for human beings should be 'listening to a sad piece of music', rather than feeling sad after hearing the music generated from AAC.

For the theoretical study of music structure features, Patrick Gomez et al. emphasised the relationships between the musical structures and induced emotions according to the reaction of the body based on four quadrants of valence-arousal space. This study also suggested that the three structure features of music (mode, rhythm and harmonic complexity) have larger contribution to distinguish the positive and negative emotions. In addition, Nielzen et al. investigated the relationship between the features of melody and emotional expression of music in the algorithmic composition systems and suggested that it is difficult to distinguish melody from ear. Recently, some studies focus on melody generation. The method in reference [7] proposed a hierarchical recurrent neural network for the conditional melody generation. And a chord conditioned melody generation model was composed through the Transformer network. There are also some models which are designed for generating melody sequences or audio waveform.

For music generation, considerable research efforts have been devoted. The software GhostWriter used the Herman real-time music generator to generate music with horror emotion colour. The research work of SentiMozart generated the emotional music by recognising facial expressions. Davis and Mohammed created the piano music with emotions through a rule-based technique. The method in reference [16] made a real-time music generator in the valence-arousal space. Specifically, the deep-learning based methods have been widely used in recent years. Ferreira et al. firstly proposed a symbolic music generation model by combining the sentiment analysis through the deep learning technique. And the method in reference [17] explored the method of generating music from images by using the emotion as the connection between the visual and auditory domains.

It should be noticed that, most of the current works either consider the internal structure of music or generate music in a given emotion, ignoring the relationship between them. Therefore, it is necessary to connect the affective expression of generated music with its internal structure closely.

3. Methodology

3.1 Representation of Symbolic Music

Generally speaking, music generation can be seen as a language modelling task. To do this, we parse MIDI files into the discrete sequences of notes which can represent events in a vocabulary format. Specifically, we leverage a python package Pretty-MIDI to extract several note characteristics such as pitch, velocity and duration. The MIDI data represents the music as a sequence of pitch, duration and hold tokens as shown in Fig. 1. These note events can be encoded as the music representations through the method of one-hot encoding. In addition, there are several structure characteristics corresponding to the emotion perception in music, such as melody, harmony, rhythm, timbre and so on reference [18]. In our work, we match the melody corresponding to the emotion label as the conditional generation input, which are embedded in the same way as the music representations.

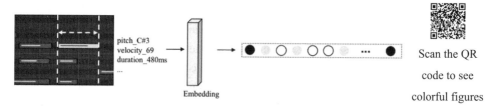

Fig. 1 The representation of symbolic music.

3.2 The Model of Emotional Music Generation

The goal of our generative model is to generate an MIDI piece in the specified emotion based on the corresponding melody characteristics. In the proposed model, we predict the MIDI event notes as the model outputs instead of generating MIDI pieces directly. Similar to the auto-regressive model, the generator is designed for predicting a distribution p over the next note sequence. In order to improve the emotional expression of the generated music, we also design a pre-trained emotional music classification network to shorten the distance between the generated and real feature distributions.

The overall framework of the proposed method is shown in Fig. 2, which consists of the music generation network and music emotion classification network. The music generation network f_G is a gated recurrent unit parameterised by weights. It can input the emotion label E and output MIDI $\hat{y}$ as $\hat{y} = f_G(E)$. And the music emotion classification network ϕ is pre-trained as the loss network to define the loss function. As shown in Fig. 2, the objective of our model is:

$$\mathcal{L} = argmin_\theta \sum \mathcal{L}_g^\phi(p, n) + \lambda \mathcal{L}_i^\phi(\hat{y}, y) \tag{1}$$

where θ denotes the parameters to be optimised, λ is a weight hyper-parameter, n represents the next MIDI event notes, and $\hat{y}$ is the generated MIDI events. Here, $\mathcal{L}_g^\phi(p, n)$ is the loss function for predicting MIDI event notes, and $\mathcal{L}_i^\phi(\hat{y}, y)$ is the perception loss for capturing emotion characteristics from real MIDI events y.

3.2.1 Music Generation Network

The details of the music generation network and music emotion classification network are shown in Fig. 3. For the music generation network as shown in Fig. 3a, we build the model in a self-supervision training mode. For the input emotion label E, we match the melody MIDI sequence corresponding to a given emotion label as the input vector of the generation network, and the output is the MIDI event sequence.

Specifically, we first encode the melody MIDI sequence as the input vector $x_i \in \mathbb{R}^{d_v}$, d_v represents the dimension of the input feature vector. Then, these input features are fed into GRU with three layers. To predict the event notes, we thus add two dense layers with a dropout rate of 0.5 to predict the distribution of the next note sequence. In this network, the loss function $\mathcal{L}_g^\phi$ is defined as follows:

$$\mathcal{L}_g^\phi(p, n) = -\log p[n], \quad p = \frac{1}{t}\sum_{i=1}^{t}\sigma(f_i[0:d_c-1]) \tag{2}$$

where t is the time-stamp of each event note, σ is a softmax function, d_c denotes the number of MIDI event notes, and f_i represents the output logits of the generation model.

3.2.2 Music Emotion Classification Network

For the music emotion classification network as shown in Fig. 3b, an emotional music classification model is pre-trained to shorten the distance between the generated and real feature distributions. The input of the network is the MIDI sequence and its output is the result of emotion classification. Specifically, the model consists of two residual blocks, three layers of GRU with 512 neurons and two dense layers. In order to reduce the number of model parameters, the residual block consists of two dilated convolution layers and a max-pooling layer. Here, the dilated convolution can increase the receptive field with a fixed kernel size, and the max-pooling layer can remove redundant information. The parameter settings of major components in the network are shown in Table I. And then we use the cross-entropy loss function to optimise the classification network.

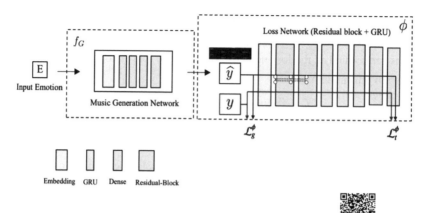

Scan the QR code to see colorful figures

Fig. 2 The framework of the proposed method. $\mathcal{L}_g^\phi$ represents the loss function of the music generation network, and $\mathcal{L}_t^\phi$ is the loss function of the music emotion classification network. We conduct the joint training for the above two loss functions, which can shorten the distance between the emotional features of the generated music and those of the target emotional music.

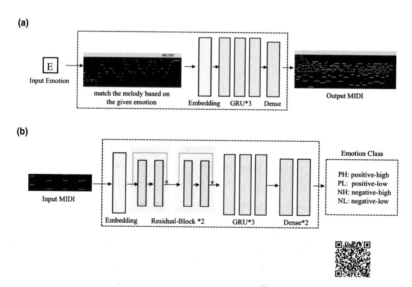

Scan the QR code to see colorful figures

Fig. 3 The details of the proposed network. (a) The description of the emotional music generation network, (b) the description of the emotional music classification network.

Table I Parameter settings of the classification model

	Units	Size	Dilated-r
Res-blk: cov1	128	3	$2^{\left[\frac{i}{3}\right]}$
Res-blk: conv2	256	3	$2^{\left[\frac{i}{3}\right]}$
GRU layers	512	–	–
Dense layer-1	64	–	–
Dense layer-2	4	–	–

3.2.3 The Perceptual Optimisation

To further improve the realism of emotion expression of the generated MIDI piece, we leverage the emotion MIDI classification network to capture different emotion characteristics. Specifically, we adopt the above pre-trained classification model as a loss network. As shown in Fig. 2, the loss network ϕ is used to define the loss function $\mathcal{L}_t^\phi$. The loss function computes a scalar value $\mathcal{L}_t^\phi(\hat{y}, y)$ by measuring the difference between the output MIDI $\hat{y}$ and the target MIDI y,

$$\mathcal{L}_t^\phi = \|\phi_1(\hat{y}) - \phi_1(y)\|_2^2 \quad (3)$$

where $\phi_1(\cdot)$ denotes the feed-forward feature extraction process of the classification network, and we extract the output logits in the last dense layer as feed-forward features as shown in Fig. 2.

Fig. 4 The dataset annotation rules

4. Experiments

4.1 Dataset Annotation

For the emotion labelling process, since the emotional expression of music is diverse and suitable for different scenes, we construct an MIDI dataset that contains four kinds of emotion labels. The annotation rules of the MIDI dataset is given in Fig. 4. We can see from Fig. 4 that the emotion space consists of two dimensions of valence (horizontal axis) and arousal (vertical axis). By considering the two dimensions jointly, we define four emotion labels corresponding to the four quarters of the Russell's model, denoted as positive-high (PH), positive-low (PL), negative-high (NH) and negative-low (NL) respectively.

For the construction of MIDI dataset, since the dataset of POP909 is constructed by the popular songs and their emotional expressions always make people comfortable, we choose to label the pieces of the POP909 dataset in the positive-valence dimension. The POP909 dataset contains 909 popular songs created by professional musicians and has multiple versions in MIDI files. On the other hand, for the negative-valence dimension, because VGMIDI is composed of music clips from game background and those music clips can stimulate the players' tension and some other negative emotions, it is selected to label in the negative-valence dimension. The VGMIDI dataset contains a total of 823 MIDI clips from video game soundtracks.

Since POP909 has a coherent emotional expression throughout each music piece while VGMIDI has large emotional we mark each piece of POP909 with an emotion label. While for each piece of the VGMIDI dataset, its each clip is labelled with an emotion label according to its emotional changes.

In the form of questionnaire, we invite 200 students on campus to judge their subjective emotional experiences of the tracks in the two datasets. About 35% of subjects have music background and the others without music background. Each subject is asked to hear 20 pieces of POP909 and 30 clips of VGMIDI randomly, and then they select one of the corresponding emotional keywords as shown in Fig. 4. Since these keywords are located in the valence-arousal emotional space, we can divide them into four emotional regions corresponding to the four emotion labels as denoted by PH, PL, NH and NL. For the result of labelling, the statistics of emotion scores are counted for each MIDI sample. In order to make the labelling result more convincing, we weigh the scores from the subjects with music background by a factor of 0.6 and those without music background by 0.4. The statistic of the labelled music from the two datasets is presented in Table II.

4.2 Implementation and Training Details

During training, we divide the datasets into training and testing sets in the ratio of 8:2. Specifically, the labelled MIDI datasets have multiple tracks as shown in Fig. 5. As mentioned in Section 3.2, the melody track is chosen for the emotional expression in our music generation procedure. Therefore, we extract the melody structure features according to the given emotion as the conditional input in our model.

Our implementation is based on Tensorflow. The hyper-parameter λ in Equation (1) is set to be 0.8. We update the parameters of the model using Adam with an initial learning rate of 0.01. Each category is trained independently. The batch size is set to be 32 and the total training iterations are about 10 000. During the training procedure, the loss values fail to decrease until more than 10 epochs.

4.3 Quantitative Experiment Results

Table III shows the loss values during training procedure and we can see that the loss value still decreases until more than 10 epochs. Some pieces of the generated emotional music by the proposed model are shown in Fig. 6. Furthermore, in order to show that our model can learn the features of music structure, we compare the input melody with the generated music as shown in Fig. 7. It can be seen clearly from Fig. 7 that the envelope lines of the input melody are consistent with those of the generated music diagrams, which suggests that the proposed method could produce proper music structure according to the music structure features of the input melody.

For the objective evaluation of the proposed method, we design an experiment on the emotion classification of the generated music. In the experiment, we use the accuracy of music emotion classification as the metric, and compare the accuracy of emotional music classification. The experimental results on four kinds of emotional melodies are given in Table IV. It can be seen from Table IV that compared with the classification results of the generated music without music structure, the introduction of the music structure features can effectively increase the accuracy of music emotion classification.

Table II Amount of MIDI samples annotated with four emotion labels

Emotion label	Valence-arousal	Dataset	Amount
PH	Positive-high	POP909	423
PL	Positive-low	POP909	486
NH	Negative-high	VGMIDI	400
NL	Negative-low	VGMIDI	403

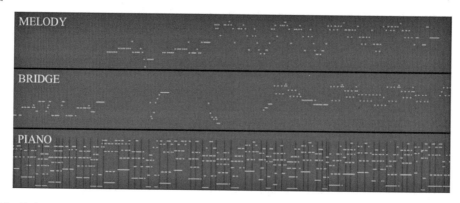

Fig. 5 An example of the MIDI file in a piano row, each piece contains three different tracks (melody, bridge, piano).

Table III The details of loss value for the training procedure

Epochs	Loss- $\mathcal{L}_g^\phi$	Loss- $\mathcal{L}_t^\phi$	Loss- $\mathcal{L}$
2	1.92	0.64	2.43
4	1.73	0.65	2.25
6	1.53	0.74	2.12
8	1.48	0.75	2.08
10	**1.23**	**0.93**	**1.97**
12	1.23	0.91	1.96
14	1.21	0.93	1.96

Note: Bold values are signifies the numerical value of the optimal convergence of the model.

4.4 Qualitative Experiment Results

In order to verify the validity of the proposed method, we also conduct the subjective evaluation tests on the generated music with different emotions. In terms of subjective evaluation, we test on four groups of the generated music correlated to four emotions of PH, PL, NH and NL, each of which containing eight samples. The questionnaire is used to collect the emotional experiences of the subjects on the generated music samples. The final statistical results are shown in Fig. 8. It can be seen from Fig. 8 that the average values of emotional scores in PH/PL emotion are higher than those in the NH/NL emotion. It indicates that the generated music samples in the PH/PL emotion are more likely to resonate with the subjects than those in the NH/NL emotion. This maybe because the influence of music on our emotion is not the emotional attribute of music itself, but from the kind of emotion we listen to songs with. Human beings usually have positive emotional state, and thus they tend to give positive emotional scores when listening to the generated music samples.

In the next experiment, we compare the proposed emotional music generation method with the existing work of reference [1], which employs the Genetic Algorithm to generate the sentiment music. On one hand, we make the comparison on the quality of the generated emotional music with two kinds of emotion labels as positive and negative. On the other hand, the comparison is also made on the emotional score of the generated music with four emotion labels as PH, PL, NH and NL.

For comparison on the quality of the generated emotional music, since the method of reference [1] generates emotional music with two kinds of emotion labels as positive and negative, we merge the PH and PL labels into the positive category, and the NH and NL into the negative category. In the experiment, we firstly re-implement the method of reference [1] with their available codes on the collected training set. And then our model is also re-trained with two kinds of emotion labels. The comparison results of the generated music examples are given in Fig. 9. It can be seen from Fig. 9 that for both of the positive and negative MIDI examples, our model exhibits better continuance and note density without break than the method in reference [1].

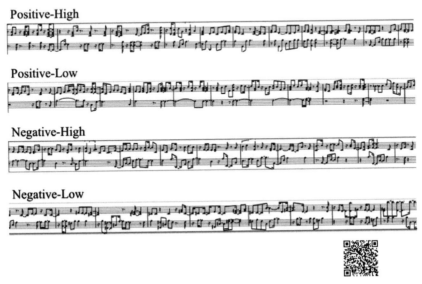

Scan the QR code to see colorful figures

Fig. 6 The generated MIDI samples through the proposed model with four emotion labels.

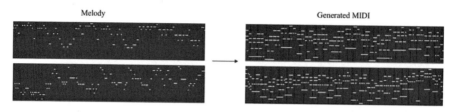

Fig. 7 Influence of melody as the conditional input on the generation of music structure.

Table IV The classification results of emotion for the generated music with/without music structure on four kinds of emotions

Emotion label	Without music structure	With music structure
PH	0.79	0.87
PL	0.73	0.83
NH	0.80	0.86
NL	0.77	0.84

For comparison on the emotional score of the generated music, we modify the emotion classifier of reference [1] correlated to four emotion labels of PH, PL, NH and NL, and retrain it on the collected training set. And then we design a questionnaire to evaluate the emotional expressions of the generated music by the two methods in terms of fluency, facticity, integrity and emotional adequacy. In the experiment, we test on four groups of generated music correlated to four emotions of PH, PL, NH and NL, each of which contains 2 samples. The questionnaire is used to collect the emotional experiences of the subjects on the generated music samples. The scores ranged from 1 to 5 for the aspects of fluency, facticity, integrity and emotional adequacy, which represent the accuracy of emotional expression compared with the suggested emotion. Finally, we gather 100 valid questionnaires and the results of the average scores are shown in Fig. 10. It can be seen from Fig. 10 that the generated music samples by our model can obtain higher scores in terms of fluency, facticity, integrity and emotional adequacy than those generated by the method of reference [1]. Furthermore, the results of the emotion scores show that the proposed method can achieve better performance in PH and PL than those in NH and NL. The experimental results are consistent with those obtained in the first subject experiment of Section 4.4 under the similar analysis.

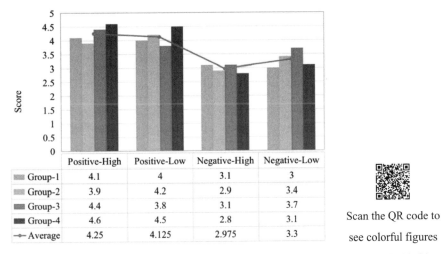

Fig. 8 The statistical results of the emotional scores correlated to four emotions of PH, PL, NH and NL.

Scan the QR code to see colorful figures

(a)

(b)

Fig. 9 The comparison results of the generated music examples by the proposed model and compared model in [1]. (a) Positive music examples. The left is generated by our model and the right is generated by the compared model, (b) Negative music examples. The left is generated by our model and the right is generated by the compared model.

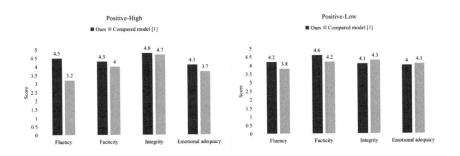

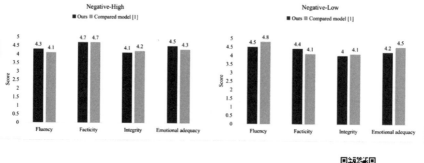

Scan the QR code to see colorful figures

Fig. 10 The comparisons on average score of the generated music by our model and the method in [1] correlated to four emotions in terms of fluency, facticity, integrity and emotional adquacy.

5. Conclusion

In this paper, we propose an emotional music generation model to generate music by considering the structure features along with its emotion label. In the proposed method, a conditional generative GRU model is used for generating music in an auto-regressive manner. By building a music emotion classification model, we jointly optimise a perceptual loss with cross-entropy loss during the training procedure, which can refine the emotion expression of the generated MIDI samples closely to the real ones. Both the subjective and objective experiments prove that our model can generate emotional music correlated to the specified emotion and music structures.

As an extension of the existing work, our future work will mainly focus on the following two parts. Firstly, we need to analyse how the combination of different music structure features affect the emotional expression of the generated music. Secondly, we will improve the quality of the generated music from the perspective of algorithm, and further establish a music emotional evaluation system.

Funding Information

National Natural Science Foundation of China, Grant/Award Numbers: 61631016, 61971383, 62001432; the Fundamental Research Funds for the Central Universities, Grant/Award Number: CUC21GZ007.

Acknowledgements

This work is supported by the National Natural Science Foundation of China under Grant Nos. 61971383, 61631016 and 62001432, and the Fundamental Research Funds for the Central Universities under Grant No. CUC21GZ007.

Conflict of Interest

We declare that we do not have any commercial or associative interest that represents a conflict of interest in connection with the work submitted.

Data Availability Statement

Research data are not shared.

ORCID

Lin Ma https://orcid.org/0000-0002-1338-2146

References

[1] FERREIRA L N, WHITEHEAD J. Learning to generate music with sentiment[J]. arxiv preprint arxiv: 2021, 2103.06125.

[2] KOH E, DUBNOV S. Comparison and analysis of deep audio embeddings for music emotion recognition[J]. arxiv preprint arxiv: 2021, 2104.06517.

[3] GOMEZ P, DANUSER B. Relationships between musical structure and psychophysiological measures of emotion[J]. Emotion, 2007, 7(2): 377-387.

[4] HEVNER K. Experimental studies of the elements of expression in music[J]. The American journal of psychology, 1936, 48(2): 246-268.

[5] GUNDLACH R H. Factors determining the characterization of musical phrases[J]. The american journal of psychology, 1935, 47(4): 624-643.

[6] NIELZEN S, CESAREC Z. Emotional experience of music as a function of musical structure[J]. Psychology of music, 1982, 10(2): 7-17.

[7] GUO Z, DIMOS M, DORIEN H. Hierarchical recurrent neural networks for conditional melody generation with long-term structure[J]. arxiv preprint arxiv: 2021, 2102.09794.

[8] CHOI K, PARK J, HEO W, et al. Chord conditioned melody generation with transformer based decoders[J]. IEEE access, 2021, 9: 42071-42080.

[9] YANG L C, CHOU S Y, YANG Y H. Midinet: a convolutional generative adversarial network for symbolic-domain music generation[C]//18th International Society for Music Information Retrieval Conference, October 23-27, 2017: 324-331.

[10] PAINE T L, KHORRAMI P, CHANG S, et al. Fast wavenet generation algorithm[J]. arxiv preprint arxiv: 2016, 1611.09482.

[11] VAN DEN OORD A, DIELEMAN S, ZEN H, et al. Wavenet: a generative model for raw audio[J]. arxiv preprint arxiv:1609.03499, 2016, 12.

[12] DHARIWAL P, JUN H, PAYNE C, et al. Jukebox: a generative model for music[J]. arxiv preprint arxiv: 2020, 2005.00341.

[13] JIAN J, LEI W, QI K. Research on the method of emotional music generation[J].

Microcomputer applications, 2019, 35(02): 52-55.

[14] MADHOK R, GOEL S, GARG S. SentiMozart: Music generation based on emotions[C]//ICAART (2). 2018: 501-506.

[15] DAVIS H, MOHAMMAD S. Generating Music from Literature[C]//Proceedings of the 3rd Workshop on Computational Linguistics for Literature (CLFL). 2014: 1-10.

[16] WALLIS I, INGALLS T, CAMPANA E. Computer-generating emotional music: the design of an affective music algorithm[J]. DAFx, espoo, finland, 2008, 712: 7-12.

[17] TAN X, ANTONY M, KONG H. Automated music generation for visual art through emotion[C]//ICCC. 2020: 247-250.

[18] THOMPSON W F, ROBITAILLE B. Can composers express emotions through music?[J]. Empirical studies of the arts, 1992, 10(1): 79-89.

[19] RUSSELL J A. A circumplex model of affect[J]. Journal of personality and social psychology, 1980, 39(6): 1161.

[20] ABADI M, AGARWAL A, BARHAM P, et al. TensorFlow: large-scale machine learning on heterogeneous distributed systems[J]. ArXiv, 2016, 1603. 04467: 1-19.

[21] KINGMA D P, BA J. Adam: a method for stochastic optimization[J]. arxiv preprint arxiv: 2014, 1412.6980.

Visually Aligned Sound Generation via Sound-Producing Motion Parsing*

1. Introduction

When watching videos, the sound events tend to occur together with visual motions, since the sounds always come from the corresponding visual perceptibly motions. For example, a person can naturally associate the dog bark with its opening mouth. Successful learning of the correct mapping from visual to sound (VTS) is beneficial to many real-world applications, such as dubbing film, video foley editing, and restoring sound for old films.

In recent years, some efforts have been made for employing the deep neural networks to produce visually aligned sounds. The objective of Zhou is to directly explore the correlation between the audio waveform and video streams. But due to the large gap in sampling rate, it fails to capture fine-grained visual information when applied to the long-term audio data. Chen et al. made up visually unavailable sound produced beyond the image to build pure mapping from visual to relevant sound. In addition to reference [2], the concurrent work aims at improving the sample fidelity and is capable of generating multi-class sound without several separated models. However, these methods overlook the importance of discrimination between visually sound-producing motion and visual still, and thus shows the limitations in applying to the motion sensitive category, i.e., dog barking.

The task of generating natural sounds from videos is still challenging because the generated sounds should be highly temporal-wise aligned with visual motions. Here comes a question: why is the temporal alignment especially challenging for VTS? In this paper, we try to uncover a possible explanation from the following three aspects.

(1) Restricted to the limited training data, the previous works often perform the pre-trained model to extract the video frame features as visual embedding, and then do not update its parameters during the translation procedure. And thus the discrimination of obtained visual embedding is entirely depended on the capacity of pre-trained networks.

---

* The paper was originally published in *Neurocomputing*, 2022, and has since been revised with new information. It was co–authored by Xin Ma, Wei Zhong, Long Ye, and Qin Zhang.

(2) The visual motions do not have a strict boundary of time, and thus their dynamic and tempo-scales are hard to characterize. A large temporal receptive field will lead to overlap the consecutive transient motions, in opposite that a short one will lose to obtain the whole action cues. So that effectively modeling such tempos of visual motions will facilitate their recognition.

(3) Little work has paid close attention to discussing the role of visual embedding and evaluate how such embedding performs in the VTS translation procedure. On one hand, most relevant CNN visualization works focus on the spatial activation map of the input image, while the VTS task instead requires more temporal information. Recent approaches leveraged the attention mechanism to model the correlation between the visual motions and corresponding sound. However, the attention module relies on the paired data as input, while one of the modalities is lost in the translation task. Therefore, it is still challenging to provide intuitive explanations for the VTS translation procedure without any attention module. On the other hand, uncertainty is one of the main issues in developing the deep learning methods. As for the VTS task, proposing uncertainty quantification method on the translation model will improve the robust of predictions, which has also not been considered in the related works.

As analyzed above, it can be found that the existing methods have not stressed the discrimination of obtained visual embedding within the visual modality. Restricted to the limited modeling capacity and missing interpretability in the perspective of visual motion, the visually sound-producing motion cannot be distinguished effectively from the visual still, and the resulting problem of visual embedding confusion may mislead the model to learn an incorrect mapping across the modality translation, which largely cripples the alignment performance. Therefore, it is necessary to emphasize the discrimination of visual embedding and better distinguish between visually sound-producing motion and visual still, facilitating the translation from visually sound-producing motion to the corresponding sound.

Based on this observation, we propose a method of sound-producing motion parsing for the visually aligned sound generation. By parsing the sound-producing motion in the task of VTS, the obtained visual embedding should not only distinguish the sound-producing motion from still, but also have sensitivity to the various tempo-scales of visual motion. In the proposed method, we make full use of the capacity of the pre-trained model to enlarge the embedding discrimination between visually sound-producing motion and visual still. Concretely, the reduced sampling stride is applied to perceive visual motion cues with half overlapped temporal field at every timestamp, which provides sufficient local motion-discriminative information for capturing transient motion. Subsequently, a series of temporal regulators are performed to capture the various tempo-scales of visual motions, which are composed of consecutive convolution layers with dropout in temporal connections. In order to resolve the mismatch issue of cross-modality temporal scales, a cross-modal temporal aligner is also introduced to align the temporal scales of two modalities in the shared space. Finally, the visual embedding is fed into the generation modules to synthesize sound.

Furthermore to make the model more interpretable, we leverage the gradient flowing into the bottleneck layer to produce a coarse temporal activation map, highlighting the important temporal cues in the visual embedding for the sound prediction. With this explanation, our parsing hypothesis can be verified on the VTS translation task, which can be summed up the discriminative visually sound-producing motion embedding aligns and projects to the corresponding sound. Extensive evaluation results demonstrate that the proposed method outperforms the existing algorithms in terms of the temporal-wise alignment and sound quality.

Our main contributions can be summarized as follows:

We consider the VTS task from the perspective of sound-producing motion parsing and propose to explore the discrimination of extracted visual embedding on distinguishing the visually sound-producing motion from visual still. With the well separated visual motion, the translation model is facilitated to map the visual embeddings to the corresponding sound ones correctly.

We perform a series of temporal regulators to capture the various tempo-scales of visual motions, and also introduce the cross-modal temporal aligner to resolve the mismatch issue of cross-modality temporal scales.

We expand the Grad-CAM to produce the temporal activation maps by leveraging the gradient flowing into the bottleneck layer, making the translation model more transparent and explainable. As an addition, we show the robustness of our translation model through Monte Carlo Dropout to further estimate the epistemic uncertainty for our predictions.

The rest of the paper is organized as follows. Section 2 gives a brief review of the related works in recent years. The problem of visual embedding confusion is formulated in Section 3, and then our methodology is proposed with module configuration details in Section 4. The experimental settings, results and discussions are presented in Section 5. Finally, the conclusion is drawn in Section 6.

2. Related Works

The existing visual to sound methods generally involve three aspects: visual to sound translation, visual motion embedding extraction and audio-visual correlation. The task of visual to sound translation is solved by synchronized mapping from the visual motions to the corresponding sound, which largely depends on the extraction of discriminative visual motion embedding. While the audio-visual correlation is designed for the purpose of learning a consistent representation between two modalities.

2.1 Visual to Sound Translation

Synthesizing sound for a video has recently drawn large attention. On one hand, the translation procedure can be seen as a retrieval task. Owens et al. first introduced the deep neural networks to predict the sound produced by the object interaction with a drumstick and then replaced it with a closed sound example indexed from the dataset. Afterward, the works in references [13,14] mainly focused on the sound classes to adopt key variations and accompany with average spectrum as a prior belonging to the same class to predict sound, which are highly relied on the temporal periodic variation. On the other hand, the related works aim to build a mapping function between two modalities. The methods in references [15-17] exploited the conditional generative networks to conduct musical playing from performance to music. They have achieved high- fidelity results, benefited from the standard digital vocoder of MIDI. Unlike the above works focusing on human musical playing action, the approaches explored the correlation between static instrument and sound. Different from the high-quality music data, recent researchers accumulated a nature sound dataset from AudioSet that includes several types of sound recorded in the wild and proposed a generative method to directly transform visual features to specific sound representation such as waveform or spectrogram.

Although the studies in references [1-3] can synthesize the considerable sound, it disregards the inner-discrimination within visual modality, which may lead to poor temporal synchronization caused by the confusion of visually sound-producing motion. Our work stresses the discriminative ability in distinguishing between visually sound-producing motion and visual still. As a result, the network can easily learn the mapping from well-separated visual motions to the corresponding sound, which may achieve high temporal alignment.

2.2 Visual Motion Embedding Extraction

Deep convolution neural networks have achieved considerable success in visual temporal modeling. In general, the previous work can be classified into three kinds of architectures, such as two-stream, 3D convolution and sequence-based frameworks. The works in references [21, 22] presented a two-stream network to model the appearance and temporal cues, respectively. Tran et al. designed a 3D convolution network to jointly learn the spatial-temporal filters. By splitting the temporal and spatial operations, several recent works attempted to apply the consecutive pseudo-3D filters to achieve better nonlinearity. Inspired by successfully applying the attention mechanism in natural language processing, the relevant approaches tried to obtain temporally coherent video embeddings with self-attention mechanism. Other works demonstrated that the temporal supervision is critical for temporal learning.

For the VTS task, the previous approaches only use visually motion-insensitive embedding, which ignore the discrimination between motion and still within the local temporal field. In this work, we apply the time-redundant visual embedding to improve the discrimination on transient sound-producing motion.

2.3 Audio-Visual Correlation

Recently, some efforts have been made for learning audio-visual representations based on nature alignment between the paired visual and sound inputs. These methods can be divided into three categories according to the size of the region of interest, such as patch-level, frame-level and clip-level. The patch-level correlation between visual regions and sound events leads to investigate the tasks of sound localization and sound separation, which introduced the visual-spatial information into sound expression. Besides, the sound event localization posed a frame-level association structure for the purpose of the fine-grained temporal event recognition. And the clip-level methods were used to identify whether the paired inputs are matched or not. Normally, the self-supervised signals could be performed in joint learning or distilling from one modality to another for audio-visual representation learning. We also learn the correlation between visual and sound embeddings and introduce the interactions between two modalities to better align them in the temporal scale.

3. Problem Formulation

In this section, we will formulate the embedding confusion problem arisen in the task of generating visually aligned sound. The visually aligned sound generation can be set up as a sequence to sequence problem. Taking a sequence of video frames as the inputs, the model is trained to translate from the visual frame features to audio sequence representations. Specifically, we denote (V_n, A_n) as a visual-audio pair. Here V_n represents the visual embeddings of n-th video clip, while A_n is the audio embeddings for the corresponding n-th audio sample. The translation procedure can be formulated as,

$$A_n = f(V_n), n \in [1, N]. \tag{1}$$

We consider this translation task needs to align the visual motions with corresponding sound. To achieve this goal, we firstly formulate the challenge of embedding confusion that will lead to a series of alignment problems, such as sound missing and sound redundancy. In each visual-audio pair (V_n, A_n), it involves T-frame visual and audio embeddings respectively, and the composition of those embeddings can be expressed by,

$$V_n = V_{n\_p} + V_{n\_s},\ A_n = A_{n\_p} + A_{n\_q},\ n_m + n_s = n_p + n_q = T, \qquad (2)$$

where the visual embeddings V_n can be decomposed into the n_m-frame motion embeddings V_{n_m} and n_s-frame still ones V_{n_s}, while the audio embeddings A_n are composed of the n_p-frame sound-producing embeddings A_{n_p} and n_q-frame sound quiet ones A_{n_q} as well. For the purpose of generating visually aligned sounds, the correct mapping only exists between the sound-producing motion and sound producing (V_{n_m} to A_{n_p}) as well as visual still and sound quiet (V_{n_s} to A_{n_q}). On the contrary, the irrelevant mapping from visual still to sound producing (V_{n_s} to A_{n_p}) as well as from the sound-producing motion to sound quiet (V_{n_m} to A_{n_q}) will bring the sound redundancy and sound missing, respectively.

Taking the dog barking as an example, Fig. 1 illustrates the correct mapping from the visual embeddings to corresponding audio ones. More concretely, we expect the distance between visual barks and the corresponding sound barks should be smaller, as well as the distance between visual still and the corresponding sound still. While for visual modality, the distance between visual motion and still is expected to be larger. The existing works are to learn an embedding space through a manner of self-supervised learning, and build the mapping function from visual barks to the corresponding temporal-aligned sound barks, making the visual embedding gradually move to its corresponding sound one in the embedding space. We consider that it follows a manner of implicit learning to obtain the visual discrimination through the discrimination between sound barks and sound still. However, there is an obvious problem in this learning objective. If the initial discrimination of visual embedding is not enough, that is, the visual bark and visual still cannot be distinguished well, the model will be confused under this optimization objective where two similar visual embeddings will be projected into two different sound-producing states, sound barks or sound still. This problem may mislead the model to learn an incorrect mapping across modality translation.

In addition, since the visual motions do not have a strict boundary of time, they may occupy one or more frames, and thus their dynamic and tempo-scales are hard to characterize. In order to better model the temporal variations of visual motions, we also stress out the importance of multiple tempo-scale modeling on visual motions. With the proper temporal modeling for various tempo-scales of visual motions, the model can effectively characterize the boundary of motion dynamics. As a result, we refer to the above issues as the embedding confusion challenge.

Considering this problem, we propose a hypothesis. If we well distinguish visual motion from visual still at the beginning, will the performance of translation model be improved? Based on this assumption, we propose to consider the VTS task from the perspective of sound-producing motion parsing. By parsing the sound-producing motion in the task of VTS, the obtained visual embedding should not only distinguish the sound-producing motion from still, but also have sensitivity to the various tempo-scales of visual motion. With the well separated visual motion, the translation model is facilitated to map the visual embeddings to the corresponding sound ones correctly. That is to say, the distance between visual motion and still has been enlarged, while the distance between visual and the corresponding sound embeddings has been shortened.

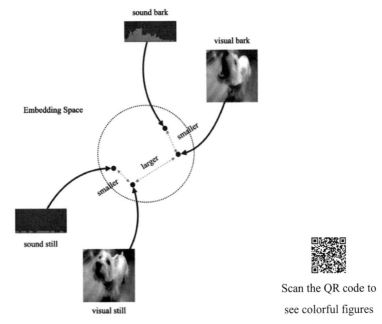

Scan the QR code to see colorful figures

Fig. 1 Illustration of the discrimination on motion and still facilitating correct mapping from the visual embeddings to corresponding audio ones.

4. The Proposed Method

In this section, we will propose the method of sound-producing motion parsing for visually aligned sound generation. By parsing the sound-producing motion in the task of VTS, the obtained visual embedding should not only distinguish the sound-producing motion from still, but also have sensitivity to the various tempo-scales of visual motion. The overall framework of the proposed method is shown in Fig. 2.

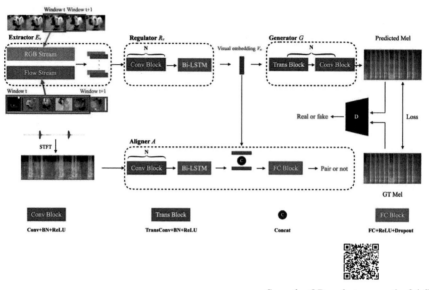

Scan the QR code to see colorful figures

Fig. 2 Overall framework of the proposed method.

As shown in Fig. 2, the proposed framework mainly consists of four modules: the visual embedding extractor E_v, regulator R_v, the cross-modal temporal aligner A and the generator G. For the visual embedding extractor E_v, the pre-trained model takes the RGB-frames and their dense flows as input x. In order to enlarge the discrimination between visual motion and visual still, we design a time-redundant sliding window to sample the flow images. And thus the flow-stream model can obtain a fine-grained temporally perceptive field to distinguish the transition motion from complex context. By concatenating with the object embedding from RGB-stream, the multiple tempo-scales modeling is conducted on the concatenated embeddings through temporal regulator R_v, where the module stacks a series of temporal convolutions with dropout on the temporal connections. After that, we can obtain the well-parsing visual embeddings $V_n \{V_n = R_v(E_v(x))\} \in \mathbb{R}^{D^*T}$.

With the visual embeddings extracted, the cross-modal temporal aligner A takes the visual embeddings V_n and their corresponding audio embeddings A_n as the input and distinguishes whether the audio-visual embeddings are matched at the same timestamp. The difference with the existing paradigm is the careful modeling of the temporal variation in both visual and sound domains, which can be regarded as a temporal aligner to align cross-modal embeddings in temporal scale. As will be discussed in the ablation study, with careful temporal capacity tuning in the specific module, the model can provide sufficient temporal information for the tempo-scales modeling. The temporal aligner is trained in a contrastive manner, where the positive pairs of (A_n, V_n) are from the same video clip and the audio input in negative pairs of (A_m, V_n) is sampled from another video clip. And thus, the alignment loss can be computed as follows:

$$L_{align} = -E\left[\log(A(A_n,V_n))\right] - E\left[\log(1-A(A_m,V_m))\right]. \qquad (3)$$

At last, The generator G predicts A_n only with the visual embeddings V_n as input. To make the generated sound more realistic, the adversarial loss is also conducted. The discriminator D minimizes the following loss:

$$L_D = -E\left[\log(D(A_n))\right] - E\left[\log(1-D(G(V_n)))\right]. \qquad (4)$$

During training, the generator tries to minimize the reconstruction loss computed by mean square error (MSE) between the predicted sound and real one along with adversarial loss:

$$L_G = \alpha^* \| A_n - G(V_n) \|_2 - \beta^* E\left[\log(D(G(V_n)))\right]. \qquad (5)$$

In total, the whole loss can be summarized as follows:

$$L_{total} = L_G + L_D + \gamma^* L_{align}. \qquad (6)$$

In the next subsections, we will provide the module configuration details of the proposed framework.

4.1 Discriminative Visual Embedding Parsing Module

We firstly design an overlapped temporal sliding window for a certain temporal extent which covers the successive five optical flow frames centered on each timestamp to extract motion features, while sampling one RGB frame to obtain the appearance features. Specifically, a two-stream BN-Inception model is utilized as a backbone to extract the visual embeddings in RGB and flow streams at each timestamp, where the model is pre-trained on UCF101 and freezes parameters during the training procedure. The purpose of two stream is to capture fine low-level motion cue which is critical in our cases, and we choose the BN-inception as the backbone due to its good balance between effectiveness and efficiency. And then, the concatenation of T-frame RGB and flow embeddings is fed to the regulator R_v, which consists of three convolutional blocks and a two-layer bidirectional LSTM (Bi-LSTM). Each convolutional block is made up of a 1D convolution layer, batch normalization (BN) layer, rectified linear unit (ReLU) and dropout layer. With careful temporal capacity tuning, the module can obtain the sensitivity to various tempo-scales of visual motion.

4.2 Cross-Modal Temporal Aligner

We propose a cross-modal temporal aligner A to compensate the mismatch in temporal scale during training. The temporal aligner can be seen as a binary classifier, and outputs 1 if the audio stream and video frames are derived from the same video sample. On the contrary, it will output 0 if the audio is sampled from another video. Here the temporal aligner is proposed to find a joint embedding space with temporally synchronized sensitivity.

In this joint space, the visual embedding V_n is expected to be close to its paired audio one A_n. The reason why we choose to use the negative pair from the same category is to force the network to recognize temporal synchronization as opposed to mere semantic association between the audio and visual inputs.

Specifically, the ground-truth spectrogram is firstly processed by an audio encoder, which consists of five consecutive convolutional blocks with each making up of 1D convolution layer, BN layer and ReLU. To match the temporal dimension size of visual embedding V_n, we down-sample the audio feature map by stride-convolution and generate the $D \times T$ output. Then a two-layer Bi-LSTM takes the feature processed by convolutional blocks as input and outputs a final audio embedding A_n. Note that the convolution layers in audio encoder and regulator R_v can be regarded as the temporal capacity adjuster and we control its temporal capacity by regulating its kernel size and dropout rate. Finally, the concatenated dual-modality features are fed into a two-layer feed-forward networks to output the binary classification result.

4.3 Generation Modules

We construct a generator G to translate the encoded video embeddings to sound spectrogram, which is made up of two parts. One part is the learnable upsampling module consisted of transposed convolutional blocks. And the other part is the convolutional blocks to recover the details in temporal level. Specifically, the convolution layer along with BN layer and ReLU constitute the basic components of a block. Moreover, the adversarial training is introduced to make the generated sound more realistic. The discriminator D needs to distinguish whether a spectrogram comes from the real or synthesized by machine. Instead of standard GAN, the PatchGANs is applied to preserve the high resolution structure in temporal scale. Specifically, we use the convolutional blocks similar to the audio encoder mentioned above. Finally, we leverage WaveNet to convert a synthesized spectrogram to waveform. Each WaveNet model is independently available for each category.

4.4 Temporal Activation Map

In order to highlight the important temporal cues in the visual embedding for the sound prediction, we need to obtain the temporal activation map of time t for the generated sound y. In particular, we firstly compute the gradient of the score for prediction y with respect to the temporal feature map activations in k-th channel V^k of a bottleneck layer (before generator module), i.e. $\frac{\partial y}{\partial V^k}$. Then, these gradients are averaged over the time dimensions (indexed by t) to obtain the channel importance weights α_k,

$$\alpha_k = \frac{1}{T}\sum_t \frac{\partial y}{\partial V_t^k}. \tag{7}$$

During the computation of α_k, while the gradients are propagated to the candidate layer, the actual computation amounts to successive weight matrix products. Hence, this weight α_k captures the 'importance' of feature map V^k for a target prediction y. We perform a weighted combination of temporal feature maps, and apply a ReLU to obtain the final temporal activation maps,

$$L_{TAM} = RELU\left(\sum_k \alpha_k V^k\right). \tag{8}$$

The ReLU function can select the features that have a positive influence on the sound prediction. Through the straightforward temporal alignment along the time axis between the highly activated visual cues and corresponding sound, it makes the translation model more explainable.

4.5 Uncertainty Quantification

The epistemic uncertainty is computed by using the weight's posterior $p(W \mid X, Y)$, which is intractable. However, the works in references [54-56] have shown that the dropout operation can be used to transform the deterministic model into a stochastic format as an approximation to the intractable posterior. In practice, we employ a Monte Carlo sampling during inference: we run n trials and compute the mean value of predictions y_i ($i \in [1, n]$) as our final output $\bar{y}$. Then we calculate the variance as the uncertainty estimation $\sigma_{epistemic}$ by

$$\sigma_{epistemic} = \frac{1}{n}\sum_{i=1}^{n}(y_i - \bar{y})^2. \tag{9}$$

Hence, the variance of the predictions can be regarded as the uncertainty degree for the final prediction $\bar{y}$, and acts as an uncertainty estimation of the translation model.

5. Experiments

In this section, we first introduce the datasets and implementation details involved in the experiments. And then the variants of visual embeddings are explored, along with tempo-scales modeling within our module. Based on these, our best variant model is compared with the three state-of-the-art methods from the aspects of semantic-wise alignment, sample fidelity and temporal-wise alignment metrics. In addition, we also conduct several human evaluations to estimate the generated results subjectively. Finally as a supplementary study, we figure out the explanation of visual embedding and visualize it in the temporal activation maps, along with the visualization experiment on uncertainty quantification.

5.1 Datasets

As for the experiment dataset, we first choose the mostly used VAS dataset given in [2] to compare our model with the three state-of-the-art methods. On the other hand, to further evaluate the temporal-wise alignment performance, we also collect three categories from the VEGAS dataset that the audio contents are temporally high sensitive to the visual signals, namely Drum, Dog and Fireworks. We define them as the Temporal-VEGAS dataset.

VAS: The VAS dataset is the mostly used dataset for the task of visually aligned sound generation. It consists of 12.5k video clips for 8 classes, including 4 sound classes from the VEGAS dataset, namely Fireworks, Dog, Drum, Baby crying, and 4 sound classes from the Audio-Set dataset, namely Cough, Hammer, Gun, Sneeze.

Temporal-VEGAS: In order to further evaluate the temporal-wise alignment of our approach, we collect three categories that the audio contents are temporally high sensitive to the visual signals. Such sensitivity is firstly introduced in the "visual relevant" experiment of the method. Inspired by their experiments, we design a task in which the sound category is temporal-sensitive with the visual content. Specifically, each test video is combined with two different audio samples. One of the audios is original from the same video, while the other is randomly selected from another video of the same class. And then the task asks the volunteers to pick the better corresponding video-audio pair. If the volunteers can perceive this difference, it means that the sound class is highly temporal-sensitive.

The human evaluation results on the visual relevance task have been shown in the Table I. It can be seen from Table I that the three sound categories namely Drum, Dog and Fireworks obtain higher scores on temporal sensitivity, while the remaining seven categories show lower temporal sensitivity which means that they do not have a clear sound-producing boundary for human to notice. Since the categories with high temporal sensitivity can better evaluate the temporal-wise alignment performance, we select three sound categories with higher sensitivity scores from the VEGAS dataset, such as Drum, Dog and Fireworks to compose the Temporal-VEGAS dataset for comparing our model with the competing methods. On average, each category contains 2835 videos with each ranging from 2-10 s. For the fairness of comparison, the last 128 videos in each category are used for testing and the remaining for training.

Table I Human evaluation results on the visual relevance task. The volunteers need to choose one of pairs which is more synchronized. Percentages indicate the frequency of a pair being judged as more synchronized, and the sensitivity represents the perception difference between the original pair and random pair.

	Original Pair	Random pair	Sensitivity
Chainsaw	63.06%	36.94%	26.12%

Table I Continued

	Original Pair	Random pair	Sensitivity
Water flowing	58.65%	41.35%	17.30%
Rail transport	60.68%	39.32%	21.36%
Printer	59.14%	40.86%	18.28%
Helicopter	56.38%	43.62%	12.76%
Snoring	59.34%	40.66%	18.68%
Baby crying	64.26%	35.74%	28.52%
Drum	73.67%	26.33%	**47.34%**
Fireworks	75.65%	23.35%	**52.30%**
Dog	78.49%	21.51%	**56.98%**

Furthermore, in order to avoid confusion with noise in onset detection evaluation, we also collect a mini-Dog dataset from Dog test dataset for testing on temporal alignment which contains least noise and least sound out of visual scene.

5.2 Implementation Details
5.2.1 Data Preprocessing

The audio samples are resampled at 22,050 Hz. And the audio representations can be computed by the following steps. We firstly obtain a spectrogram via the Short Time Fourier Transform by using a window length of 1024 and a hop size of 256. And thus the time dimension of the spectrogram is 860. Then the spectrogram is transformed to Mel space using an 80 channel Mel filter-bank. At last, the Mel-spectrograms are normalized to between 0 and 1.

For the training videos, the video frames are resampled at 21.5 fps and have been padded to the same length of 10 s by duplicating and concatenating. And thus a video reaches a total frames of 215. For capturing the fine low-level motion cues, we extract the dense optical flows by performing the TVL1 algorithm. During the cross-modality translation procedure, the size of visual embedding V_n is set to be 215*1024. Meanwhile, the sound bottleneck feature map is with the same size of 215*1024, where 215 is the compressed time dimension with a downsampling rate of 4.

5.2.2 Model Hyperparameters

Here we list the hyperparameters used in regulator module, aligner module, generator module and discriminator module of our model shown in Fig. 2. For the regulator module, we stack 3 convolutional blocks where each one consists of a 1D convolution layer, BN layer, ReLU and dropout layer. Here the kernel size of 1D convolution layer is chosen to be 5 with the hidden state size equal to 1024. And the dropout rate is 0.5. In addition, we also use a bidirectional two-layer-LSTM structure with the hidden state size being 512 to model the whole context of the visual features. The aligner module stacks 5 convolutional blocks with each having the similar structure with that of regulator to perceive longer temporal field in audio samples. The hidden state size of 1D convolution layer in each block increases from 80 to 1024 gradually. In the generator module, the two transposed convolutional block project the features from 1024 to 80 dimensions, which keep the same dimension with input spectrogram. Each transposed convolutional block is followed by one convolutional block to recover the details in temporal level. Here the kernel size of convolution layer is also chosen to be 5 for the transposed convolutional and convolutional blocks. Finally, the discriminator module is built with the same architecture as the aligner mentioned above and the same hyperparameters are also used.

5.2.3 Training Details

Our implementation is based on Pytorch. The hyper parameters α, β and γ in Eqs. (5) and (6) are set to be 10000, 1 and 1000, respectively. We update the parameters of the model using Adam with an initial learning rate of 0.0002. Each category is independently trained. The batch size is set to be 64 and the total training iterations are about 14000.

5.2.4 Competing Methods

In the experiments, we choose three state-of-the-art methods namely V2S, RegNet and SpecVQGAN as the competing methods to make comparisons. The V2S directly explored the correlation between the audio waveform and video streams and can be regarded as a benchmark for the VTS task. The RegNet is a strong baseline for generating natural sound from the visual input, which applies the audio forwarding regularizer to leverage sufficient information to learn the correlation between visual and sound signals. In addition to reference [2], SpecVQGAN aims at improving the sample fidelity and is capable of generating multi-class sound without several separated models, which is the concurrent state-of-the-art method. For a fair comparison, we use the publicly available codes and pre-trained models provided by the authors of references [2, 3]. Since the V2S has no publicly available code, we reproduce it as a baseline.

5.2.5 Onset Detection Evaluation

Since the peaks of amplitude are largely corresponding to the onset of barks, we detect the amplitude peaks to perform the onset detection evaluation on the preprocessed mini-Dog dataset. Specifically, we firstly compute the short-time energy and zero-crossing rate of the audio as density for detection. Then we threshold the amplitude of short-time energy to find an initial set of peaks, and merge the nearby peaks with the threshold of zero-crossing rate. Finally, we treat the selected frame list for each peak as the onset detection results.

To quantitatively evaluate the detection accuracy as a measure of temporal alignment, we compare the timing of peaks between the generated sound and real one. A bark is considered to be detected if the start frame of a predicted peak occurs within 0.15s towards one of real peaks and is longer than 0.15s with adjacent peaks. For instance, in a video, we define Corr (correct) as the number of matched peaks. And Redt (redundant) represents the redundant peaks of the predicted sound, which can be computed by the total number of predicted peaks subtracting Corr. While Miss (missing) can be calculated by the amount of the ground-truth peaks subtracting Corr. Finally, we compute the statistic of the above three metrics on the mini-Dog dataset namely Corr-p, Redt-p and Miss-p.

5.3 Exploration on Temporal Variants and Temporal Aligner

With the appropriate regulation of the temporal receptive field, the model tends to align better. In this subsection, we will show how we control the temporal capacity and how it makes impacts on the temporal alignment.

Temporal variants We adjust the temporal capacity by changing its kernel size K and dropout rate P. By default, K and P are set to be 5 and 0.5 respectively, and the corresponding variant is denoted as K5P0.5. While we represent the variant without temporal aligner by K5P0.5w/oA as a comparative term to verify the effectiveness of temporal aligner. For a larger temporal field with more information passed through the kernel, we design two variants K19P0 and K19P0:5, whose kernel reaches a size of 19. Conversely, we restrict K to be 1 as a small variant K1P0.

Setup We train the model variants with different temporal capacities on the Dog dataset. Besides, we conduct the onset detection experiments on the mini-Dog dataset and leverage the Corr-p, Redt-p and Miss-p metrics to evaluate the temporal-wise alignment of the generated audio from different model variants.

Results In Table II, we show the onset detection results of different model variants on the mini-Dog dataset. It can be seen from Table II that, the model with temporal aligner K5P0:5 is fully superior to K5P0:5w=oA which indicates that the temporal-wise alignment benefits from the temporal aligner module. For the temporal variations, the variant K5P0:5 shows the best performance on all of the three alignment metrics. Specifically, the 57.25% barks are correctly detected, compared to 38.28% and 53.87% for the large variants of K19P0 and K19P0:5 as well as 36.01% for the small variant of K1P0. This is because the kernel with a large temporal field provides redundant information about the target motion inside the receptive field and the generator uses these redundant information for reconstruction which leads to the timing diffusion effect and misaligns the correct producing timing. While for the small variant of K1P0, the temporal field is too small to conclude the variations of tempo-scales and thus hard to learn the correct mapping from visual to sound. These findings indicate that the appropriate temporal receptive field can achieve the optimal result.

Furthermore, in order to observe the influence of dropout clearly, we choose the large variants to conduct the comparison between K19P0 and K19P0:5. Surprisingly, when adding a dropout rate of 0.5 to the convolution layer, it shows 15.59% gains on Corr-p. We consider that the dropout operation has randomly closed some connections in temporal field, which introduces the hidden noise in temporal connections to force the model to learn necessary mapping from a broken pipe of visual cues to the corresponding sound.

Table II Onset detection comparison on the mini-Dog dataset for different temporal variations of our models. Higher Core-p score is better, while lower Redt-p and Miss-p scores are better. Best results are highlighted in bold.

Various	Corr–p ↑	Redt–p ↓	Miss–p ↓
K1P0	36.01%	63.99%	52.27%
K19P0	38.28%	61.72%	56.36%
K19P0.5	53.87%	46.13%	53.82%
K5P0.5w/oA	53.83%	46.17%	56.47%
K5P0.5	**57.25%**	**42.75%**	**52.74%**

5.4 Exploration on the Discrimination of Visual Embeddings

In this subsection, we aim to explore the discrimination of different pre-trained visual embeddings, and how well they perform on the alignment evaluation.

Setup We extract several types of embeddings from the pre-trained visual models, namely MC3, time-dependent and time-redundant. A candidate is to see if the MC3 embedding proposed in reference [24] that keeps the temporal resolution of the visual embeddings can maintain the fine motion details. While the time-dependent embedding represents the flow computed by the adjacent two frames which can be regarded as the nonoverlap temporal feature. In contrast, the proposed time-redundant embedding is extracted by using the overlapped temporal sliding window with a receptive field of five flow frames.

For exploring the discrimination of different pre-trained visual embeddings, we first manually mark a list of dog barking frame indices on a Dog video example for testing. With the above three pre-trained visual models performed on each frame, the t-SNE is applied to cluster the obtained visual embeddings and measure the discrimination between the bark and quiet frames among different types of visual embeddings. Furthermore, we study how well the model performs on the temporal alignment when it takes different types of visual embeddings as input. Specifically, three models are trained on the Dog dataset by using different embeddings as visual input respectively. For each trained model, we synthesize the audio on the mini-Dog dataset. Then the onset detection is conducted on the generated audios to measure the temporal alignment of different visual embeddings.

Results We illustrate the qualitative t-SNE for different visual embeddings in Fig. 3 and the numerical onset detection results are shown in Table III. It can be seen from Fig. 3 that, compared to MC3 and time-dependent embeddings, our time-redundant visual embeddings have a clearer boundary between visual motion and still, showing stronger discrimination. The weak-discrimination of the time-dependent embedding lies in the fact that it may have a lack of discrimination on the detailed visual motion cues. While for the MC3, it performs the enough temporal receptive field for taking motion information as consideration but fails to handle it. We demonstrate that MC3 tends to aggregate the temporal visual cues inside the temporal receptive field instead of maintaining details. Furthermore we can also find from Table III that, our time-redundant visual embeddings achieve the best results on the temporal alignment metrics, which demonstrates the well-discriminative visual embeddings can facilitate the correct mapping from visual to sound. Besides, this testifies that our model performs better on the motion-sensitive categories such as Dog.

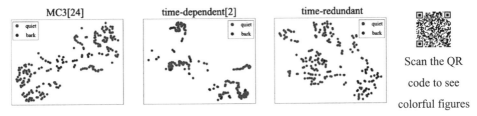

Scan the QR code to see colorful figures

Fig. 3 2-D t-SNE cluster visualization of different visual embeddings adopted by our models. Color code according to the cluster labels is assigned by human label. The quiet frame is marked with red, while the bark frame is blue. The clustering algorithm identifies the discrimination between visual motion and still among different visual embeddings.

Table III Comparison of three metrics on onset detection with mini-Dog dataset among different visual embeddings used in our models. Higher Core-p score is better, while lower Redt-p and Miss-p scores are better. The best results are highlighted in bold.

Various	Corr–p ↑	Redt–p ↓	Miss–p ↓
MC3	40.45%	59.55%	62.74%
time-dependent	45.42%	54.58%	66.32%
time-redundant	**57.25%**	**42.75%**	**52.74%**

5.5 Comparison with State-of-the-Art Methods

In this subsection, we compare the proposed model with three state-of-the-art methods on the VAS dataset, which is a mostly used dataset for the task of visually aligned sound generation. For a fair comparison, we employ the same config with the competing methods. During the training process, we perform the same strategies followed by V2S and RegNet, which train a separate model for each class. While SpecVQGAN also includes a class label into the transformer conditioning sequence modeling, allowing the model to learn to separate the subspaces for each class. As for the codebook of SpecVQGAN, we choose the pre-trained codebook on VAS dataset.

Setup For the experimental analysis, we perform the comparisons between the proposed model with the above three state-of-the-art methods in the aspects of semantic-wise alignment, sample fidelity and temporal-wise alignment metrics. For the semantic-wise alignment, we conduct the evaluation with the help of two experiments. The first one is the prediction task which uses a SVM classifier to verify if a generated sound can capture the semantic information of the specific category. And the second one is the retrieval task that retrieves the nearest sample from real test set to see whether the generated samples can keep the semantic-wise coherence with the real sample in category level. We leverage the same pre-trained audio feature extractor for both experiments. For the aspect of sample fidelity, we first employ the Fréchet Distance (FID) between the fake and real sample features to perform the evaluation. Here we adopt a Melception architecture given by reference [3] to extract the sample features. Since the FID metric relies on dataset-level distributions, it is not suitable for the conditional generation. And thus we also adopt Melception-based KL-divergence (MKL) metric proposed in reference [3] that individually compares the distances between output distributions of the fake and real samples with the visual input as a condition. For the temporal-wise alignment, we collect 20 samples with less noise for each category from VAS test set, and conduct the onset detection experiment on this subset. The Corr-p metric is used to evaluate the temporal-wise alignment performance.

Results Table IV shows the comparison results between the proposed model with three state-of-the-art methods in the aspects of semantic-wise alignment, sample fidelity and temporal-wise alignment metrics. It can be seen from Table IV that, the proposed model achieves the highest classification and retrieval accuracy, suggesting our model can produce better semantic-wise alignment performance. Benefit from the discrimination within the extracted visual embedding, the translation model can also bring this discrimination into the corresponding audio domain. And thus the obtained discrimination not only makes the synthesized audio more consistent within the same sound category, but also maintains a clear boundary among different sound-producing objects.

For the aspect of sample fidelity, the proposed model achieves better FID score than V2S and RegNet, but worse than SpecVQGAN. This maybe because SpecVQGAN adopts a non-autoregressive MelGAN as their vocoder and a multi-class co-learning objective to access the diverse data distribution. As for the conditional generation evaluation, we can also see from Table IV that our model obtains the best MKL score, because the proposed method produces more discriminative visual motion embedding as a meaningful condition for the translation model.

Moreover, our model achieves better temporal-wise alignment performance than the competing methods in terms of Corr-p metric. We consider that the traditional autoregressive method only receives the visual context as the initial hidden state to the SampleRNN, and thus it is hard to keep the synchronization with visual modality during the long sequence translation procedure, which would crop the temporal-wise alignment performance. Although RegNet and SpecVQGAN leverage more advanced modules to attend all visual contexts during translation, the weak discrimination in visual modality will lose the sensitivity on visual motion, leading to worse temporal-wise alignment performance. Contrast to the competing methods, our model provides sufficient locally motion-discriminative information for capturing transient motion, along with temporal multi-scale modeling capacity for different durations of sound-producing motion.

5.6 Qualitative Visualization

In order to transparently compare the performance of temporal-wise alignment, we visualize the generated Mel-spectrogram results from the proposed model, V2S, RegNet and SpecVQ-GAN as well as the ground-truth ones in Fig. 4. The results of three categories are shown from left to right: Drum, Dog and Fire-works. For each demo example, we show the spectrograms from the real audio, V2S, RegNet, SpecVQGAN and our model from top to bottom. It can be seen from Fig. 4 that compared to the competing methods, our results of dog demos show better temporal-wise alignment that successfully detect more accurate barking timing, which temporally fit the opening mouth of dog. While RegNet fails to generate any sound in some barking timing, this situation is even worse for V2S and SpecVQGAN. The same results can also be seen in the case of Fireworks.

For the category of Drum, although the amplitude of the target sound is hard to be clearly distinguished from the ambient sounds and accompaniments, our model can also perceive the specific beat rhythm of the performance. Specifically, the drum and cymbal sounds appear alternately in a regular way through observing the specific frequency ranges as shown in the red dotted boxes of the first column in Fig. 4.

It can be concluded that for the categories with higher temporal sensitivity such as Drum, Dog and Fireworks, we consider our model provides sufficient locally motion-discriminative information for capturing more accurate transient motion, facilitating to improve the temporal-wise alignment performance. Furthermore, more interesting samples will be shown in the supplementary files.

Table IV The comparison results between the proposed model with three state-of-the-art methods in the aspects of semantic-wise alignment, sample fidelity and temporal-wise alignment metrics. The Acc represents the classification accuracy on the test set, Top1 and Top5 are the retrieval accuracy. While the FID and MKL adopted by [3] are also used here as the sample fidelity metrics. The Corr-p is the temporal-wise alignment metric.

	Acc ↑	Top1 ↑	Top5 ↑	FID ↓	MKL ↓	Corr−p ↑
V2S	23.60%	14.50%	33.00%	74.15	8.57	12.35%
RegNet	26.10%	27.00%	59.80%	65.83	6.17	28.46%
SpecVQGAN	39.00%	32.90%	61.00%	**33.38**	6.38	15.19%
Ours	**46.20%**	**42.10%**	**67.40%**	55.79	**6.07**	**34.33%**

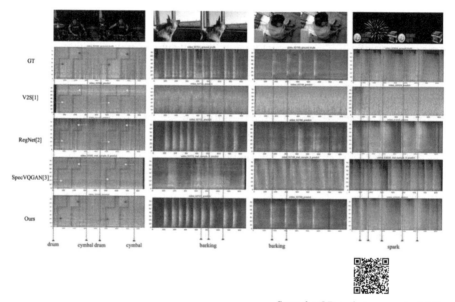

Scan the QR code to see colorful figures

Fig. 4 Temporal-wise alignment visualization comparison of sound samples from original video, V2S, RegNet, SpecVQGAN and our model on the Temporal-VEGAS dataset through Mel-spectrogram representation. The results of three categories are shown from left to right: Drum, Dog and Fireworks. Horizontal and vertical axes are showing time and frequency respectively.

5.7 Sound Retrieval Experiment

Since the direct quality evaluation is quite difficult for the synthesized sound, we conduct a good proxy with the help of two experiments as listed below. The first one is the retrieval task that retrieves the nearest sample from real set to see whether the generated samples can keep the semantic-wise coherence with the real sample category. And the second one is the prediction task which will be given in the next subsection.

Audio to audio retrieval In order to evaluate the quality of synthesized sound in the retrieval task, we firstly extract the normalized audio features of real and synthesized sounds by the model given in reference [61], which is a CNN-based network pre-trained on the AudioSet. Then we consider a retrieval experiment where the feature of synthesized audio is used as query and the real audio with the maximum cosine similarity is retrieved. Here the real audios are selected from each category of our Temporal-VEGAS test set and composed into a dataset of 384 audios. The retrieval performance is measured in statistics by Top1 and Top5. If our model has the capability of reasonable mapping, the query audio should belong to the same category as the retrieved one. Table V shows the Top1 and Top5 retrieval accuracy of competing methods and our model on the category level. Note that the task can be very challenging because the videos recorded in the wild may contain very similar irrelevant ambient noises such as people talking, screaming and so on. It can be observed from Table V that, our model achieves the best results with the average retrieval accuracy of 71.20% in Top1 and 89.50% in Top5, which indicates that the proposed model can maintain the semantic-wise coherence with the real sound. The discriminative visual embedding results in all synthesized audios in the same class maintaining higher similarity, while keeping a clear boundary among different sound categories, which facilitates to the category-level retrieval.

Visual to audio retrieval Following the idea of reference [64], we perform the additional cross-modal retrieval experiment on the Temporal-VEGAS dataset to see if the proposed method can learn a shared feature space through the way of cross-modality learning. Similar to the related work, our aligner module is also trying to project the cross-modal embedding into a shared feature space. Through the learned shared space, we can easily evaluate the distance of the cross-modal inputs.

In the designed cross-modal retrieval experiment, the paired audio-visual samples are used as the inputs of the temporal aligner. Here a synthesized audio sample is respectively combined with 384 test videos, composing 384 input pairs. For 384 synthesized audio samples, the final test set counts 384*384 audio-visual pairs in total. If our model can learn a reasonable mapping, the real audio-visual pair should have a higher score than the others. Note that this task can be very challenging since the videos in the same class may contain very similar contents. For each test video, we retrieve the audio with the highest matched score. The experimental result shows that the proposed method achieves the retrieval accuracy of 14.03% in the instance level significantly higher than that of chance being 0.26%, which indicates our method can learn a meaningful cross-modal embedding.

Table V Audio retrieval comparison of sound samples from V2S, RegNet, SpecVQGAN and our model on the Temporal–VEGAS dataset. Higher Top1 and Top5 scores are better.

		V2S	RegNet	RegNet	Ours
Dog	Top1	21.90%	51.60%	46.90%	**73.40%**
	Top5	54.60%	**92.20%**	75.80%	89.80%
Drum	Top1	12.80%	23.20%	**52.30%**	49.60%
	Top5	31.20%	42.20%	72.70%	**78.70%**
Fireworks	Top1	88.30%	84.40%	64.80%	**90.60%**
	Top5	100%	99.20%	85.30%	**100%**
Average	Top1	41.00%	53.20%	54.60%	**71.20%**
	Top5	61.90%	77.80%	77.90%	**89.50%**

Table VI Audio classification comparison of sound samples from V2S, RegNet, SpecVQGAN, our model and original videos on the Temporal–VEGAS dataset.

	V2S	RegNet	RegNet	Ours	Real
Dog	28.41%	53.91%	49.62%	65.62%	94.53%
Drum	2.65%	3.91%	32.68%	20.47%	98.43%
Fireworks	96.90%	100%	92.90%	100%	97.66%
Average	42.65%	52.60%	58.40%	62.03%	96.87%

5.8 Sound Prediction Experiment

In this subsection, we further evaluate the quality of the synthesized sound with the help of the prediction task. This task focuses on determining whether the translation model can produce a sound which captures the semantic information of the specific category.

Setup For this evaluation, instead of training a classification model on audio waveforms, we leverage the same audio feature extractor as in the retrieval experiment due to its summarizing ability of compacting high dimensional signals with less necessary for large amounts of data. Based on this, a SVM is trained on these audio features to predict the sound event category. Here the training audios are randomly chosen 1000 samples from each category of our Temporal-VEGAS training set. After training, the features of synthesized audios are fed to the SVM classifier model for predicting.

Results The classification accuracy of the sounds generated from competing methods and our model is shown in Table VI. To evaluate the performance of the classification model, we also test the trained SVM on the original audios from the test set and obtain an average prediction accuracy of 96.87%. It can be seen from Table VI that, our model achieves the average prediction accuracy of 62.03% on the three categories of generated sounds, which outperforms the competing methods. While for the category of Fireworks, the classification accuracy of synthesized audios are even higher than that of real ones, perhaps because most of the original audio tracks are recorded in the wild which may exist the dominated-misleading ambient sounds.

In addition, we also notice from Table V and Table VI that the generated sounds for the category of Drum show poor quality in both of the sound retrieval and prediction experiments, while the other two sound categories perform quite well. This may be because these two categories of Dog and Fireworks have relatively monotonous sound distribution, telling the model what tones should be generated. However, the drum record tracks contain many types of instruments with both inner-class and intra-class differences. Due to the mode collapse and extremely rich instrument timbre, the training distribution is difficult to match, which seems to lead to the poor quality of the generated sounds.

Table VII Audio instance quality comparison of sound samples from V2S, RegNet, SpecVQGAN and our model on the Temporal–VEGAS dataset. The lower FID and MKL scores are better.

	V2S	RegNet	RegNet	Ours
FID	83.33	62.75	**33.57**	53.48
MKL	7.73	5.13	5.49	**5.12**

5.9 Sound Instance Quality Analysis

The above evaluations mainly focus on the semantic-wise alignment with the real sounds, which may ignore the diversity of the generated sounds. That means even if all the generated sounds tend to be the same one which keeps the class-specific semantic information that the experiment can also reach an unreasonable good result. We hence choose to further evaluate the diversity of the generated sound waveforms.

Setup In this experiment, we use the metrics of FID and MKL for measuring the perceptual quality (fidelity) and diversity with the features extracted from pre-trained inception model. In order to keep the same settings of sample fidelity experiment with reference [3], we adopt the pre-trained Melception model used in reference [3] as the feature extractor. Specifically, we first calculate the FID scores to measure how much similarity exists between the generated and original audio signals on the whole test set. Since the FID metric relies on the dataset-level distributions, we also adopt MKL metric given in reference [3] that individually compares the distances between output distributions of the fake and real samples with the visual input as a condition. The lower FID and MKL scores indicate better performance.

Results Table VII presents the FID and MKL scores of the sounds generated from the competing methods and our model on the Temporal-VEGAS dataset. It can be found from Table VII that, the proposed model achieves better FID score than V2S and RegNet, but worse than SpecVQGAN. Compared to SpecVQGAN, we also notice that the FID scores of the other models are still relatively high. Here we consider the following two possible reasons. One is driven from the autoregressive sampling vocoders. Specifically, the WaveNet used in our model and RegNet and SampleRNN used in V2S tend to generate the same type of amplitudes for a given Mel-spectrogram condition, resulting in the lack of diversity. Beyond that, SpecVQGAN obtains higher FID score mainly benefiting from the multi-class co-learning objective to access the diverse class distributions during training procedure. As for the conditional generation results, we can also see from Table VII that the proposed method obtains the best MKL score. This is because our model can produce more discriminative visual motion embedding as a meaningful condition for the translation model.

5.10 Human Evaluation on the Alignment

Although many heuristic evaluation metrics have been proposed, the perceptual quality of generated sounds is still difficult to measure in the automatic evaluation. For instance, the generated audio may still sound like a reasonable match to the original video, although it may not be exactly similar to the real one. This is mainly because the overall pattern is more important than the specific frequencies, and the pattern is difficult to detect by applying the distance metrics. Therefore, we conduct a human study among the local college volunteers to effectively measure the true quality of the synthetic sounds.

Setup In this subsection, we directly compare the generated sounds from our model and competing methods through the human evaluation on the alignment. In this experiment, the subjects are presented with a video combined with two different audio samples, one is generated by our model and the other is randomly chosen from one of competing methods. For each test video, the subjects are asked to select the video with better performance on the temporal alignment. Finally, we aggregate the votes conducted by three subjects to get the final results.

Specifically, the subjects are required to evaluate the results by the following two criteria. 1) Sound-Missing: select the less missing sound that the model fails to generate when a visually sound-producing motion arises. 2) Sound-Redundant: select the less redundant sound that the model generates when no visual motion that may reveal sound occurs.

Results We compare our model with competing methods and show the average preference rate in Table VIII on the Dog category. It can be found from Table VIII that our approach significantly outperforms the competing methods, which indicates that our model can better capture the correlation between the visual motion and the corresponding sound. Specifically, the volunteers consider that approximately 67.06% of our synthesized sounds contain less missing and redundant parts, which benefits from characterizing the discrimination and dynamic of visually sound-producing motion by the means of constructing the time-redundant motion features and carefully modeling the temporal field.

Table VIII Evaluation of alignment performance by human judgments, where the results are preferred rate based on two temporal alignment criteria. Higher score is better.

	Competing Methods	Ours
Sound-Missing	31.43%	**68.57%**
Sound-Redundant	34.45%	**65.55%**
Average	32.94%	**67.06%**

Table IX Human evaluation results on the real-or-fake task, where people judge whether a video-audio pair is real or generated. Percentages indicate the frequency of a pair being judged as real.

	V2S	RegNet	RegNet	Ours
Dog	9.25%	29.68%	18.71%	**39.84%**
Drum	15.46%	55.06%	45.68%	**69.13%**
Fireworks	58.79%	88.67%	84.13%	**89.84%**
Average	27.83%	57.80%	49.51%	**66.27%**

5.11 Human Evaluation on the Real-or-Fake Task

Setup For this task, we would like to see whether the generated sound is able to be considered as real. Specifically, besides the results generated from the proposed method, we also combine the video with the audio generated from competing methods as the additional comparison. Then we require the volunteers to make their own decisions. The criteria of being fake can be bad synchronization or poor quality in hearing. The overall results are shown in Table IX.

Results It can be seen from Table IX that, 66.27% of the sounds synthesized by our model successfully to fool the volunteers into thinking that the generated audios are real, which outperforms the second place RegNet by 8.47%. Meanwhile, we also show the results split into three categories in Table IX. Unsurprisingly, the competing methods achieve the decent results on the category of Fireworks which is insensitive to visual motion, but their results are much worse than our method on the categories that are sensitive to visual motion, such as Dog and Drum. For the category of Fireworks, the participants are almost fooled by the synthesized sounds. It is mainly because the sound of fireworks is simple and prototypical, and the sound-producing frames can be easily distinguished by the appearance difference between darkness and spark. As for the category of Drum, the participants are always insensitive to the fast drumstick hit. This is because instead of the expectation of hearing a precise timing of the hit, the overall music pattern is more perceptible. The dog barking, for example, is highly varied in the tempo-scales and fine-grained visual motions so that it is hard to predict from a silent video. For instance, when the dog is opening its mouth, if the model fails to generate any bark sound soon at that time, the human can easily confirm that the sound is fake.

5.12 Supplementary Study for Visualizing the Temporal Activation Map

In this subsection, we aim to consider the problem related to the visually sound-producing motion localization, that is "which video frames are making sounds". Based on this, we can figure out the motion-activation maps by varying different temporal variants and visual embeddings in addition, highlighting the important temporal cues in the visual embeddings for the sound prediction.

Setup We conduct the visualization experiment on the mini-Dog dataset and leverage the gradient flowing back to the bottle-neck layer to visualize the temporal activation maps of the generated audios from different model variants.

Results We visualize the temporal heat-maps of our trained models with different temporal variants in Fig. 5, and those of different visual embeddings in Fig. 6. In general, it can be shown that, the temporal activation maps of different video demos relate to the corresponding visually sound-producing frames as shown in the red boxes. For instance, the models aim to highlight the opening mouth and leaning-forward body of a dog in the video frames, which intuitively resembles how humans interpret the sound-producing events with the visual content. Specifically in Fig. 5, we try to visualize how the temporal variants with different kernel size perform on the temporal activation maps. It can be seen from Fig. 5 that, similarly to the results of SubSection 5.3, the variant of K19P0 with a large temporal field provides the redundant attention on the visual cues, which leads to the timing diffusion effects on heat-maps. While for the small variant of K1P0, the temporal field is too small to conclude the whole motion cues and thus can only detect few short-time motion cues. In contrast, the variant of K5P0:5 with appropriate temporal field works well to almost discover all informative motion cues over time.

Furthermore, it can be found from Fig. 6 that, the discrimination of visual embeddings also appears on the temporal attention maps, which is consistent with the results of SubSection 5.4. More concretely, the time-dependent embeddings hardly locate the visually sound-producing frames in the video due to its lack of discrimination on the detailed visual motion cues. While the MC3 aggregates the adjacent temporal visual cues, failing to maintain the details along time. By comparison, our time-redundant embeddings achieve the best results, which keep the discrimination on the motion details and facilitate correct translating from visual to sound.

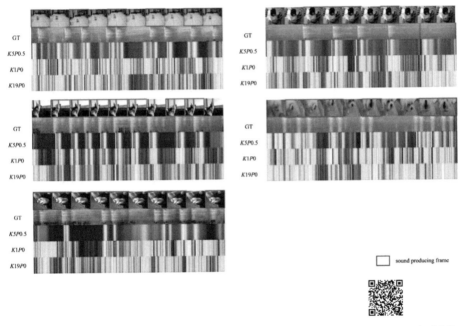

Scan the QR code to see colorful figures

Fig. 5 The temporal activation maps obtained by different temporal variants used in our models with mini-Dog dataset. Color code according to the colormap is assigned by the Jet mapping algorithm.

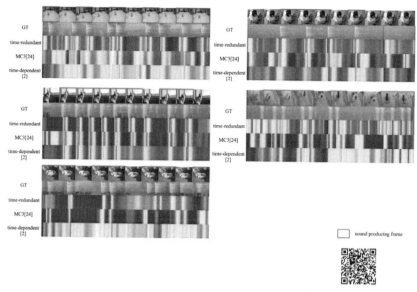

Scan the QR code to see colorful figures

Fig. 6 The temporal activation maps obtained by different visual embeddings adopted by our models with mini-Dog dataset. Color code according to the colormap is assigned by the Jet mapping algorithm.

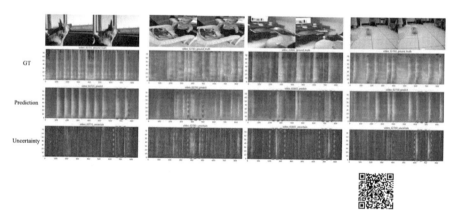

Scan the QR code to see colorful figures

Fig. 7 The uncertainty estimation maps obtained by applying Monte Carlo Dropout in our model on the Dog category.

5.13 Supplementary Study for Uncertainty Quantification

In this subsection, we aim to consider the robustness of translation model for the VTS task, that is "which part of sound predictions is unreliable". In order to compute the uncertainty, we try to leverage the uncertainty quantification method to highlight the uncertainty degree for the sound predictions.

We conduct the visualization experiment on the Dog category and apply a Monte Carlo sampling during inference through the dropout layer to obtain the uncertainty estimation for the sound predictions. For the qualitative evaluation, Fig. 7 visualizes the Mel-spectrograms of real and predicted audios for some video samples in the Dog category, as well as the estimated epistemic uncertainty values projected on the Mel-spectrogram for the sake of clarity. For the uncertainty maps shown in the third row of Fig. 7, the light points indicate the higher uncertainty predictions. It can be observed from Fig. 7 that, the high epistemic uncertainty mainly appears around the regions with more blurry predictions, which may represent the unseen sound frequencies in the training set.

6. Conclusion

In this work, we introduce a perspective of limited modeling capacity and missing interpretability in visual motion to explain why the synchronization remains challenging for generating sound from the silent video. To solve this challenge, we propose a method of sound-producing motion parsing for the visually aligned sound generation. Here the well-parsing motion embedding should not only distinguish the sound-producing motion from still, but also have sensitivity to the various tempo-scales of sound-producing motion. With sufficiently leveraging the discriminative visual motion information, our model is able to build a correct mapping from the visual motion to the corresponding sound. Moreover, a visualized explanation by using the temporal activation map is also given to better understand the whole translation procedure. Both the results in the automated metrics and human evaluations show that our synthesized sounds are more temporal-aligned to the corresponding videos than the state-of-the-art approach.

For the future research directions, a major challenge facing cross-modal translation methods is that they are very difficult to evaluate through an automatic metric. The mostly used way to evaluate a cross-modal translation task is through human judgments. However, they are time consuming and laborious work. Although a number of automatic metrics have already been proposed, they have been shown to only weakly correspond to human judgments. We believe that addressing the approximate evaluation issue will be crucial for further study of cross-modal translation. This will benefit not only for better comparison between different approaches, but also for a better learning objective to optimize.

Credit authorship contribution statement

Xin Ma: Conceptualization, Methodology, Software, Writing - original draft. **Wei Zhong:** Data curation, Writing - review & editing. **Long Ye:** Visualization, Investigation.

Qin Zhang: Supervision, Validation.

Declaration of Competing Interest

The authors declare that they have no known competing financial interests or personal relationships that could have appeared to influence the work reported in this paper.

Acknowledgements

This work is supported by the National Natural Science Foundation of China under Grant Nos. 61631016 and 61971383, and the Fundamental Research Funds for the Central Universities under Grant No. CUC19ZD006.

Appendix A. Supplementary data

Supplementary data associated with this article can be found, in the online version, at https://doi.org/10.1016/j.neucom.2022.04.018.

References

[1] ZHOU Y, WANG Z, FANG C, et al. Visual to sound: generating natural sound for videos in the wild[C]//Proceedings of the IEEE conference on computer vision and pattern recognition. 2018: 3550-3558.

[2] CHEN P, ZHANG Y, TAN M, et al. Generating visually aligned sound from videos[J]. IEEE transactions on image processing, 2020, 29: 8292-8302.

[3] ASHIN V, RAHTU E. Taming Visually Guided Sound Generation[J]. ArXiv, 2021, 2110.08791: 1-36.

[4] YANG C, XU Y, SHI J, et al. Temporal pyramid network for action recognition[C]//2020 IEEE/CVF Conference on Computer Vision and Pattern Recognition (CVPR). IEEE, 2020: 588-597.

[5] ZHOU B, KHOSLA A, LAPEDRIZA A, et al. Learning deep features for discriminative localization[C]//Proceedings of the IEEE conference on computer vision and pattern recognition. 2016: 2921-2929.

[6] SELVARAJU R R, COGSWELL M, DAS A, et al. Grad-CAM: visual explanations from deep networks via gradient-based localization[J]. International journal of computer vision, 2019, 128(2): 336-359.

[7] ZHAO H, GAN C, ROUDITCHENKO A, et al. The Sound of Pixels[J]. arxiv preprint arxiv:1804.03160, 2018.

[8] ZHAO H, GAN C, MA W C, et al. The sound of motions[C]//Proceedings of the IEEE/CVF International Conference on Computer Vision. 2019: 1735-1744.

[9] CHEN Y, HUMMEL T, KOEPKE A, et al. Where and when: space-time attention for audio-visual explanations[J]. arxiv preprint arxiv:2105.01517, 2021.

[10] ABDAR M, POURPANAH F, HUSSAIN S, et al. A review of uncertainty quantification in deep learning: techniques, applications and challenges[J]. Information fusion, 2021, 76: 243-297.

[11] WANG Y, ROCKOVÁ V. Uncertainty quantification for sparse deep learning[C]//

International Conference on Artificial Intelligence and Statistics. PMLR, 2020: 298-308.

[12] OWENS A, ISOLA P, MCDERMOTT J, et al. Visually indicated sounds[C]//Proceedings of the IEEE conference on computer vision and pattern recognition. 2016: 2405-2413.

[13] CHEN K, ZHANG C, FANG C, et al. Visually indicated sound generation by perceptually optimized classification[C]//Proceedings of the European Conference on Computer Vision (ECCV) Workshops. 2018.

[14] GHOSE S, PREVOST J J. Autofoley: artificial synthesis of synchronized sound tracks for silent videos with deep learning[J]. IEEE transactions on multimedia, 2020, 23: 1895-1907.

[15] GAN C, HUANG D, CHEN P, et al. Foley music: learning to generate music from videos[C]//Computer Vision–ECCV 2020: 16th European Conference, Glasgow, UK, August 23–28, 2020, Proceedings, Part XI 16. Springer International Publishing, 2020: 758-775.

[16] SU K, LIU X, SHLIZERMAN E. Audeo: audio generation for a silent performance video[J]. arxiv preprint arxiv:2006.14348, 2020.

[17] KOEPKE A S, WILES O, MOSES Y, et al. Sight to sound: an end-to-end approach for visual piano transcription[C]//ICASSP 2020-2020 IEEE International Conference on Acoustics, Speech and Signal Processing (ICASSP). IEEE, 2020: 1838-1842.

[18] CHEN L, SRIVASTAVA S, DUAN Z, et al. Deep cross-modal audio-visual generation[C]//Proceedings of the on Thematic Workshops of ACM Multimedia 2017. 2017: 349-357.

[19] HAO W, ZHANG Z, GUAN H. Cmcgan: A uniform framework for cross-modal visual-audio mutual generation[C]//Proceedings of the AAAI conference on artificial intelligence. 2018, 32(1).

[20] GEMMEKE J F, ELLIS D P W, FREEDMAN D, et al. Audio set: an ontology and human-labeled dataset for audio events[C]//2017 IEEE international conference on acoustics, speech and signal processing (ICASSP). IEEE, 2017: 776-780.

[21] SIMONYAN K, ZISSERMAN A. Two-stream Convolutional Networks for Action Recognition in Videos[C]//Proceedings of the 27th International Conference on Neural Information Processing Systems. MIT Press, Cambridge, MA, USA, 2014, 1: 568-576.

[22] WANG L, XIONG Y, WANG Z, et al. Temporal segment networks: towards good practices for deep action recognition[J]. arxiv e-prints, 2016: arxiv: 1608.00859.

[23] TRAN D, BOURDEV L, FERGUS R, et al. Learning spatiotemporal features with 3D convolutional networks[C]//Proceedings of the IEEE international conference on computer vision. 2015: 4489-4497.

[24] TRAN D, WANG H, TORRESANI L, et al. A closer look at spatiotemporal convolutions for action recognition[C]//Proceedings of the IEEE conference on Computer Vision and Pattern Recognition. 2018: 6450-6459.

[25] TRAN D, RAY J, SHOU Z, et al. Convnet architecture search for spatiotemporal feature learning[J]. arxiv preprint arxiv:1708.05038, 2017.

[26] QIU Z, YAO T, MEI T. Learning spatio-temporal representation with pseudo-3D residual networks[C]//2017 IEEE International Conference on Computer Vision (ICCV). IEEE Computer Society, 2017: 5534-5542.

[27] GIRDHAR R, CARREIRA J, DOERSCH C, et al. Video action transformer network[C]//

Proceedings of the IEEE/CVF conference on computer vision and pattern recognition. 2019: 244-253.

[28] LIU X, LEE J Y, JIN H. Learning video representations from correspondence proposals[C]//2019 IEEE/CVF Conference on Computer Vision and Pattern Recognition (CVPR). IEEE Computer Society, 2019: 4268-4276.

[29] WANG X, GIRSHICK R, GUPTA A, et al. Non-local neural networks[C]//Proceedings of the IEEE conference on computer vision and pattern recognition. 2018: 7794-7803.

[30] ZHOU B, ANDONIAN A, OLIVA A, et al. Temporal relational reasoning in videos[J]. arxiv e-prints, 2017: arxiv: 1711.08496.

[31] HUANG D A, RAMANATHAN V, MAHAJAN D, et al. What makes a video a video: analyzing temporal information in video understanding models and datasets[C]//Proceedings of the IEEE Conference on Computer Vision and Pattern Recognition. 2018: 7366-7375.

[32] SEVILLA-LARA L, ZHA S, YAN Z, et al. Only time can tell: discovering temporal data for temporal modeling[C]//Proceedings of the IEEE/CVF winter conference on applications of computer vision. 2021: 535-544.

[33] OYA T, IWASE S, NATSUME R, et al. Do we need sound for sound source localization?[J]. arxiv preprint arxiv:2007.05722, 2020.

[34] YANG K, RUSSELL B, SALAMON J. Telling left from right: learning spatial correspondence of sight and sound[C]//2020 IEEE/CVF Conference on Computer Vision and Pattern Recognition (CVPR). IEEE Computer Society, 2020: 9929-9938.

[35] GAO R, GRAUMAN K. 2.5D visual sound[C]//Proceedings of the IEEE/CVF Conference on Computer Vision and Pattern Recognition. 2019: 324-333.

[36] CHATTERJEE M, LE ROUX J, AHUJA N, et al. Visual scene graphs for audio source separation[J]. arxiv e-prints, 2021: arxiv: 2109.11955.

[37] GAO R, GRAUMAN K. Visualvoice: audio-visual speech separation with cross-modal consistency[C]//2021 IEEE/CVF Conference on Computer Vision and Pattern Recognition (CVPR). IEEE, 2021: 15490-15500.

[38] TIAN Y, SHI J, LI B, et al. Audio-visual event localization in unconstrained videos[C]//Proceedings of the European conference on computer vision (ECCV). 2018: 247-263.

[39] WU Y, ZHU L, YAN Y, et al. Dual attention matching for audio-visual event localization[C]//2019 IEEE/CVF International Conference on Computer Vision (ICCV). IEEE, 2019: 6291-6299.

[40] AFOURAS T, OWENS A, CHUNG J S, et al. Self-supervised learning of audio-visual objects from video[J]. arxiv preprint arxiv:2008.04237, 2020.

[41] ARANDJELOVIC R, ZISSERMAN A. Objects that sound[C]//Proceedings of the European conference on computer vision. 2018: 435-451.

[42] CHUNG J S, ZISSERMAN A. Out of time: automated lip sync in the wild[C]//Computer Vision–ACCV 2016 Workshops: ACCV 2016 International Workshops, Taipei, Taiwan, November 20-24, 2016, Revised Selected Papers, Part II 13. Springer International Publishing, 2017: 251-263.

[43] LIU J Y, YANG Y H, JENG S K. Weakly-supervised visual instrument-playing action

detection in videos[J]. IEEE transactions on multimedia, 2018, 21(4): 887-901.

[44] AYTAR Y, VONDRICK C, TORRALBA A. Soundnet: Learning Sound Representations from Unlabeled Video[C]//Proceedings of the 30th International Conference on Neural Information Processing Systems. Curran Associates Inc., Red Hook, NY, USA, 2016: 892-900.

[45] AFOURAS T, CHUNG J S, ZISSERMAN A. Asr is all you need. cross-modal distillation for lip reading[C]//ICASSP 2020-2020 IEEE International Conference on Acoustics, Speech and Signal Processing. IEEE, 2020: 2143-2147.

[46] SZEGEDY C, VANHOUCKE V, IOFFE S, et al. Rethinking the inception architecture for computer vision[C]//Proceedings of the IEEE conference on computer vision and pattern recognition. 2016: 2818-2826.

[47] HOCHREITER S, SCHMIDHUBER J. Long short-term memory[J]. Neural computation, 1997, 9(8): 1735-1780.

[48] IOFFE S, SZEGEDY C. Batch normalization: accelerating deep network training by reducing internal covariate shift[J]. arxiv preprint arxiv:1502.03167, 2015.

[49] NAIR V, HINTON G E. Rectified linear units improve restricted boltzmann machines[C]// Proceedings of the 27th international conference on machine learning (ICML-10). 2010: 807-814.

[50] SRIVASTAVA N, HINTON G, KRIZHEVSKY A, et al. Dropout: a simple way to prevent neural networks from overfitting[J]. The journal of machine learning research, 2014, 15(1): 1929-1958.

[51] GOODFELLOW I, POUGET-ABADIE J, MIRZA M, et al. Generative Adversarial Nets[C]//Proceedings of the 27th International Conference on Neural Information Processing Systems. MIT Press, Cambridge, MA, USA, 2014, 2: 2672-2680.

[52] ISOLA P, ZHU J Y, ZHOU T, et al. Image-to-image translation with conditional adversarial networks[C]//2017 IEEE Conference on Computer Vision and Pattern Recognition. IEEE Computer Society, 2017: 5967-5976.

[53] VAN DEN OORD A, DIELEMAN S, ZEN H, et al. Wavenet: a generative model for raw audio[J]. arXiv preprint arXiv:1609.03499, 2016, 12.

[54] ABDAR M, SALARI S, QAHREMANI S, et al. UncertaintyFuseNet: robust uncertainty-aware hierarchical feature fusion model with ensemble monte carlo dropout for COVID-19 detection[J]. arxiv preprint arxiv:2105.08590, 2021.

[55] CORTINHAL T, TZELEPIS G, ERDAL AKSOY E. Salsanext: fast, uncertainty-aware semantic segmentation of lidar point clouds[C]//Advances in Visual Computing: 15th International Symposium, ISVC 2020, San Diego, CA, USA, October 5–7, 2020, Proceedings, Part II 15. Springer International Publishing, 2020: 207-222.

[56] ABDAR M, FAHAMI M A, CHAKRABARTI S, et al. BARF: a new direct and cross-based binary residual feature fusion with uncertainty-aware module for medical image classification[J]. Information sciences, 2021, 577: 353-378.

[57] ZACH C, POCK T, BISCHOF H. A duality based approach for realtime TV-L1 optical flow[C]//Pattern Recognition: 29th DAGM Symposium, Heidelberg, Germany, September 12-14, 2007. Proceedings 29. Springer Berlin Heidelberg, 2007: 214-223.

[58] PASZKE A, GROSS S, MASSA F, et al. PyTorch: an Imperative Style, High-Performance Deep Learning Library[C]//Proceedings of the 33rd International Conference on Neural Information Processing Systems. Curran Associates Inc., Red Hook, NY, USA, 2019: 8026-8037.

[59] KINGMA D P. Adam: a method for stochastic optimization[J]. arxiv: 1412.6980, 2015.

[60] VAN DER MAATEN L, HINTON G. Visualizing data using t-SNE[J]. Journal of machine learning research, 2008, 9: 2579-2605.

[61] KUMAR A, KHADKEVICH M, FÜGEN C. Knowledge transfer from weakly labeled audio using convolutional neural network for sound events and scenes[C]//2018 IEEE International Conference on Acoustics, Speech and Signal Processing. IEEE, 2018: 326-330.

[62] HEUSEL M, RAMSAUER H, UNTERTHINER T, et al. GANs Trained by a Two Time-Scale Update Rule Converge to a Local Nash Equilibrium[C]//Proceedings of the 31st International Conference on Neural Information Processing Systems. Curran Associates Inc., Red Hook, NY, USA, 2017: 6629-6640.

[63] MEHRI S, KUMAR K, GULRAJANI I, et al. SampleRNN: an unconditional end-to-end neural audio generation model[J]. arxiv preprint arxiv:1612.07837, 2016.

[64] JING L, VAHDANI E, TAN J, et al. Cross-modal center loss for 3D Cross-modal retrieval[C]//2021 IEEE/CVF Conference on Computer Vision and Pattern Recognition. IEEE, 2021: 3141-3150.

MovieREP: a New Movie Reproduction Framework for Film Soundtrack*

1. Introduction

It is an important principle for movie restoration to keep the film as closely as possible to its original presentation. There are several disadvantages existing in traditional reproduction method.

1. Traditional reproduction methods deeply rely on manual restoration, which costs a lot of manpower and material resources.

2. The traditional reproduction method may injure the film soundtrack. The process requires a stable light source to irradiate the soundtrack surface, which may cause great fire hazards to the nitrate films.

3. The traditional methods ignore the fact that the soundtrack is recorded on the image form and restore the soundtrack merely on audio domain.

To solve the problems mentioned above, we propose a film soundtrack reproduction framework, which combines film soundtrack information capturing, reproduction and restoration together. This paper is the first systematical study on the research problem of restoring the film soundtrack on image level and constructing a whole reproduction framework. The main contributions of this work can be summarized as three-fold:

- In order to reduce the physical damage caused by film projector on film soundtrack, we propose an optical imaging-based film capturing system.
- To restore the film soundtrack on image domain, we propose a film soundtrack reproduction framework.
- We propose the MovieAD dataset, which provides 19125 high resolution film optical soundtrack images and the correspond original sound reproduced by film projector serving as ground truth.

---

* The paper was originally published in *ACM MM*, 2021, 40 (35), and has since been revised with new information. It was co-authored by Ruiqi Wang, Long Ye, and Qin Zhang.

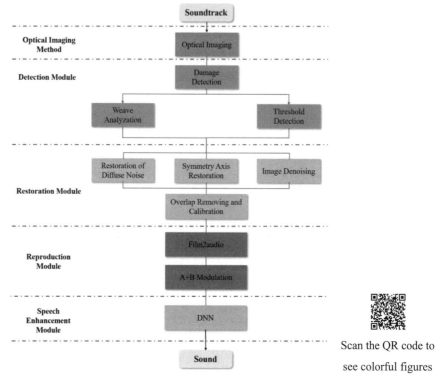

Fig. 1 The reproduction framework based on the optical imaging.

Scan the QR code to see colorful figures

2. Reproduction Framework

This section proposes an optical imaging based film soundtrack reproduction framework (shown in Fig. 1).

2.1 The Optical Imaging Method

In our framework, instead of using traditional method's laser devices radiating scanning lights, we use optical imaging method to capture the film soundtrack information to avoid the irreversible damage on the film soundtrack.

So far, the optical imaging method can obtain the film image with the resolution up to 5760*3240. The method can reproduce audio of 77.8kHz sampling rate, which is much higher (162.08%) than the movie industry standard (48kHz). Moreover, the reproduction speed is enhanced to twice the speed of the state-of-art film projector Arriscan.

2.2 Damage Detection Module

Film soundtrack images are converted to Damage Detection Module after captured by the optical imaging system. In this module, the location and type (mildew, spot, diffuse noise, film weave, scratch) of the damage is detected. Based on the grayscale histogram threshold, the global threshold detection method is introduced to the module. In the method, an unbroken film frame is set to be the reference frame. Then, a light transmission rate based method combining with the RGB channel values adopted method is performed in this module.

2.3 Restoration Module

The function of restoration module is to restore soundtrack damage on image level. Considering the characteristics of the optical soundtrack, a symmetry axis based method is developed to restore the damage. Through the adaptive threshold method, most of the diffuse noises can be removed. The film weave problem and overlap problem can be modeled and solved by the nonconvex optimization problem.

2.4 Audio Reproduction Module

This module is designed to reproduce the restored film soundtrack image to movie sound. The reproduction method simulates the photoelectrulation process on the basis of the luminous flux integration. Since the proposed reproduction module obeys similar principals to the traditional reproduction system, we simulated the A-chain and B-chain characteristics curve in the modulation system.

2.5 Speech Enhancement Module

A DNN-based speech enhancement method is adopted in the framework, which can greatly reduce the false image of music noise and enhance the speech quality.

3. Experiments and Demo

The film soundtrack picture dataset, MovieAD, is built using optical imaging method. The MovieAD dataset contains 19125 high resolution film soundtrack pictures captured by optical imaging system. 12960 frames have their corresponding audio generated by film projector and restored by professional manually serving as ground truth.

Restoration experiments have been performed using MovieAD dataset. The results of damage detection are shown in Fig. 2. The results of soundtrack weave calibration are shown in Fig. 3.

Sound evaluation experiments have been performed according to the audio type. Compared with the traditional method results, sound reproduced by MovieREP has improved by 8.2% in average in DNSMOS scores while maintaining twice reproduction speed. The detailed results are shown in Figure 4.

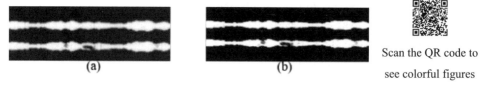

Scan the QR code to see colorful figures

Fig. 2 Damage detection. (a) original image. (b) detection results.

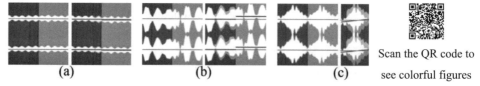

Scan the QR code to see colorful figures

Fig. 3 The results of soundtrack calibration. (a) mild weave. (b) moderate weave. (c) heavy weave.

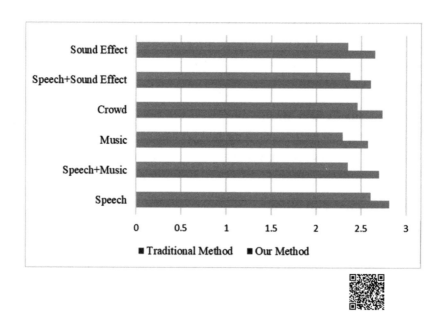

Scan the QR code to see colorful figures

Fig. 4 The objective evaluation results on different audio types.

Acknowledgments

This work is supported by the National Natural Science Foundation of China under Grant Nos. 61631016 and 61971383, and the Fundamental Research Funds for the Central Universities.

References

[1] BORDWELL D, THOMPSON K, SMITH J. Film art: an introduction[M]. New York: McGraw-Hill, 2004.

[2] CHAN C. Memory-efficient and fast implementation of local adaptive binarization methods[J]. arxiv preprint arXiv:1905.13038, 2019.

[3] GU S, ZHANG L, ZUO W, et al. Weighted nuclear norm minimization with application to image denoising[C]//Proceedings of the IEEE conference on computer vision and pattern recognition. 2014: 2862-2869.

[4] JOYEUX L, BUISSON O, BESSERER B, et al. Detection and removal of line scratches in motion picture films[C]//Proceedings. 1999 IEEE Computer Society Conference on Computer Vision and Pattern Recognition (Cat. No PR00149). IEEE, 1999, 1: 548-553.

[5] KUMAR A, FLORENCIO D. Speech enhancement in multiple-noise conditions using deep neural networks[J]. arxiv preprint arxiv:1605.02427, 2016.

[6] READ P, MEYER M P. Restoration of motion picture film[M]. Amsterdam: Elsevier, 2000.

[7] REDDY C K A, GOPAL V, CUTLER R. DNSMOS: A non-intrusive perceptual objective speech quality metric to evaluate noise suppressors[C]//IEEE International Conference on Acoustics, Speech and Signal Processing. IEEE, 2021: 6493-6497.

[8] TAN K S, ISA N A M. Color image segmentation using histogram thresholding-Fuzzy C-means hybrid approach[J]. Pattern recognition, 2011, 44(1): 1-15.

第三部分
视觉处理

Distributed Markov Chain Monte Carlo Kernel Based Particle Filtering for Object Tracking*

1. Introduction

Object tracking is a task for many computer vision applications, such as surveillance and navigation. Particle filtering (PF), which uses state and observation models to locate and track, is a stochastic algorithm, which solves non-linear and non-Gaussian tracking problems; it is more effective than extended Kalman filter (EKF) and other nonlinear filters, due to its accuracy in occluded and cluttered environments, and ease of fusion of observations.

Particle filtering works in a recursive Bayesian estimation framework; it generates a set of random samples (the particles) that approximate the unknown posterior states. Object tracking uses particle sets to predict the location of objects from a state model; deviations are corrected by maintaining multiple hypotheses over time. However, when the scene is complex, for example, the observation noise as very small variance, the filtering performance decreased clearly unless a large number of particles are used to represent the posterior distribution of object state. It results in heavy computational cost and complexity, so as to increase execution time. On the other hand, the inherent drawback of PF-loss of diversity after resampling, which means most samples may have low likelihood and their contribution to the posterior distribution are negligible-affects efficient tracking performance. Both two weaknesses limit generic particle filtering to be applied to real-time systems in practice.

In previous works, improved algorithms and hardware implementations have been proposed to address aforementioned problems: intensive computation and degeneracy problems. Bolic et al. identify resampling as the main computation cost and presented parallel resampling algorithms and architectures. Kotecha and Djuric designed a Gaussian particle filter-a fully parallel algorithm that avoided resampling. Maskell et al. proposed a

---

* The paper was originally published in *Multimedia Tools and Applications*, 2012, 56, and has since been revised with new information. It was co–authored by Danling Wang, Qin Zhang, and John Morris.

SIMD particle filter that uses n processors to process n particles. Sutharsan et al. proposed an application to multi-target tracking using a SIMD particle filter. The auxiliary particle filtering (APF), unscented particle filtering (UPF), and genetic particle filtering (GPF) are the modifications to overcome degeneracy phenomenon.

Particle filtering mainly involves three modules: generate particles, compute weights and their normalization, and resampling step. Normally, several thousand particles have to be estimated in each iteration in order to achieve satisfactory tracking accuracy. Due to computational intensity and complexity in the modules of weight computation (likelihood) and resampling step, thus, they create main restricts in the real-time implementations. Hence, we study an efficient parallel implementation which estimates the posterior distribution with kernel function, allocates the particles with mean shift, pipelines data with Markov Chain Monte Carlo (MCMC) sampling in a cluster environment to overcome the drawbacks of generic particle filtering.

In the paper, we present two schemes for implementing pipelined particle filtering-a local parallel scheme and a global one based on kernel function with an alternative MCMC resampling. In Section 2, generic particle filtering (SIS/R), as well as MCMC based resampling are described. In Section 3, kernel based PF with MCMC resampling is presented. Two parallel PF schemes are described in Section 4. Results and analysis are followed in Section 5. Finally, we conclude in Section 6.

2. Generic Particle Filtering

2.1 Sequential Importance Sampling

Sequential Importance Sampling (SIS) is a generic particle filtering technique; it is basically a sequential Monte Carlo technique. SIS tries to approximate the posterior probability density (PDF) of the states using a set of M random particles with associated weights:

$$p(x_k | z_{1:k}) \approx \sum_{i=1}^{M} \omega_k^{(i)} \delta(x_k - x_k^{(i)}) \quad (1)$$

where $\omega_k^{(i)}$ is the normalized weight of a particle $\left(\omega_k^{(i)} / \sum_{i=1}^{M} \omega_k^{(i)}\right)$, i, in iteration, $\omega_k^{(i)}$ is updated using

$$\omega_k^{(i)} = \omega_{k-1}^{(i-1)} \frac{p(z_k | x_k^{(i)}) p(x_k^{(i)} | x_{k-1}^{(i)})}{q(x_k^{(i)} | x_{k-1}^{(i)}, z_{1:k})} \quad (2)$$

$p(z_k | x_k^{(i)})$ is the observation likelihood density, which could be described by an observation model $z_k = h_k(x_k, v_k)$ where v_k is a noise process.

$p\left(x_k^{(i)} \mid x_{k-1}^{(i)}\right)$ is the state transition density, which could be described with a state model $x_k = f_k(x_{k-1}, u_k)$ where u_k is a noise process, independent of v_k.

$q\left(x_k^{(i)} \mid x_{k-1}^{(i)}, z_{1:k}\right)$ is the importance sampling function, which should ideally be close to the posterior density.

$q\left(x_k^{(i)} \mid x_{k-1}^{(i)}, z_{1:k}\right) = p\left(x_k \mid x_{k-1}^{(i)}\right)$ is often chosen to be optimal. Thus, (2) could be rewritten:

$$\omega_k^{(i)} \propto \omega_{k-1}^{(i)} p\left(z_k \mid x_k^{(i)}\right) \tag{3}$$

Equations (1), (2) and (3) describe generic particle filtering.

2.2 Resampling Algorithm

Particles may become degenerate after several iterations-a drawback of SIS filtering. Resampling avoids the degeneracy by replicating particles with large weights and discarding ones with small weights. A general resampling, also known as binary search resampling algorithm, may be summarized (Fig. 1).

PURPOSE: Generate a set of new particles $\left\{x_k^{(i)}\right\}_{i=1}^{M}$ at time $k, k > 0$

INPUT: A set of particles with weights $\left\{x_{k-1}^{(i)}, \omega_{k-1}^{(i)}\right\}_{i=1}^{M}$

1: $u_i \sim U(0,1]$, u_i is a uniform random in $(0,1]$, $i = 1, 2, \ldots, N$

2: $C_j = \sum_{i=0}^{M} \omega^{(i)}$

3: **for** $i = 1:M$ **do**

4: **for** $j = 1:M$ **do**

5: **if** $C_{j-1} < u_i \leq C_j$ **then**

6: $x_k^{(i)} = x_{k-1}^{(j)}$

7: **end if**

8: **end for**

9: **end for**

Fig. 1 Algorithm of binary search resampling.

PURPOSE: Generate a set of particles to approximate the posterior density probability

INPUT: The first element of a Markov chain with arbitrary value $x^{(0)} = x_0$

1: **for** $i = 1 : N-1$ **do**

2:   $u \sim U[0,1]$

3:   $x^* \sim q(x^* | x^{(i)})$

4: **if** $U < \mathcal{A}(x^{(i)}, x^*) = \min\left\{1, \dfrac{p(x^*)q(x^{(i)|x^*})}{p(x^{(i)})q(x^* | x^{(i)})}\right\}$ **then**

5:   $x^{(i+1)} = x^*$

6: **else**

7:   $x^{(i+1)} = x^i$

8: **end if**

9: **end for**

Fig. 2 Algorithm of MH resampling.

2.3 Markov Chain Monte Carlo Sampling

The previous resampling resolves degeneracy, but a new problem is introduced: multiple particles are generated from ones with large weights. As a result, the new particle set may lose diversity. Here, Markov Chain Monte Carlo (MCMC) sampling replaces the general resampling (a) to prevent loss of diversity and (b) to provide a resampling framework, which is amenable to pipeline processing.

The Metropolis–Hasting (MH) algorithm is one MCMC method, which uses periodic MCMC steps to diversify particles in SIS filtering. MH sampling still uses an importance density $q(.)$ to generate a Markov chain $\{x^{(i)}, i = 1, 2, ..., N\}$ with density $p(.)$. The algorithm is presented in Fig. 2. With this algorithm, we see that MCMC accepts a move from a state $x^{(i)}$ to $x^{(i+1)}$ with probability $\mathcal{A}(x^{(i)}, x^*)$; otherwise, it sets $x^{(i+1)} = x^{(i)}$.

3. Kernel Based Particle Filtering with Markov Chain Monte Carlo Resampling

Kernel particle filter (KPF) have been proposed in the literature [2] for visual tracking. In this work, we extended KPF with an alternative MCMC resampling so as to adapt for pipelined implementation. A kernel function on each particle in the KPF is used to approximate the posterior distribution, then the mean shift procedure is applied to facilitate the particles to high weight areas. Thus, the posterior distribution $p(x_k | z_{1:k})$ with kernel K and window radius (bandwidth) h in the d-dimensional space R^d, is formulated as follows

$$\hat{p}(x_k | z_{1:k}) = \frac{1}{M} \sum_{i=1}^{M} K_h(x_k - x_k^{(i)}) \omega_k^{(i)} \qquad (4)$$

where

$$K_h(x_k - x_k^{(i)}) = \frac{1}{h^d} K\left(\frac{x_k - x_k^{(i)}}{h}\right) \qquad (5)$$

PURPOSE: Obtain the estimates of posterior distribution $\hat{X}_k$

INPUT: The observation z_k, initial distribution $p(x_0)$

1: **for** $i = 1 : M$ **do**
2: $\quad x_0^{(i)} \sim p(x_0)$
3: **end for**
4: **for** $k = 1 : K$ **do**
5: $\quad$ **for** $i = 1 : M$ **do**
6: $\quad\quad x_k^{(i)} = \text{generate}(x_{k-1}^{(i)}, \omega_{k-1}^{(i)})$
7: $\quad\quad \omega_k^{(i)} = \text{weight}(x_k^{(i)})$
8: $\quad\quad \tilde{x}_k^{(i)} = \text{mean-shift}(x_k^{(i)}, \omega_k^{(i)})$
9: $\quad\quad \tilde{\omega}_k^{(i)} = \text{re-weight}(\tilde{x}_k^{(i)})$
10: $\quad$ **end for**
11: $\quad \hat{X}_k = \sum_{i=1}^{M} \tilde{x}_k^{(i)} \tilde{\omega}_k^{(i)}$
12: $\quad (x_k^{(i)}, 1/M) = \text{MCMCresampling}(\tilde{x}_k^{(i)}, \tilde{\omega}_k^{(i)})$
13: **end for**

Fig. 3 Algorithm of kernel based PF with MH resampling.

According to the mean shift theory, given the posterior estimation (4), a mean shift procedure is defined recursively by computing the mean shift vector $m(x_k)$ and translating the center of kernel H_h to search the steepest direction toward the modes of the posterior distribution. In this procedure, each particle mean shift vector is determined by

$$m(x_k) - x_k = \frac{\sum_{i=1}^{M} H_h\left(x_k - x_k^{(i)}\right) \omega_k^{(i)} x_k^{(i)}}{\sum_{i=1}^{M} H_h\left(x_k - x_k^{(i)}\right) \omega_k^{(i)}} - x_k \quad (6)$$

After a couple of iterations, the mean shift vector $m(x_k) - x_k$ would be in the gradient direction of (4). Equation (6) also shows that particles change their position, which means new particle don't follow the posterior distribution anymore. Therefore, the weight needs to be re-computed after each mean shift procedure when the new posterior distribution, denoted by $p(\tilde{x}_k | z_k)$, is estimated, adjusted by

$$\omega_k^{(i)} = \frac{p\left(\tilde{x}_k^{(i)} | z_k\right)}{q\left(\tilde{x}_k^{(i)} | \tilde{x}_{k-1}^{(i)}\right)} \quad (7)$$

where $q\left(\tilde{x}_k^{(i)} | \tilde{x}_{k-1}^{(i)}\right) = \sum_{i=1}^{M} K_h\left(x_k - \tilde{x}_k^{(i)}\right)$ is the new proposal distribution. Kernel based particle filtering using MCMC resampling algorithm is shown in Fig. 3.

4. Parallel Implementation

Computing weights and the resampling algorithm are the two main bottlenecks preventing efficient implementation of particle filtering in parallel since the weight computation step has a high computational complexity and resampling requires not only intensive computations but is a sequential step. Thus, we suggest two parallel schemes to resolve the bottlenecks that prevent reduction in overall execution time.

4.1 Local Parallel Scheme

Fig. 4 shows a sequentially generic particle filtering, which generates one particle at a time, so M particles require M steps. Since there is no data dependency in particle generation and computation weight steps, we can map the first two steps into a distributed scheme directly. The scheme uses a master node and several slaves. The generation steps for M particles are distributed to P nodes (including the master): each node performs the same operations on different particles from $\lfloor iM/P \rfloor$ to $\lfloor (i+1)M/P \rfloor - 1$, where i is the node index. The master node computes normalized weights from local weight sums, resampling and communications (scatter and gather operations). The slave nodes generate local particles, compute and sum local weights. Equation (1) can be re-written:

$$p(x_k \mid z_{1:k}) = \frac{\sum_{j=1}^{P}\sum_{i=1}^{n_k^j} \omega_k^{(i,j)} \delta(x_k^{(i)} - x_k^{(i,j)})}{\sum_{j=1}^{P} W_k^{(j)}} \tag{8}$$

where n_k^j is the number of particles on node index j at time k; $W_k^{(j)} = \sum_{i}^{n_k^j} \omega_k^{(i,j)}$ is the sum of local normalized weights on each node at time k. The local parallel scheme is illustrated in Fig. 5.

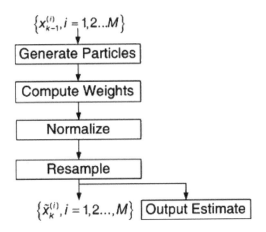

Fig. 4 Sequential scheme.

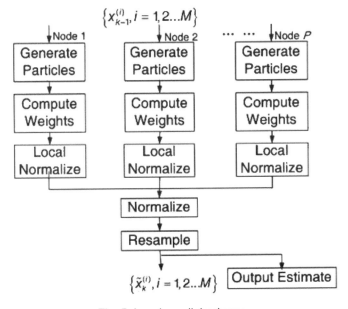

Fig. 5 Local parallel scheme.

4.2 Global Parallel Scheme

In the local parallel scheme, we note that resampling can not begin until normalized weights are calculated. So the resampling step can not be paralleled with particle generation and weight computation. Moreover, resampling remains sequential due to the need to continuously sum weights as larger weight particles are replicated.

An alternative MH resampling and kernel function were added to generic filtering to improve tracking performance, reduce execution time. In particle filtering, the importance sampling function, $q\left(\tilde{x}_k^{(i)} \mid \tilde{x}_{k-1}^{(i)}\right)$, is independent of the current state, $q\left(\tilde{x}_k^* \mid x_k^{(i)}\right) = q\left(\tilde{x}_k^*\right)$. So, the acceptance probability simplifies to

$$\mathcal{A}\left(\tilde{x}_k^{(i)}, \tilde{x}_k^*\right) = \min\left\{1, \frac{p\left(\tilde{x}_k^*\right)q\left(\tilde{x}_k^{(i)}\right)}{p\left(\tilde{x}_k^{(i)}\right)q\left(\tilde{x}_k^*\right)}\right\} = \min\left\{1, \frac{\tilde{\omega}_k^*}{\tilde{\omega}_k^{(i)}}\right\} \tag{9}$$

$$\tilde{\omega}_k = p(z_k \mid \tilde{x}_k)\frac{\sum_{i=1}^{M} p\left(\tilde{x}_k \mid \tilde{x}_{k-1}^{(i)}\right)}{\sum_{i=1}^{M} q\left(\tilde{x}_k \mid \tilde{x}_{k-1}^{(i)}\right)} \tag{10}$$

Thus, (9) and (10) describe a kernel based particle filtering with MCMC resampling. We also note that normalization is not required; we only need to know the state density up to a proportionality constant. Another advantage of MCMC sampling is that it is easy to simulate multiple independent resampling operations in parallel because there is no data dependence. The global parallel scheme is shown in Fig. 6.

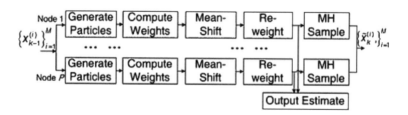

Fig. 6 Global parallel scheme.

If the state transition density, $p\left(\tilde{x}_k \mid \tilde{x}_{k-1}^{(i)}\right)$, is chosen to be the important sampling density, then the acceptance probability becomes

$$\mathcal{A}\left(\tilde{x}_k^{(i)}, \tilde{x}_k^*\right) = \min\left\{1, \frac{p\left(z_k \mid \tilde{x}_k^*\right)}{p\left(z_k \mid \tilde{x}_k^{(i)}\right)}\right\} \tag{11}$$

Thus, the acceptance probability of particles in a Markov Chain is simplified again.

In the global parallel scheme, particle generation and weight computation, as well as independent MH sampling, all run in parallel on each node. The master node scatters and gathers particles only once each iteration. The communication cost is thus significantly reduced, compared with the local parallel scheme, which needs at least three scatter or gather operations in each iteration.

5. Results

In this section, we present results obtained with message passing interface (MPICH 2 variant) implementations of the two schemes for various numbers of nodes and particles. The simulations used a cluster of eight computational nodes with 3.8 GHz CPUs and 4 GB memory, running Redhat Linux. Data block decomposition was used to partition the data, group communication patterns were used to send particles from and to master node.

Firstly, we applied the two schemes to Huang's nonlinear models. The state model is defined by

$$x_k = 0.5 x_{k-1} + \frac{25 x_{k-1}}{1+x_{k-1}^2} + 8\cos(1.2(k-1)) + \omega_k \qquad (12)$$

The observation model is given by

$$z_k = \frac{x_k^2}{20} + v_k \qquad (13)$$

where $\omega_k \sim N(0,10)$ and $v_k \sim N(0,1)$ are mutually independent white Gaussian noise. A run of ten times repeatedly with $M = 2000$ particles on $P = 2$ nodes over 50 time steps $(k = 1, 2, \ldots, 50)$ was used to evaluate the accuracy of the distributed schemes, where the error is defined by

$$RMSE = \left[\frac{1}{T} \sum_{k=1}^{T} (\hat{x}_k - x_k)^2 \right]^{1/2}, \quad \hat{x}_k = \frac{1}{M} \sum_{i=1}^{M} x_k^i \qquad (14)$$

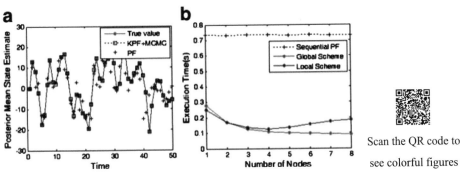

Scan the QR code to see colorful figures

Fig. 7 Posterior estimate and execution time: a Posterior mean state estimate for two schemes. b Execution time with increasing of nodes, compared with streamline scheme.

Table I RMSE and execution time

Nodes(P)/RMSE	Execution time (s)	
	Local parallel scheme	Global parallel scheme
RMSE	5.20	2.75
P=1	0.25	0.28
2	0.17	0.17
3	0.14	0.12
4	0.13	0.11
5	0.14	0.10
6	0.16	0.10
8	0.18	0.09

Fig. 7a shows estimate of posterior state probability using SIS/R and KPF-MCMC on two nodes. KPF-MCMC is closer to the true value than SIS/R since kernel function moves particles toward high distribution areas, meanwhile MCMC sampling avoids loss of particle diversity. So, the filtering performance has been obviously improved. RMSE values and time performance on different nodes are given in Table I, which shows that the global scheme significantly reduces execution time. Fig. 7b presents the execution time for the streamline schemes and two parallel ones.

Fig. 8a shows speedup for the two parallel schemes. The global scheme performs much better since it reduces communication among nodes and sequential resampling—the main bottlenecks in the local scheme which reaches maximum speedup with only four nodes, whereas the global scheme's speedup is still increasing for eight nodes. The elapsed time on resampling and communication in the local scheme is shown in Fig. 8b. In the global scheme, resampling with an independent MH algorithm is pipelined, and the volume of data communicated among nodes is reduced, leading to its clearly better performance.

Also, the algorithms using the local scheme and global one with 800 particles and color distribution are used to process 20 frames of a video sequence, tracking human, respectively. The tracked human is modeled as an ellipse with a fixed region, so that the state can be determined by two parameters, $X_k = (x_k, y_k)$, where (x_k, y_k) is center of the ellipse at time k. Fig. 9 shows the tracking results obtained from the local parallel scheme (left side), the global one (right side). As we expected, tracking performance of KPF-MCMC outperforms generic PF's.

Distributed Markov Chain Monte Carlo Kernel Based Particle Filtering for Object Tracking

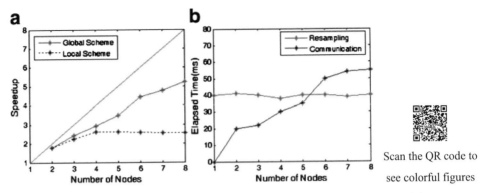

Scan the QR code to see colorful figures

Fig. 8 Speedup and elapsed time: a Speedup for two schemes. b Elapsed time on resampling and communication in the local scheme.

Scan the QR code to see colorful figures

Fig. 9 Simulation results from frame #3, #8, #12.

219

6. Conclusion

Two parallel distributed particle filtering schemes were examined. The simple local scheme is a partially pipelined strategy, which directly maps particle generation and weight computation steps onto clusters. The results show that it suffers from excessive communication and its sequential resampling steps. In contrast, the global scheme is a full parallelization, which shows reduced execution time up to more than 8 nodes. Tracking accuracy was also increased by likelihood with a kernel function, and resampling with an independent MH algorithm. Thus the global scheme has a dual benefit.

Acknowledgement

This research was supported by the National Science Foundation of China under Grants 60572041 and 60832004.

References

[1] BOLIC M, DJURIC PM, HONG S. Resampling algorithms for particle filters: a computational complexity perspective[J]. EURASIP journal on applied signal processing, 2004, 44(1–2): 2267-2277.

[2] CHANG C, ANSARI R. Kernel particle filter for visual tracking[J]. IEEE signal processing letters, 2005, 12(3): 242-245.

[3] CHIB S, GREENBERG E. Understanding the metropolis-hastings algorithm[J]. Journal of the american statistical association, 1995, 49: 327-335.

[4] CHO JU, JIN SH, PHAM XD, et al. Multiple objects tracking circuit using particle filters with multiple features[C]//IEEE International Conference on Robotics and Automation, April 10-14, 2007, Roma, Italy. IEEE, c2007: 4639-4644.

[5] DOUCET A, GODSILL S, ANDREIEU C. On sequential Monte Carlo sampling methods for Bayesian filtering[J]. Journal of statistical computation and simulation, 2000, 10: 197-208.

[6] HIGUCHI T. Monte Carlo filter using the genetic algorithm operators[J]. Journal of statistical computation and simulation, 1997, 59(1): 1-23

[7] HONG S, CHIN S, DJURIC P. Design and implementation of flexible resampling mechanism for high-speed parallel particle filters[J]. Journal of VLSI signal processing, 2006, 44: 47-62.

[8] HUANG AJ. A tutorial on Bayesian estimation and tracking techniques applicable to nonlinear and non-Gaussian processes[R]. Virginia: MIRTE, 2005.

[9] KOTECHA J, DJURIC P. Gaussian particle filtering[J]. IEEE transactions on signal processing, 2003, 51(10): 2592-2601

[10] LIU JS, CHEN R, LOGVINENKO T. A theoretical framework for sequential importance sampling and resampling[M]//Sequential Monte Carlo Methods in Practice. New York:

Springer, 2001: 225-246.

[11] MASKELL S, ALUN B, MACLEOD M. A single instruction multiple data particle filter[C]//IEEE Nonlinear Statistical Signal Processing Workshop, September 13-15, 2006, Cambridge, UK. IEEE, c2006: 51-54.

[12] MIGUEZ J. Analysis of parallelizable resampling algorithms for particle filtering[J]. IEEE transactions on signal processing, 2007, 87: 3155-3174.

[13] MUSA ZB, WATADA J. Motion tracking using particle filter.[C]//The 12th International Conference on Knowledge-Based Intelligent Information and Engineering Systems, September. Springer, 2008: 119-126.

[14] PEREZ P, VERMAAK V, BLAKE A. Data fusion for visual tracking with particles[J]. Proceedings of the IEEE, 2004, 92: 495-513.

[15] PITT M, SHEPARD N. Filtering via simulation: auxiliary particle filters[J]. Journal of the american statistical association, 1999, 94(446): 590-599.

[16] QUINN M J. Parallel programming in C with MPI and OpenMP[M]. New York: McGraw Hill, 2003.

[17] RUI Y, CHEN Y. Better proposal distributions: object tracking using unscented particle filter[C]//IEEE Conference on Computer Vision and Pattern Recognition, December 08-14, 2001, Kauai, HI, USA. IEEE, c2001: 786-793.

[18] SUTHARSAN S, SINHA A, KIRUBARAJAN T, et al. An optimization based parallel particle filter for multitarget tracking[J]. IEEE transactions on aerospace and electronic systems, 2005, 5913: 87-98.

[19] ZHOU S K, CHELLAPPA R, MOGHADDAM B. Visual tracking and recognition using appearance adaptive models in particle filters[J]. IEEE transactions on image processing, 2004, 13(11): 1434-1456.

Use Hierarchical Genetic Particle Filter to Figure Articulated Human Tracking*

1. Introduction

Articulated human movement tracking is an active research area since it can prepare data for pose estimation and action recognition, with the basic idea that estimate the Degree of Freedom (DoF) of human body by the observation of image features. Because of occlusion and cluttered background, the distributions during filtering are usually multi-modal, which highly restrict the use of Kalman filter. Particle filter (PF) can provide an effective solution to the multi-modal filtering problem as approximating the posterior density by a finite set of normalized weighed particles. However, a major concern in PF is that the number of particles increases exponentially with the dimension of state space. As the dimension in articulated human tracking is usually high, then how to draw samples in DoF space becomes a big problem.

One way to improve the sample effectiveness is using human motion model. With the guide of motion model, the search space can be constrained or the state space dimension can be reduced. D.Ormoneit *et al.* use PCA to present human walking in a low-dimensional linear subspace and then estimate the natural variations among people and activities. Rohr reduces the dimension of the problem to the phase of walking cycle, and estimates the phase and body location using Kalman filter. However, motion model based human tracking has a problem called model invalidation which indicates the phenomenon that motion model fails to constrain the particles in the region with a high likelihood, so the motion model can be seen as a double-edged sword. Another way is to improve the samples with the knowledge of current observation. Deutscher *et al.* use *annealed particle filter* to track an articulated model in a high dimensional space. M. Bray *et al.* figure out high-dimensional tracking by smart particle filter, which uses "*Stochastic Meta-Descent*" as a sample optimizer. It has been realized that improving the sampling and global optimization is more decisive to the

---

* The paper was originally published in *ICME*, 2008, and has since been revised with new information. It was co-authored by Long Ye, Qin Zhang, and Ling Guan.

success of articulated model based human motion tracking with less particle number. Our work also belongs to this category.

In our previous work, *Genetic Resampling Particle Filter* (GRPF) is adopted to improve the sampling with the idea of solving the particle degeneration problem by evolutionary method. However, the fixed-length binary representation of GA highly restricts the use of GRPF in high-dimensional estimation. Evolutionary Computation (EC) and Differential Evolution (DE) extend GA in a more generic sense but take no account of the hierarchic rank in high-dimensional data. Consequently, not only the precision of optimization has been lowered but also the computation efficiency is decreased.

In this paper, we will present a novel optimization method named Hierarchical Genetic Optimizer (HGO), and apply it to the sampling step in particle filter. Finally, we use the Hierarchical Genetic Particle Filter (HGPF) to do 2D articulated human movement tracking. Result demonstrates the effectiveness of the new framework in dealing with the self-occlusion and cluttered background which is unavoidable in 2D monocular human tracking.

2. Hierarchical Genetic Optimizer

As a Bayesian estimation problem, human motion tracking consists of two main steps: prediction and update. Prediction uses Chapman-Kolmogorov (C-K) function to obtain target movement prior probability $p(x_n | y_{1:n-1})$. In update stage, the likelihood in Bayes' theorem is used to modify the priori PDF. Particle filter approximates the posterior density as a finite set of normalized weighed particles, as:

$$p(x_n | y_{1:n}) \approx \sum_{i=1}^{Ns} \omega_n^{(i)} \delta(x_n - x_n^{(i)}) \tag{1}$$

where $\omega^{(i)}$ is the normalized weight of the particle and it can be updated by Eq.2, $q(\cdot)$ is importance sampling function and often be chosen as $p(x_t | x_{t-1})$. Then Eq.2 becomes Eq.3.

$$\omega_n^{(i)} \propto \omega_{n-1}^{(i)} \frac{p(y_n | x_n^{(i)}) p(x_n^{(i)} | x_{n-1}^{(i)})}{q(x_n | x_n^i, y_n)} \tag{2}$$

$$\omega_n^i \propto \omega_{n-1}^i p(\mathbf{y}_n | \mathbf{x}_n^{(i)}) \tag{3}$$

Eq.3 shows that in traditional particle filter, particles are sampled without any knowledge of the observation, and greatly increase the risk to converge the particles in a low likelihood region. To cope with this problem, a hierarchical genetic optimizer is adopted in this paper to increase the probability of finding global optima.

Evolutionary Computation takes more generic sense as extending the GA in the float processing. In EC, crossover and mutation operator are showed as Eq.4 and Eq.5, where X

denotes the particle/chromosome in a population; ε is a random parameter between 0 and 1; η represents the normal distribution.

$$\text{Crossover}: \begin{cases} X^{newA} = \varepsilon X^A + (1-\varepsilon)X^B \\ X^{newB} = \varepsilon X^B + (1-\varepsilon)X^A \end{cases} \qquad (4)$$

$$\text{Mutation}: X^{newA} = X^A + \eta \qquad (5)$$

Assume X_{min} and X_{max} present the vector composed by the minimums and maximums of all the state parameter, Crossover can provide a good local search result in state space $S = [X_{min} \; X_{max}]$, and mutation has the ability to do the neighborhood search around S. With these two operators, EC can search the space S by crossover, simultaneity change the region shape of S by mutation to find global optima. One concern in EC and other evolutionary algorithms is that it takes no account of the hierarchic rank in high-dimensional data which have been demonstrated as an effective way to reduce the configuration space.

Hierarchical search usually can be seen as tree-based estimation which has three useful assumptions: 1) Parameters on the same level are independent therefore these parameter partitions can be executed independently; 2) In the same branch, the observation model for high level parameter is independent of the low level parameter; 3) The parameters in the same branch have a mutual observation model. Combining the three assumptions above, M branches and N level tree-based particle state can be estimated by Eq.6, where $x_{m,n,r}$ denote the parameter state in a tree-based data, m is the branch index, n is the level index, and r indexes the freedom in a same node.

$$p(x_{1:M,1:N} \mid y) = \prod_{m=1}^{M} p(x_{m,N_m} \mid x_{m,1:N_m-1}, y_{m,1:N_m}) p(x_{m,1:N_m-1} \mid y_{m,1:N_m-1})$$

$$= \prod_{m=1}^{M} p(x_{m,N_m} \mid x_{m,1:N_m-1}, y_{m,1:N_m}) p(x_{m,N_m-1} \mid x_{m,1:N_m-2}, y_{m,1:N_m-1}) \cdots p(x_{m,1} \mid x_0, y_{m,1}) \qquad (6)$$

$$\propto \prod_{m=1}^{M} p(y_{m,1:N_m} \mid x_{m,1:N_m-1}^{(i)}, x_{m,N_m}^{(i)}) p(y_{m,1:N_m-1}, x_{m,N_m-2}^{(i)} \mid x_{m,N_m-1}^{(i)}) \cdots p(y_{m,1}^{(i)} \mid x_{m,1}^{(i)})$$

Eq.6 shows the basis of hierarchical search which can be summarized as: The global optimum of each branch is up to the accuracy of parameter level by level. This is also the basic idea of HGO which uses hierarchical mutation to find the search space, and then uses crossover operator to find local optima.

2.1 Hierarchical Mutation

In the mutation step, mutation number and intensity η are presented by mutation probability p_m and the covariance vector σ_η. Like gene mutation usually happens in bad environment, the value of p_m and σ_η are also determined by the observation fitness. In the hierarchical search problem, according to Eq.6, the global fitness can be ensured by optimizing the state parameters level by level. In other words, if we let u_l denote the fitness of level l, then the global optima can be achieved by convergence from $l=1$ to $l=leaves$ level. In each level, $p_m = \lambda_m \mu_l$, and η is 0 mean normal distribution with covariance vector $\sigma_\eta = [\lambda_\eta u_1, I]$, where I is an h-k dimensional zero vector and h is the dimension of the state space, k is the parameters number of level $1:l$.

2.2 Local Crossover

After l_m reaches the leaf level, Hierarchical Mutation steers the particles into a higher likelihood, increasing the opportunity to find global optima. Then the local optimization is achieved by crossover. We define our crossover operator as Eq.8. Compared with the definition in Eq.4, our method adds a path reform point with the consideration of maintaining the part property of parent chromosome. X' and X'' in Eq.7 are two parts of a chromosome divided by a random path reform point.

$$\begin{cases} X^{newA} = \left[\varepsilon X'^A + (1-\varepsilon) X'^B \quad (1-\varepsilon) X''^A + \varepsilon X''^B \right] \\ X^{newB} = \left[(1-\varepsilon) X'^A + \varepsilon X'^B \quad \varepsilon X''^A + (1-\varepsilon) X''^B \right] \end{cases} \quad (7)$$

Combining Hierarchical Mutation and Local Crossover, the detail of HGO is as follows.

- Choose initial population with N individuals
- Evaluate the individual fitness μ_i in the population and the global fitness μ_G of the population as:

$$\mu_G = \frac{1}{N} \sum_{i=1}^{N} u_i$$

- Repeat

if $\mu_G > \mu_{T2}$ (Global Optimization)

Select best-ranking individuals to reproduce

1) Computer p_c by:

$$p_c = \lambda_c \mu_G$$

2) Breed new generation through crossover (Eq. 7) and give birth to offspring;

else (Hierarchical Optimization)

1) Evaluate level fitness μ_l from the top level until $\mu_l > \mu_{T3}$ and set mutation level $l_m =$ current level;

2) Computer p_m and σ_η by:

$$p_m = \lambda_m \mu_{lm}; \sigma_\eta = \lambda_\eta \mu_{lm}$$

3) Breed new generation through mutation (Eq.5) and give birth to offspring;

end

Evaluate the individual fitness of the offspring;

Replace worst ranked part of population with offspring

- Convergence if

$$\mu_G > \mu_{T1}$$

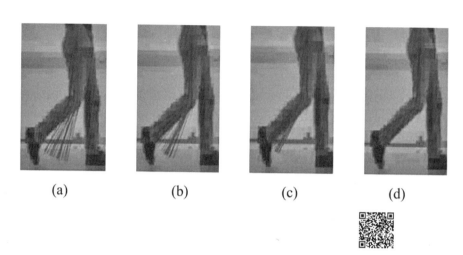

(a)　　　　(b)　　　　(c)　　　　(d)

Scan the QR code to see colorful figures

Fig. 1 The sample state after prediction and three kinds of convergences (a) after prediction; (b) level 1 (thigh) convergence; (c) region convergence, search space conformed, local optimization begin; and (d) global convergence.

In HGO, there are three important thresholds $\mu_{T3}, \mu_{T2}, \mu_{T1}$ which are levels of convergence, region convergence and global convergence. Fig. 1 shows how the three thresholds effect the optimization by a sample leg matching experiment.

3. Articulated Human Tarcking

3.1 Initialization

Before the tracking, we need to do two initializations: 1) build human body model and 2) initialize sample states. Initialization plays an important role in tracking as it may apply more useful prior information for tracking and draw many researches on [10]. To make it easy and accurate, we adopt a manual initialization process, shown in figure 2. With the guide of 'marker', the size of human model and initialization pose can be fixed as the prior information.

3.2 Image Likelihood

The image likelihood or observation model is used to measure the fitness of the particles. It is built by calculating the feature difference between the model and current frame. Edge and region color are two mainly used image features. In order to get these features, it is necessary to do the video segmentation to obtain the foreground, so the accuracy of segmentation has a great impact on the tracking performance. In our work, we extract the edge and region features directly, eliminating the effect of the segmentation.

(a) (b) (c)

Scan the QR code to see colorful figures

Fig. 2 (a) Initialized Human Model, (b) Smoothed Gradient Image, and (c) Cropped Region.

Edge: A gradient based edge detection mask introduced in the reference [5] is used to extract edge feature. Firstly, a gradient image is built and then smoothed with a Gaussian mask and remapped between [0 1], as shown in figure 2 (b). It is clear that in the smoothed image, the more the pixels close to the edge, the more the pixel value close to 1. So the edge feature can be calculated with a sum-squared difference function as in Eq.8, where P_{edge} is the value of the edge pixel taken along the model's boundary and N is the number of edge pixels.

$$E_{edge} = \frac{1}{N}\sum_{i=1}^{N}\left(1 - P_{edge}^{(i)}\right)^2 \quad (8)$$

Region color: In the initializing process, we can get color histogram of each body segment, as in figure 2 (c). For the 2D tracking, it is feasible to consider that the color statistics change slightly and can be used for region feature, as shown in Eq.9, where is the value of bin in histogram and N is the number of bins.

$$E_{region} = \frac{1}{M}\sum_{j=1}^{M}\left(C_{reference}^{(j)} - C_{estimate}^{(j)}\right)^2 \quad (9)$$

3.3 HGPF for Tracking

Combining all the ideas above, the human tracking framework based on HGPF can be summarized as follows.

- Manual initialization, then draw the N particles $\{X_0, \omega_0\}_{i=1:N}$ around the initialized pose, set $V_0 = 0$;

For $k = 1$ to the frame number
- Prediction: $X_k = X_{k-1} + V_{k-1}$
- Measure ω_k with Eq.10 and do optimization with HGO;

$$w_k^{(i)} = exp\left(-\left(E_{edge} + E_{region}\right)\right) \quad (10)$$

- MAP output $X_{out,k}$ and set $V_k = X_{out,k} - X_{out,k-1}$.

4. Experiments

Besides high-dimensional sampling, another difficulty in 2D human tracking is the information missing or information illegibility caused by self-occlusion and cluttered background. The difference between HGO and other hierarchical search programs is that our method does not output the high level state when it has converged, but considers them as the potential nodes to extend to next level, as shown in Fig. 3 (a). This idea is similar to the viterbi algorithm and we use hierarchical mutation to estimate the path probability level by level, and use crossover to trace back the output. As the viterbi algorithm can figure the FEC decoding, our method also has the capability in dealing with the information illegibility.

We carry out experiments on two 2D monocular human motion sequences, among which one has the cluttered background (figure 4 (a)), and the other one has self-occlusion (figure 4 (b)). We use 20 particles in both cases.

For gait tracking, the serious self-occlusion happens when the human body is on a stand state. Fig. 3 (b) shows how HGO can probability deal with the problem.

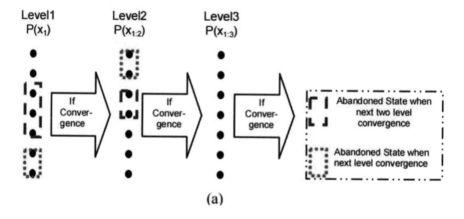

(a)

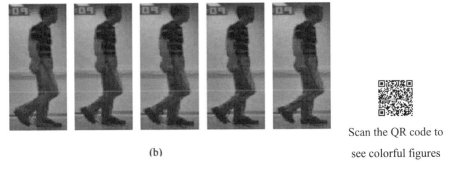

Scan the QR code to see colorful figures

(b)

Fig. 3 (a) Illustrate of Convergence Process, (b) convergence process with 0,1,2,4,6 iterations.

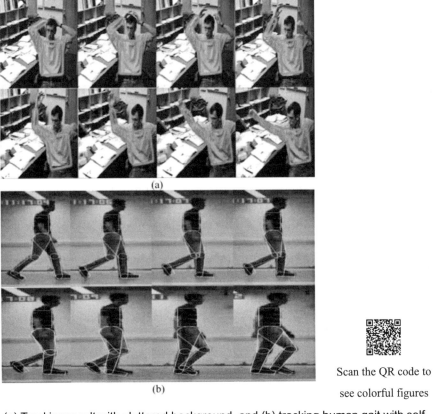

Scan the QR code to see colorful figures

(b)

Fig. 4 (a) Tracking result with cluttered background, and (b) tracking human gait with self-occlusion.

5. Conclusion

In this paper, we proposed a novel hierarchical search method, namely hierarchical genetic optimization. Then a framework for articulated human tracking based on HGO and particle filter is used to track 2D human movement. The experiments result showed that our method performs well when the objects are disordered by occlusion and cluttered background.

In the future, we will extend our method to 3D human tracking, in which we need to take more consideration on the correlation between parameters.

References

[1] HUANG A J. A tutorial on bayesian estimation and tracking techniques applicable to non-linear and non-gaussian process[R]. Virginia: MITRE, 2005.

[2] ORMONEIT D, BLACK M J, HASTIE T. Representing cyclic human motion using functional analysis[J]. Image and vision computing, 2005, 23(14): 1264-1276.

[3] ROHR K. Human movement analysis based on explicit motion models[M]//Motion-Based Recognition. Dordrecht: Springer, 1997: 171-198.

[4] DEUTSCHER J, BLAKE A, REID I. Articulated body motion capture by annealed particle filtering[C]//IEEE Conference on Computer Vision and Pattern Recognition, June 13-15, 2000, Hilton Head, SC, USA. IEEE, c2000: 126-133.

[5] BRAY M, KOLLER-MEIER E, GOOL L V. Smart particle filtering for high-dimensional tracking[J]. Computer vision and image understanding, 2007, 106: 116-129..

[6] YE L, WANG J, ZHANG Q. Genetic resampling particle filter[J]. ACTA automatica sinica, 2007, 33: 885-887.

[7] BACK T, FOGEL D B, MICHALEWICZ. Handbook of evolutionary computation[M]. Oxford: University Press, 1997.

[8] DU M, GUAN L. Monocular human motion tracking with the DE-MC particle filter[C]//IEEE International Conference on Acoustics, Speech and Signal Processing, May 14-19, 2006, Toulouse, France. IEEE, c2006: 205-208.

[9] MACCORMICK J P, ISARD M. Partitioned sampling, articulated objects, and interface-quality hand tracking[C]//European Conference on Computer Vision, June 26-July 1, Dublin, Ireland. Springer, c2000: 3-19.

[10] NING H, TAN T, WANG L et al. Kinematics-based tracking of human walking in monocular video sequences[J]. Image and vision computing, 2004, 22(5): 429-441.

Human Action Recognition Using Multi-Velocity STIPs and Motion Energy Orientation Histogram*

1. Introduction

Recognizing human activities from video sequences has become a hot topic in computer vision because of its wide range of applications such as event analysis, action understanding, and intelligent surveillance. In recent years, many methods have shown good performance. These methods can be classified into two groups according to features: methods based on global features and methods based on local features. However, these two categories face the same challenge induced by complex and dynamic backgrounds.

To solve this problem, methods based on global features generally introduce a preprocessing step to mark the action region or segment the foreground from the background according to the content of video sequences. Bregonzio et al. adopted frame differences and pedestrian tracking for the Weizmann and KTH datasets, respectively. For datasets with complex backgrounds (e.g., UCF sports), human detector was employed because the action region is difficult to mark. Global features such as motion descriptors, combined local-global optic flow, and space-time shape were then extracted to describe actions. These methods have high accuracy when processing video sequences with static and simple backgrounds. However, a good result of foreground segmentation is important and difficult to recognize. Therefore, manual calibration is necessary to cope with complex conditions such as multiple views, camera motions, and dynamic backgrounds.

Compared with the above methods, methods based on local features are robust and generalized for action recognition with the help of interest point detection. However, the recognition accuracy of methods based on local features is usually lower than the recognition accuracy of methods based on global features.

In this paper, we propose a novel and robust action recognition approach based on

---

* The paper was originally published in *Journal of Information Science and Engineering*, 2014, 30, and has since been revised with new information. It was co-authored by Chuanzhen Li, Bailiang Su, Jingling Wang, Hui Wang, and Qin Zhang.

multi-velocity spatiotemporal interest points (MVSTIPs) and a novel local descriptor called motion energy (ME) orientation histogram (MEOH). First, we utilize multi-direction ME filters that are tuned to different velocities to detect significant changes in space and time at the pixel level. Second, a surround suppression model is employed to rectify the ME deviation caused by camera motion and complicated backgrounds. Third, unlike previously proposed methods, MVSTIPs at various speeds are detected by local maximum (LM) filtering. Thereafter, we develop MEOH descriptor to capture motion features in local regions around MVSTIPs. Finally, a bag-of-words (BoW) framework is applied to implement the training and testing for human action recognition.

The main contributions of our work are summarized as follows:

(1) Biologically motivated motion detection algorithms (i.e., filtering with multi-direction ME filters at different speeds) are introduced to detect spatio-temporal events.

(2) The surround suppression model is introduced to reduce redundant and false detection points in the texture region. Our spatio-temporal interest point (STIP) detector performs better than existing STIP detectors, such as the 1D Gabor detector.

(3) MEOH descriptor is developed to capture distinguishing motion orientations in local regions around spatio-temporal interest points.

Based on the above work, the proposed human action recognition method is adaptive to various conditions, such as low contrasts, moving cameras, and dynamic backgrounds. The robustness and generality of the proposed method can be proven by obtaining high performance on the three public human action datasets, namely, KTH, Weizmann, and UCF sports.

The rest of the paper is organized as follows. Section 2 introduces the related works. Section 3 presents the MVSTIP detection method and MEOH descriptor. Section 4 presents our experimental results. Finally, Section 5 concludes.

2. Related Works

Generally, methods based on global features are more accurate under simple conditions with static backgrounds and fixed viewpoints. However, perfect foreground segmentation cannot be implemented easily in reality. By contrast, methods based on local features are robust and generalized for action recognition because they focus on the local information of actions. Thus, actions are analyzed by a collection of local patterns instead of global features such as shape, appearance, or silhouette.

Methods based on local features detect interest points or regions that are informative and descriptive. These methods capture local features before learning a recognition model. These methods contain four principal steps: interest point detection, feature description, model learning, and classification.

2.1 Interest Point Detection

A large number of STIP detection methods have been proposed. Laptev extended the Harris corner detector into 3D and extracted scale-invariant STIPs. However, the computation is expensive, and the results are sparse. Chakraborty combined the Harris detector and surround suppression to detect STIPs. This combined method is selective; thus, most informative STIPs are retained. Dollar developed a method that uses a pair of 1D Gabor filters on the temporal domain. This method is very popular in action recognition because of its high performance and low computation. Brengonzio introduced a 2D Gabor filter-based detector. However, this method works well only for simple and static backgrounds. Inspired by a 2D speeded-up robust features (SURF) detector, Willems proposed a dense and scale-invariant detector based on 3D Hessian matrix.

2.2 Feature Description

In most methods based on local features, volumetric descriptors have been emphasized to improve computational efficiency and increase recognition accuracy. Dollar designed and tested many volumetric descriptors including normalized brightness, gradient, and windowed optical flow. Niebles calculated the brightness gradients of volumetric descriptors in the x, y, and t directions instead of more complex descriptors to obtain a descriptor for each spatiotemporal cube without increasing computational complexity. Laptev detected local structures in space-time, where image values have significant local variations in both space and time, estimated the spatiotemporal extents of detected events, and computed the scale-invariant spatiotemporal descriptors of such events. Given the success of the scale-invariant feature transform (SIFT), Scovanner proposed a new descriptor named 3D SIFT to capture vital temporal information that cannot be obtained by 2D representations. Klaser presented a 3D HoG-like descriptor that can be computed for arbitrary scales and is concentrated on the sub-histograms of 3D gradients in local regions. Knopp extended the SURF descriptor into 3D domains that can be considered speeded-up versions of 3D SIFT.

2.3 Model Learning and Classification

The BoW model has been widely used in methods based on local features. Many other recognizing models have also been successfully applied. Naïve-Bayes mutual information maximization is used as a discriminative pattern matching criterion in Yuan's action detection and recognition system. Guha used the orthogonal matching pursuit and K-singular value decomposition to investigate three overcomplete dictionary learning frameworks. Rich and compact dictionaries are learned by a series of local descriptors.

The support vector machine (SVM), hidden Markov model, and k-nearest neighbors are widely used because of their efficiency and quickness in the classification stage. Topic model-based methods, such as probabilistic latent semantic analysis and latent Dirichlet allocation, are also applied to categorize action because of the high flexibility of these generative graphical models.

3. MVSTIP Detection and Feature Description

The BoW model is adopted to implement human action recognition, learn different action classes in training video sequences, and apply the learned model to perform action categorization in query video sequences. In this paper, we focus on the spatio-temporal interest point detector and descriptor because of their importance in BoW methods. On one hand, STIPs are extracted as the most descriptive points with significant local variations in time and space. On the other hand, the description of STIPs should be meaningful enough to distinguish different motions. Our purpose is to avoid false detection and redundant detection, as well as maintain appropriate descriptions.

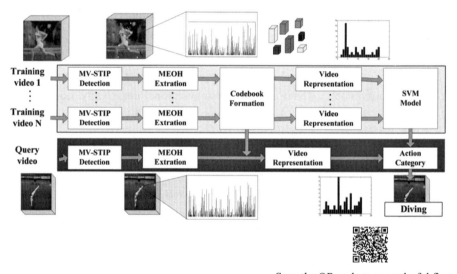

Scan the QR code to see colorful figures

Fig. 1 Flowchart of the proposed approach.

The flowchart of our approach is shown in Fig. 1. The proposed method is composed of two steps: training and testing. We separate video sequences into training and testing sets. MVSTIPs are detected for different types of actions in the training set, and MEOH features are exacted. Thereafter, we cluster the training data (MEOH features) by using the k-means algorithm and Euclidean distance as the clustering metrics. Each clustering center is called a code-word, and the collection of all centers forms a codebook (or vocabulary). Thereafter, each MVSTIP is assigned an index of the unique closest code-word to represent a training video as a collection (i.e., histogram of indices) of spatio-temporal words from the codebook. Each testing video can be processed in the same manner and can be represented as a collection of spatio-temporal words from the codebook. Finally, we use LibSVM with RBF kernel as a classifier to train and recognize videos that contain actions. The histograms are normalized before the training and testing of the model. When a query video arrives, we can input the video representation into LibSVM to obtain the action category.

In the following, we describe the MVSTIP detector (Sections 3.1-3.3) and MEOH feature descriptor (Section 3.4) in detail.

3.1 ME Responses

The proposed detector is inspired by Petkov's research, which focuses on the orientation and speed tuning properties of the spatio-temporal 3D Gabor and ME filters as models of time-dependent receptive fields of simple and complex cells in the primary visual cortex (V1). We believe that our visual perception of human actions in video sequences also follows the same mechanism because we observe that the pixels move at multi-speeds in various directions in the human region. For example, the limbs move faster than the head and body and the hands shake up and down while walking. Thus, we need a bank of filters at various speeds in different directions to detect the motion.

A 3D Gabor filter function is shown as follows:

$$g_{v,\theta,\varphi}(x,y,t) = \frac{\gamma}{2\pi\sigma^2} \exp\left(\frac{-\left((x'+v_c t)^2 + \gamma^2 y'^2\right)}{2\sigma^2}\right) \cdot \cos\left(\frac{2\pi}{\lambda}(x'+vt)+\varphi\right) \cdot$$

$$\frac{1}{\sqrt{2\pi}\tau} \exp\left(-\frac{(t-\mu_t)^2}{2\tau^2}\right) \cdot U(t)$$

$$\begin{cases} x' = x\cos(\theta) + y\sin(\theta) \\ y' = -x\sin(\theta) + y\cos(\theta) \end{cases}$$

$$U(t) = \begin{cases} 1 & \text{if } t \geq 0, \\ 0 & \text{others} \end{cases} \tag{1}$$

where v is the preferred speed, λ is the preferred spatial wavelength, θ is the preferred direction of the motion and preferred spatial orientation of the filter, and φ is the phase offset that determines whether the kernel is symmetric or anti-symmetric. γ is the spatial aspect ratio that determines the spatial symmetry of the function, σ is the deviation in spatial domain, and v_c is the speed in which the center of the spatial Gaussian envelope moves along the x' axis. A Gaussian distribution with a mean μ_t and standard deviation τ is applied to model the change in intensities of the excitatory and inhibitory lobes of the receptive field with time. The unit step function $U(t)$ ensures that the filter is causal and considers inputs only from the past. According to literature [18], $\tau = 2.75$, $\mu_t = 1.75$, $\gamma = 0.5$, $\lambda = 2\sqrt{1+v^2}$, and $\sigma = 0.56\lambda$.

When a 3D signal $l(x, y, t)$ arrives, 3D Gabor responses are computed by convolving the signal with the 3D Gabor kernel (Eq. 2).

$$r_{v,\theta,\varphi}(x, y, t) = l(x, y, t) * g_{v,\theta,\varphi}(x, y, t). \tag{2}$$

ME denotes the phase insensitive response obtained by the quadrature pair summation of the responses of two filters with a phase difference of $\pi/2$.

$$E_{v,\theta}(x, y, t) = \sqrt{r_{v,\theta,0}^2(x, y, t) + r_{v,\theta,\pi/2}^2(x, y, t)}. \tag{3}$$

ME filters are sensitive to motions at multi-speeds in various directions in video sequences. Thus, ME filters can be used to detect spatio-temporal interest points. Fig. 3 (b) shows the ME responses normalized for visualization in different directions. We observe significant changes in video frames at all speeds.

3.2 Surround Suppression Model

Generally, we take no interest in texture areas and complex backgrounds. However, the ME responses in these areas are high, and the results are coarse and inaccurate (Fig. 3 (b)). Therefore, a subtractive linear mechanism followed by non-linear half wave rectification (named surround suppression model) is employed to overcome the above drawbacks. The inhibition term and surround suppressed ME are computed by using Eqs. (4) and (5), respectively:

$$S_{v,\theta}(x, y, t) = E_{v,\theta}(x, y, t) * \omega_{v,\theta,k_1,k_2}(x, y, t), \tag{4}$$

$$\tilde{E}_{v,\theta}(x, y, t) = \left| E_{v,\theta}(x, y, t) - \alpha S_{v,\theta}(x, y, t) \right|^+, \tag{5}$$

where α controls the strength of surround suppression. $\omega_{v,\theta,k_1,k_2}(x, y, t)$ is defined as the surround suppression weighting function, which is implemented by Eqs. (6), (7), and (8).

$$G_{v,0,k}(x, y, t) = \frac{\gamma}{2\pi k^2 \sigma^2} \exp\left(\frac{-\left((x' + v_c t)^2 + \gamma^2 y'^2 \right)}{2k^2 \sigma^2} \right) \cdot \frac{1}{\sqrt{2\pi\tau}} \exp\left(\frac{-(t-\mu_t)^2}{2\tau^2} \right) \cdot U(t), \tag{6}$$

$$I_{v,\theta,k_1,k_2}(x, y, t) = \left| G_{v,\theta,k_2}(x, y, t) - G_{v,\theta,k_1}(x, y, t) \right|^+, \tag{7}$$

$$\omega_{v,\theta,k_1,k_2}(x, y, t) = \frac{I_{v,\theta,k_1,k_2}(x, y, t)}{\left\| I_{v,0,k_1,k_2} \right\|_{L1}}, \tag{8}$$

where k_1 and k_2 determine the spatial inhibition range. The visualization of the surround suppressed ME response calculation is shown in Fig. 2.

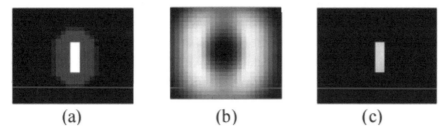

Fig. 2 Schematic of the effect of the surround suppression operator: (a) ME response; (b) inhibition term; (c) surround suppressed ME response.

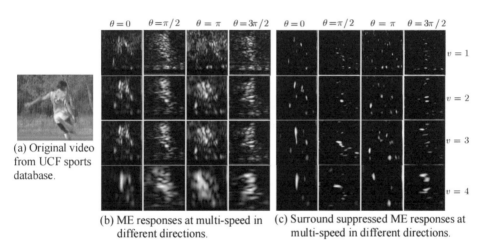

(a) Original video from UCF sports database.

(b) ME responses at multi-speed in different directions.

(c) Surround suppressed ME responses at multi-speed in different directions.

Fig. 3 ME responses and surround suppressed ME responses at multi-speed in different directions.

By using the surround suppression model, ME responses around the cubic center are reserved, whereas ME responses in farther surround are suppressed. The ME responses of the texture area are significantly decreased to separate the motions easily from the background. Surround suppression is effective for complicated or moving backgrounds to avoid false detection and redundant detection (Fig. 3 (c)).

3.3 MVSTIP Detector

The MVSTIPs are detected as follows. First, the motion energy responses at multi-speeds v in various directions θ recomputed by Eq. (3). Second, surround suppressed ME responses in various directions at each speed are computed by Eq. (5). Third, the responses in various directions at a certain speed are added to acquire the cumulative surround suppressed ME (CSSME). Figs. 4 (a) and (c) show the cumulative ME and

CSSME responses at various speeds, respectively. Finally, we remove the points with small responses by LM filtering and detect the primary STIPs at a certain speed. We repeat the above steps from $v = 1$ to 4 to obtain the collection of MVSTIPs. Figs. 4 (b) and (d) show the result of the MVSTIP detector with cumulative ME (CME) and CSSME responses, respectively. The last column shows all MVSTIPs. Significant changes in the human region can be captured by surround suppressed ME filters because the responses of the complex background region are small. However, the results of CME contain false STIPs and redundant STIPs because the ME responses are not salient in complex backgrounds (ground and woods). The density of MVSTIPs can be adjusted by setting different thresholds in LM filtering.

Our method is different from Dollar's 1D Gabor method and Bregonzio's 2D Gabor method. The STIP detector in [13] has been developed to detect pixel variations in the temporal domain and ignore the spatial relationship of pixels. The detector is very efficient for slow motion or static camera views but can be easily confused in complex backgrounds or video noises because of the lack of spatial information. Bregonzio's 2D Gabor method is exploited to overcome these drawbacks; however, foreground segmentation is needed because of the lack of temporal information. Therefore, the detector is restricted to the result of the foreground segmentation and is not robust for complex backgrounds. We conduct a comparison with the popular 1D Gabor detector on the UCF sports dataset to validate the effectiveness of the STIP detector under complex scenes. Regions of interest provided by literature [31] are used to locate the STIPs in both methods. STIPs in both methods are dense, but the false detection of 1D Gabor is denser (Fig. 5). The dynamic texture is easily confused with interest regions in 1D Gabor, but the proposed method inhibits the responses of the texture regions successfully because of the surround suppression model. The MVSTIPs of the proposed detector are also informative and descriptive because they are obtained by both spatial and temporal domains.

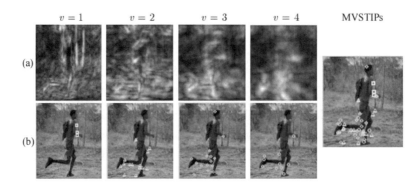

Human Action Recognition Using Multi-Velocity STIPs and Motion Energy Orientation Histogram

Fig. 4 Results of the MVSTIP detector: (a) ME responses; (b) STIPs at each speed with ME; (c) surround suppressed ME responses; (d) STIPs at each speed with CSSME. The labels denote the STIPs of $v=1$: white square, $v=2$: yellow triangle, $v=3$: pale blue rhombus, $v=4$: green circle. The last column shows all MVSTIPs obtained by using CME and CSSME, respectively.

3.4 MEOH Feature Descriptor

We develop MEOH after obtaining the MVSTIPs to obtain a descriptor for each spatio-temporal cube. The descriptor is inspired by 3D SIFT descriptor and cuboid features. First, we extract an $M \times M \times N$ brick around each interest point, and each brick is split into 64 sub-bricks with a size of $M/4 \times M/4 \times N/4$ (Fig. 6).

Fig. 5 Comparison between the proposed detector and 1D Gabor detector. The top row shows the STIPs with the 1D Gabor method. The bottom row shows the MVSTIPs with the proposed MVSTIP detector. The labels denote the same meanings and show the STIPs of $v=1$: white square, $v=2$: yellow triangle, $v=3$: pale blue rhombus, $v=4$: green circle.

The ME responses in all directions at preferred speeds are used to represent each pixel in the sub-brick. Specifically, $R_{\max}^{(B_k)}$ in Eq. (9) denotes the maximum response in the sub-brick B_k $(k = 1, 2, \cdots, 64)$. The characteristic function $f_i^{(B_k)}$ is defined in Eq. (10), i.e., if the response at a point in a given direction θ_i is larger than half of $R_{\max}^{(B_k)}$, the point moves in a direction θ_i. Then we record "1" and "0" for motion and stillness, respectively.

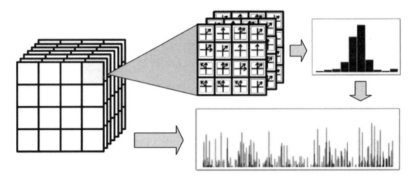

Fig. 6 Illustration of MEOH feature calculation. The center of the brick is one STIP. The brick is split into 64 sub-regions. A sub-histogram of orientations of the surround suppressed ME in a sub-region is computed. The MEOH features are then obtained by arranging 64 sub-histograms.

$$R_{\max}^{(B_k)} = \max\left\{\tilde{E}_{v_p,\theta}^{(B_k)}(x, y, t)\right\}, \tag{9}$$

$$f_i^{(B_k)}(x, y, t) = \begin{cases} 1 & \text{if } \tilde{E}_{v_p,\theta_i}^{(B_k)} \geq 0.5 R_{\max}^{(B_k)}, \\ 0 & \text{others} \end{cases} \tag{10}$$

$$s\_h_{ki} = \sum_{x=1}^{M/4}\sum_{y=1}^{M/4}\sum_{t=1}^{N/4} f_i^{(B_k)}(x, y, t). \tag{11}$$

For each pixel in the sub-bricks, the feature value is obtained by $f_i^{(B_k)}$, and a sub-histogram for each sub-brick is obtained by Eq. (11) to represent the distribution of motion in all directions. Thereafter, all sub-histograms of the sub-bricks are added to form a histogram matrix (Eq. (12)). Thereafter, the MEOH descriptor is obtained by arranging $H(\mathbf{B})$ to one column (Eq. (13)).

$$H(\mathbf{B}) = \begin{bmatrix} s\_h_{1,1} & s\_h_{1,2} & \cdots & s\_h_{1,i_{\max}} \\ s\_h_{2,1} & s\_h_{2,2} & \cdots & s\_h_{2,i_{\max}} \\ \cdots & \cdots & \cdots & \cdots \\ s\_h_{64,1} & s\_h_{64,2} & \cdots & s\_h_{64,i_{\max}} \end{bmatrix}, \tag{12}$$

$$MEOH(\mathbf{B}) = \left(s\_h_{1,1}, s\_h_{2,1}, \cdots, s\_h_{64,1}, \cdots, s\_h_{1,i_{\max}}, \cdots, s\_h_{64,i_{\max}}\right)^T. \tag{13}$$

When $i = 1, 2, \cdots, 8$, the MEOH feature vector of an interest point is computed by arranging all 64 sub-histograms into a 512-dimensional vector. The MEOH descriptor calculates the orientations of surround suppressed ME in local regions similar to the 3D SIFT or 3D Gradient. The features capture the variation of motion directions around an interest point. The features are easy to calculate because the surround suppressed ME has been obtained in the MVSTIP detection.

Fig. 7 shows the MEOH feature vectors obtained by Eqs. (9) to (13). Fig. 7 (a) shows the original frame with interest points (i.e., A, B, C, D, and E) from the UCF sports database. Fig. 7 (b) shows the numbering of directions: "1" denotes $\theta = 0$, "2" denotes $\theta = \pi/4$, and so on. Figs. 7 (c)-(g) show the MEOH feature vectors of Points A, B, C, D, and E. Some peak points appear in the corresponding edges and motion directions.

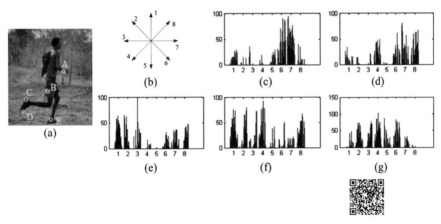

Scan the QR code to see colorful figures

Fig. 7 MEOH feature vectors: (a) original frame with interest points (*i.e.*, A, B, C, D, and E) from the UCF sports database; (b) number of directions; (c)-(g) MEOH feature vectors of Points A, B, C, D, and E, where x-axes represent directions $\theta = \{0, \pi/4, \pi/2, \cdots, 7\pi/4\}$ and y-axes represent the numbers of pixels marked by "motion" in all directions.

4. Experimental Results

In this section, we evaluate the effectiveness of our method on three public human action datasets: Weizmann, KTH, and UCF sports. With MATLAB implementation running on an Intel core i7 and 2.0 GHz computer with 8GB RAM, the proposed method (without optimization) runs with a speed of about 1 second per frame.

4.1 Experimental Results on the Weizmann Dataset

The Weizmann dataset (Fig. 8) contains 10 actions performed by 9 persons, thus giving a total of 90 videos. The action categories include "bend," "jack," "jump," "pjump," "run," "side," "skip," "walk," "wave1," and "wave2." Each video lasts for 2s to 3s with a simple background and static camera. The dataset is challenging because several actions are similar, such as "skip" and "jump."

In MVSTIP detection, we eliminate the points that have cumulative responses of less than 1.5 or 20% of the maximum. The largest 100 points are considered MVSTIPs. We then set the brick size as 20×20×12 in the MEOH description. The codebook is obtained by clustering all 90 videos.

We adopt the leave-one-person-out cross-validation to test the effectiveness, i.e., we select 10 videos performed by 1 person as the test set, whereas the other videos are used as the training set. The leave-one-person-out cross-validation is conducted 9 times until all videos are predicted. The final average recognition rate is 98.89%. The recognition rate is defined as the rate of successfully recognized actions over all test actions. The confusion matrix is shown in Fig. 9. Each column of the matrix represents the actions in a predicted class, whereas each row represents the actions in an actual class. We have obtained the ideal performance on most actions, except for "jump" and "skip." These 2 actions are similar in the way that the actors pass the video, thus resulting in similar 3D spatio-temporal cubes and therefore similar MEOH descriptors. Table I shows the existing methods based on local features performed on the Weizmann dataset. The accuracy of our approach is superior to other methods based on local features.

Fig. 8 Sample frames from the video sequences of the Weizmann dataset.

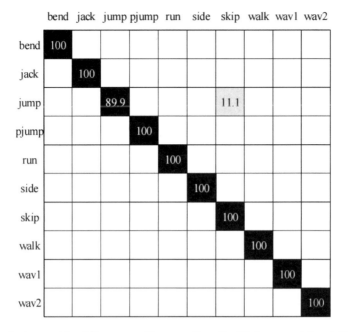

Fig. 9 Confusion matrix of the recognition result on the Weizmann dataset. The overall accuracy is 98.89%.

Table I Comparison of the recognition algorithms based on local features on the Weizmann dataset.

Ours		98.89%			
Zhang	2012	93.87%	Klaser	2008	84.30%
Guha	2012	98.90%	Niebles	2008	90.00%
Brengonzio	2009	96.67%	Scovanner	2007	82.60%
Liu	2008	90.40%			

4.2 Experimental Results on the KTH Dataset

The KTH dataset (Fig. 10) contains 598 videos of 25 different persons performing 6 actions, namely, "walking," "running," "jogging," "boxing," "handclapping," and "handwaving." The video sequences are recorded in 4 scenarios, including outdoor and indoor scenes. The KTH dataset is more difficult than the Weizmann dataset because of camera vibration, viewpoint changes, and different appearance scenarios and scales.

(a) Walking (b) Jogging (c) Running (d) Boxing (e) Handclapping (f) Handwaving

Fig. 10 Sample frames from the video sequences of the KTH dataset.

In MVSTIP detection, we eliminate the points that have cumulative responses of less than 1 or 20% of the maximum. The largest 200 points are considered MVSTIPs. We set the brick size as 24×24×12 in MEOH description. The codebook is obtained by clustering 96 videos performed by four persons.

We adopt the leave-one-person-out cross-validation to test the effectiveness. We select 24 videos performed by 1 person for the test and are run 25 times. The final average recognition rate is 95.48%. The confusion matrix is shown in Fig. 11. We have obtained reasonable performance on most of the actions, except for "run" and "jog." These two actions are similar in the way that the actors run across the video at different speeds, thus resulting in similar 3D spatiotemporal cubes and MEOH descriptors. Existing methods for human action recognition on the KTH dataset are listed in Table II. Table II shows that our approach outperforms the state-of-the-art against methods that use local descriptors.

	walk	run	jog	box	wave	clap
walk	100					
run	1	87	12			
jog	3	6	91			
box				100		
wave					98	2
clap				1.01	2.02	96.97

Fig. 11 Confusion matrix of the recognition result on the KTH dataset. The overall accuracy is 95.48%.

Table II Comparison of recognition algorithms based on local features on the KTH dataset.

Ours		95.48%			
Zhang	2012	93.50%	Brengonzio	2009	93.17%
Le	2011	93.90%	Niebles	2008	83.00%
Yuan	2011	94.00%	Laptev	2008	91.80%
Kovasha	2010	94.53%	Klaser	2008	91.40%

4.3 Experimental Results on the UCF Sports Dataset

The UCF sports dataset (Fig. 12) has been collected from broadcasting television channels, such as BBC and ESPN. This dataset contains 10 sport action classes: "diving," "golf swing," "kicking," "lifting," "horse riding," "running," "skateboarding," "bench swinging," "high-bar swinging," and "walking," thus giving a total of 150 videos. The actions are performed in various scenarios, and the camera generally moves to focus the players. The action durations and scales are different, and some videos contain 1 or more persons. This dataset is one of the most challenging datasets because of the complicated background and large intra-class variations.

The ground truth region provided by the dataset is employed to locate the action region, and we subsample the videos to save the computational cost. Four videos without ground truths are removed from the dataset. In MVSTIP detection, we eliminate the points that have cumulative responses of less than 20% of the maximum. The largest 200 points are considered MVSTIPs. We then set the brick size as 24×24×12 in the MEOH description. The codebook is obtained by clustering all videos.

Scan the QR code to see colorful figures

Fig. 12 Sample frames from video sequences of the UCF sports dataset.

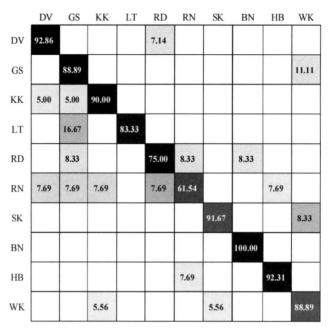

Fig. 13 Confusion matrix of the recognition result on the UCF sports dataset. The overall accuracy is 87.67%. The actions are DV=diving, GS=golf swing, KK=kicking, LT=lifting, RD=riding, RN=running, SK=skateboarding, BN=bench swing, HB=high-bar swing, and WK=walking.

We adopt the leave-one-out cross-validation to test the effectiveness, i.e., we select one video performed by 1 person as the test set and is run 146 times. The final average recognition rate is 87.67%. The confusion matrix is shown in Fig. 13. The algorithm correctly classifies most actions. Most of the mistakes are intuitively reasonable because of similar 3D spatio-temporal cubes, e.g., "skateboarding" is confused with "walking," "running" is confused with "riding," and "walking" is confused with "golf swing." Table III shows the results performed by existing methods based on local features. The proposed method is superior to other methods based on local features.

Table III Comparison of recognition algorithms based on local features on UCF sports dataset.

Ours		87.67%
Guha	2012	83.80%
Le	2011	86.50%
Kovashka	2010	87.27%
Wang	2009	85.60%

4.4 Comparison with Other Combinations

We compare our detector and descriptor with other methods in the BoW framework. Tables IV and V are the evaluation results performed by Wang. The detectors to be compared include the Harris 3D, 1D Gabor detector, and Hessian detector. The descriptors to be compared include the HOG3D, HOG, HOF, 1D Gabor, and 3D SURF. Given that our detector and descriptor cannot be separated from each other, the combination is evaluated. For the KTH dataset, the 95.48% accuracy obtained in the current paper is much better than the other various combinations. For the UCF sports dataset, we achieve the best accuracy of 87.67%. Overall, the proposed method shows significant improvements compared with other detector (Harris 3D, 1D Gabor, and Hessian) and descriptor (HOG3D, HOG, HOF, 1D Gabor, and 3D SURF) combinations.

Table IV Average accuracy of various detector/descriptor combinations on the KTH dataset.

	HOG3D	HOG/HOF	HOG	HOF	1D Gabor	3D SURF	**Proposed method**
Harris3D	89.0%	91.8%	80.9%	92.1%			
Cuboids	90.0%	88.7%	82.3%	88.2%	89.1%		**95.48%**
Hessian	84.6%	88.7%	77.7%	88.6%		81.4%	

Table V Average accuracy of various detector/descriptor combinations on the UCF dataset.

	HOG3D	HOG/HOF	HOG	HOF	1D Gabor	3D SURF	**Proposed method**
Harris3D	79.7%	78.1%	71.4%	75.4%			
Cuboids	82.9%	77.7%	72.7%	76.7%	76.6%		**87.67%**
Hessian	79.0%	79.3%	66.0%	75.3%		77.3%	

5. Conclusion

We have presented a robust human action recognition algorithm based on the MVSTIP detector and MEOH descriptor. By using the proposed detector and descriptor, the proposed method achieves high recognition accuracy and high robustness for both simple and complex backgrounds. There are mainly three reasons for this. Based on models of time-dependent receptive fields of simple and complex cells in the primary visual cortex (V1), multi-direction ME filters at different speeds detect the significant changes of points; the surround suppression model reduces the false detection points and redundant detection points caused by complex background; the MEOH descriptor of MVSTIPs captures the motion features in local regions around interest points. In future works, we will try other video representation models instead of the standard BoW model and optimize the code to reduce computation costs.

Acknowledgments

This work was supported in part by the National Natural Science Foundation of China (Grant No. 61071149), National Natural Science Foundation of China (Grant No. 61101166), and National Natural Science Foundation of China (Grant No. 61231015).

References

[1] TURAGA P, CHELLAPPA R, SUBRAHMANIAN V S, et al. Machine recognition of human activities: a survey[J]. IEEE transactions on circuits and systems for video technology, 2008, 18: 1473-1488.

[2] JIANG Z, LIN Z, DAVIS L. Recognizing human actions by learning and matching shape-motion prototype trees[J]. IEEE transactions on pattern analysis and machine intelligence, 2012, 34: 533-547.

[3] FATHI A, MORI G. Action recognition by learning mid-level motion features[C]//IEEE Conference on Computer Vision and Pattern Recognition, June 23-28, 2008, Anchorage, AK. IEEE, c2008.

[4] MIKOLAJCZYK K, UEMURA H. Action recognition with motion-appearance vocabulary forest[C]//IEEE Conference on Computer Vision and Pattern Recognition, June 23-28, 2008, Anchorage, AK. IEEE, c2008.

[5] EFROS A A, BERG A C, MORI G, et al. Recognizing action at a distance[C]//IEEE International Conference on Computer Vision, June 16-22, 2003, Madison, WI, USA. IEEE, c2003: 726-733.

[6] AHMAD M, LEE S W. Human action recognition using shape and CLG-motion flow from multi-view image sequences[J]. Pattern recognition, 2008, 41: 2237-2252.

[7] GORELICK M B L, SHECHTMAN E, IRANI M, et al. Actions as space-time shapes[J]. IEEE transactions on pattern analysis and machine intelligence, 2007, 29: 2247-2253.

[8] SCHINDLER K, VAN GOOL L. Action snippets: how many frames does human action recognition require?[C]//IEEE Conference on Computer Vision and Pattern Recognition, June 23-28, 2008, Anchorage, AK. IEEE, c2008.

[9] SCOVANNER P, ALI S, SHAH M. A 3-dimensional sift descriptor and its application to action recognition[C]//The 15th International Conference on Multimedia, September 24-29, 2007, Augsburg, Germany. ACM, c2007: 357-360.

[10] KLASER A, MARSZALEK M. A spatio-temporal descriptor based on 3D-gradients[C]// British Machine Vision Conference, September 2008, Leeds, UK. British Machine Vision Association, c2008: 995-1004.

[11] LE Q V, ZOU W Y, YEUNG S Y, et al. Learning hierarchical invariant spatio-temporal features for action recognition with independent subspace analysis[C]//IEEE Conference on Computer Vision and Pattern Recognition, June 20-25, 2011, Colorado Springs, CO, USA. IEEE, c2011: 3361-3368.

[12] LAPTEV I, MARSZALEK M, SCHMID C, et al. Learning realistic human actions from movies[C]//IEEE Conference on Computer Vision and Pattern Recognition, June 23-28, 2008, Anchorage, AK. IEEE, c2008.

[13] DOLLAR P, RABAUD V, COTTRELL G, et al. Behavior recognition via sparse spatio-temporal features[C]//The 2nd Joint IEEE International Workshop on Visual Surveillance and Performance Evaluation of Tracking and Surveillance, October 15-16, 2005, Beijing, China. IEEE, c2005: 65-72.

[14] GUHA T, WARD R K. Learning sparse representations for human action recognition[J]. IEEE transactions on pattern analysis and machine intelligence, 2012, 34: 1576-1588.

[15] NIEBLES J C, WANG H, LI F F. Unsupervised learning of human action categories using spatial-temporal words[J]. International journal of computer vision, 2008, 79: 299-318.

[16] BREGONZIO M, GONG S, XIANG T. Recognising action as clouds of space-time interest points[C]//IEEE Conference on Computer Vision and Pattern Recognition, June 20-25, 2009, Miami, FL, USA. IEEE, c2009: 1948-1955.

[17] WILLEMS G, TUYTELAARS T, VAN GOOL L. An efficient dense and scale-invariant spatio-temporal interest point detector[C]//European Conference on Computer Vision, October 12-18, 2008, Marseille, France. Springer, c2008: 650-663.

[18] PETKOV N, SUBRAMANIAN E. Motion detection, noise reduction, texture suppression, and contour enhancement by spatiotemporal Gabor filters with surround inhibition[J]. Biological cybernetics, 2007, 97: 423-439.

[19] LAPTEV I. On space-time interest points[J]. International journal of computer vision, 2005, 64: 107-123.

[20] CHAKRABORTY B, HOLTE M B, MOESLUND T B, et al. A selective spatio-temporal interest point detector for human action recognition in complex scenes[C]//IEEE International Conference on Computer Vision, November 6-13, 2011, Barcelona, Spain. IEEE, c2011: 1776-1783.

[21] BAY H, TUYTELAARS T, VAN GOOL L. Surf: speeded up robust features[C]//European Conference on Computer Vision, May 7-13, 2006, Graz, Austria. Springer, c2006: 404-417.

[22] LOWE D G. Distinctive image features from scale-invariant keypoints[J]. International journal of computer vision, 2004, 60: 91-110.

[23] KNOPP J, PRASAD M, WILLEMS G, et al. Hough transform and 3d surf for robust three dimensional classification[C]//European Conference on Computer Vision, September 5-11, 2010, Heraklion, Crete, Greece. Springer, c2010: 589-602.

[24] YUAN Z L J, WU Y. Discriminative subvolume search for efficient action detection[C]//IEEE Conference on Computer Vision and Pattern Recognition, June 20-25, 2009, Miami, FL, USA. IEEE, c2009:1-8.

[25] CHANG C C, LIN C J. LIBSVM: a library for support vector machines[J]. ACM transactions on intelligent systems and technology, 2011, 2: 27.

[26] CHAN M T, HOOGS A, SCHMIEDERER J, et al. Detecting rare events in video using semantic primitives with HMM[C]//The 17th International Conference on Pattern Recognition, August 23-26, 2004, Cambridge, UK. IEEE, c2004: 150-154.

[27] HUANG Y, LI Y. Prediction of protein subcellular locations using fuzzy k-NN method[J]. Bioinformatics, 2004, 20: 21-28.

[28] HOFMANN T. Probabilistic latent semantic indexing[C]//The 22nd annual International ACM SIGIR Conference on Research and Development in Information Retrieval, August 15-19, 1999, Berkeley, CA, USA. ACM, c1999: 50-57.

[29] BLEI D M, NG A Y, JORDAN M I. Latent dirichlet allocation[J]. Journal of machine learning research, 2003, 3: 993-1022.

[30] LIU J, ALI S, SHAH M. Recognizing human actions using multiple features[C]//IEEE Conference on Computer Vision and Pattern Recognition, June 23-28, 2008, Anchorage, AK. IEEE, c2008: 1-8.

[31] Weizmann dataset[EB/OL]. (2007-12-24)[2014-03-01]. http://www.wisdom.weizmann.ac.il/~vision/SpaceTimeActions.html.

[32] ZHANG Z, TAO D. Slow feature analysis for human action recognition[J]. IEEE transactions on pattern analysis and machine intelligence, 2012, 34: 436-450.

[33] YUAN J, LIU Z, WU Y. Discriminative video pattern search for efficient action detection[J]. IEEE transactions on pattern analysis and machine intelligence, 2011: 1728-1743.

[34] KOVASHKA A, GRAUMAN K. Learning a hierarchy of discriminative space-time neighborhood features for human action recognition[C]//IEEE Conference on Computer Vision and Pattern Recognition, June 13-18, 2010, San Francisco, CA, USA. IEEE, c2010: 2046-2053.

[35] WANG H, ULLAH M M, KLASER A, et al. Evaluation of local spatio-temporal features for action recognition[C]//British Machine Vision Conference, September 7-10, 2009, London, UK. British Machine Vision Association, c2009: 127-138.

Semantic Based Autoencoder-Attention 3D Reconstruction Network*

1. Introduction

The task of 3D reconstruction from one or several 2D images is a classic issue which can be traced back to Horn et al. It is a generally scientific problem in a wide variety of fields, such as virtual reality, autopilot, etc. Accordingly, image-based 3D reconstruction has been a focus of computer vision research for many years. The classic multi-view 3D reconstruction methods concentrate on tackling the correlation between various camera perspective parameters and 3D object model reconstruction, e.g. volumetric graph cuts, compressed sensing, structure from motion (SFM) and simultaneous localization and mapping (SLAM).

Moreover, recovering 3D geometric shape from single-view image is also an important problem which can be used in many fields. In high-level image editing, this geometric information can be used to change the lighting and material properties in the scene. Furthermore, single view reconstruction can be used as a baseline for complex modeling.

However, the single-view 3D reconstruction is an ill-posed problem due to the lack of disparity information. The traditional methods require additional prior knowledge, such as specific scene structure or geometrical constraints on the 3D structure, which limit their applications.

Owing to the remarkable achievement of learning methods and the establishment of various 3D object databases, learning methods have been gradually introduced into 3D reconstruction tasks such as Wu et al., Tatarchenko et al., Yan Xinchen et al., Choy et al., and Fan H et al.

There are three popular ways to represent a model: polygonal mesh, points and voxel. A voxel is an abbreviation for a volume element, a three-dimensional version of a pixel. In contrast to voxels, points and polygons are often explicitly represented by the coordinates

---

* The paper was originally published in *Graphical Models*, 2019, 106, and has since been revised with new information. It was co-authored by Fei Hu, Long Ye, Wei Zhong, Li Fang, Yun Tie, and Qin Zhang.

of their vertices. A direct consequence of this difference is that polygons can efficiently represent simple 3D structures with lots of empty or homogeneously filled space, while voxels excel at representing regularly sampled spaces that are non-homogeneously filled. As the neural network has regular requirements for the representation of input and output, the representation of polygonal meshes and points is not easy to apply to the learning method. Although the complexity of the voxel representation is high, the nature of the matrix representation of voxel fully satisfies the requirements of the regularity. Therefore, voxel representation is a common representation in a 3D reconstruction method based on learning. By the way, the point cloud representation of fixed points is also a compromise representation. This paper mainly studies methods based on voxel representation.

The most commonly used 3D reconstruction model architecture is autoencoder structure such as the reference [13]. However, when we reproduce the experiments of [13] and [14], we find that the output models often lack detailed information on single-view reconstruction task. We attribute the problem to the ill-conditioned nature of the single-view reconstruction task and the one-sidedness of the current networks. As parts of missing information can be found in the corresponding image, we propose to address this problem by introducing attention mechanism to the task.

Besides, we also find that some of the generated files miss obvious semantic information. For example, in the generated voxels of some bookcase as shown in Fig. 1, the generated voxels to be closer to the box because of the missing or redundant grid. We hope to use the semantic comparison module to correct this problem.

Of course, we can also explain the semantic comparison module according to the idea of GAN (Generative Adversarial Networks). The generated voxels are compared with the semantics of the original images, and the semantics comparison stage is against to the generation stage. With the adversary of these two stages, the semantic features of the generated models are more obvious. However, compared to GAN, we do not have the cross-training phase of GAN, but just complete the fine-tuning of the network on the pre-trained network. In this sense, we can think of this stage as the resemble-GAN based fine-tuning.

Based on these two ideas, we build an end-to-end system semantic based 3D AE-attention network (SAAN) for single view 3D reconstruction task. Our proposed SAAN consists of two parts. The first part AE-attention network has two branches. The upper branch learns to generate 3D rough shape of an object. The other one integrates the details of the 3D object by attention mechanism. We feed the corresponding image into modified 3D variational autoencoder reconstruction architecture to get the general volumetric occupancy. It learns to endow higher weights to the features of missing details in the image. Consequently, in the Attention Network, we can obtain volumetric occupancy which represents the details of object. Finally, we put the volumetric occupancy of these two branches together to get the full 3D shape object model. In the second semantic stage, we compare the semantic information between the input image and the output object to modify the output. Our architecture generates 3D object models which contain more vivid details and make qualitative and quantitative improvement on ShapeNet dataset compared with the references [13] and [14].

Fig. 1 A sample of semantic missing.

The rest of this paper is organized as follows. Section II briefly reviews related work on learning based single view 3D reconstruction. Section III introduces the details of our proposed method. Section IV describes the experimental setup and discusses experimental results of our proposed method, with comparison to the state-of-the-art, and our conclusions are given in Section V.

2. Related Work

This section presents a brief overview of the existing algorithms for sing-view image 3D reconstruction based on deep learning.

2.1 Related Works on Learning Based 3d Reconstruction

Wu et al. combined 2.5D depth map and 3D shape class to complete the reconstruction task. Girdhar et al. fed the image and 3D voxel grid into their TL-embedding network to train a predictable vector representation for 3D object reconstruction. Tatarchenko et al. proposed a convolutional network which predicted a depth map from an RGB image. Yan et al. proposed an encoder-decoder network with a projection loss defined by the perspective transformation which enables the unsupervised learning using 2D observation and fixed-viewpoint without explicit 3D supervision. Wu et al. proposed 3D Generative Adversarial Network (3D-GAN) which generates 3D objects from a probabilistic space. This research first introduced GAN to 3D field.

Although the above methods can be applied to perform the 3D reconstruction tasks, they have not given the subjective and objective evaluations.

Choy et al. proposed an overall framework called 3D-R2N2 for single-view and multi-view reconstruction based on LSTM which means it had high computation complexity. Fan et al. proposed a point set generation network for 3D object reconstruction from one single image. Although named point cloud, it actually learns the coordinates of the voxel grid. And the fixed number of points would not do any help to represent complex geometric structure of objects. Sun et al. solves this problem by combining other tasks. DEFORMNET takes an image input, searches the nearest shape template from a database, and deforms the template to match the query image. By the way, some papers have completed experiments on part category of the dataset. Although they have achieved good results, the results are not well generalized.

By applying a convolution on the graph spanned by the vertices and edges of the grid, the grid is first considered for discriminative 3D classification tasks. Recently, the grid has also been considered to be the output representation of 3D reconstruction. Chen et al. advocate the use of implicit fields for learning generative models of shapes and introduce an implicit field decoder called IM-NET for shape generation. Mescheder et al. try to represent the 3D surface as the continuous decision boundary of a deep neural network classifier and propose Occupancy Networks for 3D reconstruction.

2.2 Attention Mechanism

The attention mechanism is a common method in the field of image processing and natural language processing, which draws on the attention mechanism of human beings.

The visual attention mechanism is a brain signal processing mechanism unique to human vision. By quickly scanning the global image, human vision obtains the target area that needs to be focused on, which is the focus of attention. Then invests more attention resources in this area to obtain more detailed information about the target and suppress other useless information. Kelvin Xu et al. introduced attention in image caption. When generating the i-th word about the content description of the image, the reference [25] use attention to associate the area of the image with the i-th word.

In this paper, attention mechanism is divided into soft attention and hard attention. In simple terms, hard attention moves to a fixed-size area each time, and soft attention is a weighted sum of all areas at a time.

We note that the core goal of the attention mechanism is to select from a wide range of information that is more critical to the current task. For the lack of details brought by the traditional AE network, we need to complete the detail repair problem from some local features of the image. The attention mechanism allows us to capture the local features that complement the details.

Inspired by recent success in employing attention in machine translation and caption generation, we investigate the model that can attend to the details of the image while generating the volume. More details will be introduced in the Section 3.

2.3 Main Contribution

The main contributions of this paper are summarized as follows:
- We investigate models that can attend to the details of the image while generating the volume.
- We propose a semantic comparison framework to modify the model in our reconstruction task.
- We propose a reasonable combined loss function for our network.
- On the task of single-view reconstruction, our framework outperforms state-of-the-art.

3. SAAN

In this section, we propose an effective network structure to reconstruct authentic 3D object model from one single view image. It decomposes the 2D-to-3D reconstruction task into two parts (as shown in Fig. 2).

The first part we called AAN is the basic 3D reconstruction network, which is completed by two branches as shown in Fig. 3. One reconstructs rough 3D object shape conditioned on a single image which is sampled from an arbitrary view with common autoencoder structure, yielding an uncompleted volumetric occupancy. The other branch makes up the defects in uncompleted volumetric occupancy and reconstructing detailed voxel occupancy to integrate the 3D shape. Eventually, we combine these two aforementioned outputs together to get the completed 3D shape of object model. The combination is visualized and shown in Fig. 4.

Suppose the input image is x, the label of the reconstructed model is X, the AE model is f_{ae}, the attention model is $f_{attention}$, $A = \{a_1, a_2, \cdots, a_m\}, a_m \in \mathcal{R}$ are the local features of the input image, I is the weight of local feature, and the decoding of attention network is $D_{attention}$, the semantic extraction network of images and voxels is SI, SV, and our modeling process can be divided into the following three parts:

- $\min_{f_{ae}} \|X - f_{ae} \cdot x\|$,
- $\min_{V, D_{attention}} \|D_{attention} \cdot (\Sigma(a_i \odot I)) - X_{detail}\|$, where $X_{detail} = X - \widetilde{f_{ae}} \cdot x$ is the details need to be repaired, and $\widetilde{f_{ae}}$ is the optimal AE model in the first optimal problem,
- $\min_{f_{ae}, f_{attention}} \|SI \cdot x - SV \cdot (f_{ae}(I) + f_{attention}(I))\|$.

The second part is a semantic-based network structure. This part is based on the basic idea that the image should have the same semantic information as the reconstructed 3D model at the category level. We extract the information separately through two pre-trained models.

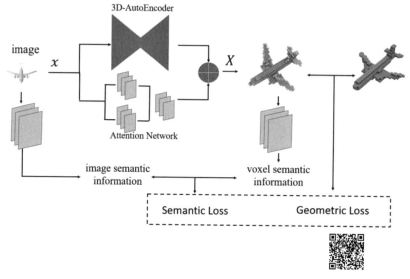

Scan the QR code to see colorful figures

Fig. 2 The general framework of our network.

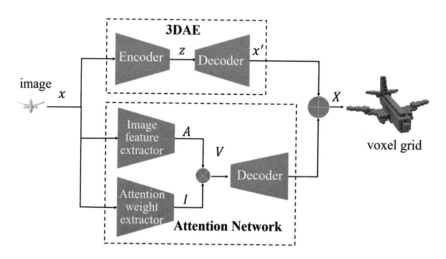

Fig. 3 The general framework of AAN.

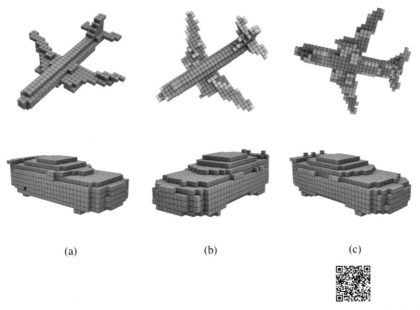

Scan the QR code to see colorful figures

Fig. 4 A sample of the two components reconstruction visualization. The first row is airplane and the second row is car. (a) is the ground truth model in the ShapeNet dataset. (b) and (c) are the predictions of AAN in different projection, in which the blue voxel girds are produced by the AE and the pink ones are made up by Attention Network.

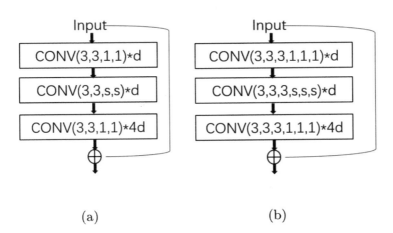

Fig. 5 (a)The basic residual unit with depth=d stride=s. (b)The basic 3D residual unit with depth=d stride=s.

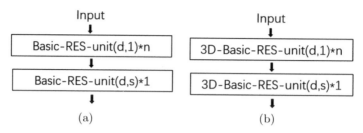

Fig. 6 (a)The residual block with depth=d, number=n, stride=s. (b)The 3D residual block with depth=d, number=n, stride=s.

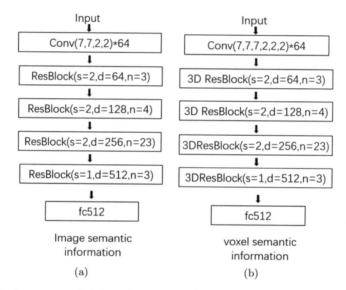

Fig. 7 (a)The image semantic information extractor. (b)The voxel semantic information extractor.

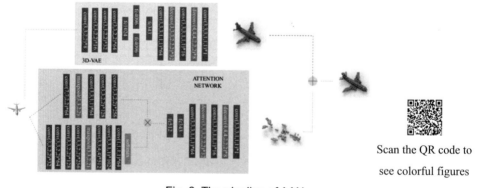

Fig. 8 The pipeline of AAN.

Scan the QR code to see colorful figures

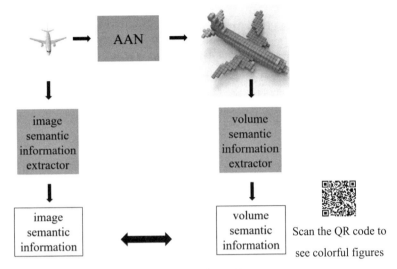

Fig. 9 The pipeline of semantic part.

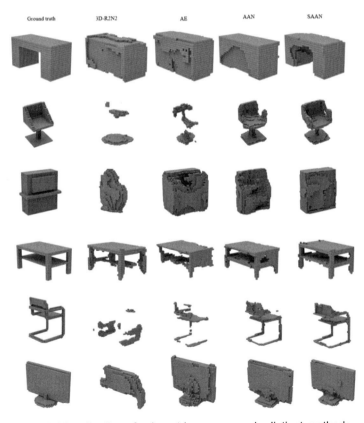

Fig. 10 Visualization of volumetric occupancy in distinct methods.

Fig. 11 The visualization of the semantic comparison process: 1) the ground truth in the ShapeNet Dataset, 2) the result with AAN, 3) the result after 25% training process 4) the result after 50% training process 5) the result after 75% training process, 6) the result with SAAN.

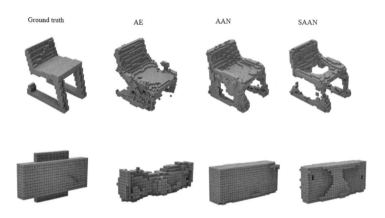

Fig. 12 Some failure samples.

3.1 Autoencoder

For our AE structure, a 2D image x sampled from an arbitrary view is fed into the 2D-image-encoder to produce low dimensional feature vector. The 3D-voxel-decoder expanses the image feature to generate the corresponding 3D volumetric occupancy. We pre-train the 3D-rough-shape generation branch with random-sampling which is a VAE structure. During the training process of upper-branch, we initialize the encoder with the pre-trained parameters and remove the random-sampling for better performance of the whole architecture.

AE is a generative model that are composed of encoder and decoder. It is used in 2D image generational tasks primitively, and we expanse it into 3D space. 3DAE utilizes single arbitrary view image to produce volumetric occupancy which can be category-wise distinguished but details lacked.

3.2 Attention Network

Inspired by the fact that the volumetric occupancy generated from AE network lacks marginal voxels often, we add an attention branch to the 3D reconstruction for completing the shape of 3D models. We design a fully convolutional Attention Network as shown in the lower half branch of Fig. 3 to establish the correspondence between missing details in volumetric occupancy and the local feature of image.

Convolutional network can extract a set of feature vectors from the 2D image. The extractor produces an m-dimensional vector $A = \{a_1, a_2, \cdots, a_m\}, a_m \in \mathcal{R}$, each element of which represent a local region of the image.

As the AE can produce a rough shape of volumetric occupancy, we can get the residual voxel occupancy which fail to reconstruct. In the convolutional Attention Network, the backpropagation algorithm will project the residual voxels back to the attention weight eigenvector. The mapping relation is recorded as I and can be obtained from an interlayer. $I = \{i_1, i_2, \cdots, i_n\}, i_n \in \mathcal{R}$ is a n-dimensional vector and each element symbolizes the pertinence between these local regions of image and the residual voxels. So that it can be regarded as importance weighted eigenvector.

As shown in the lower half branch of Fig. 3, we feed an 2D image x into the Attention Network. From two sub-branches (i.e. regarded image feature extraction as weight matrix W_f and importance weighted extraction as weight matrix W_ω) we can get a attention vector V when $m = n$. V contains the information that the different regions of image contribute to shape completion variously. After feeding into the decoder ($D_{attention}$), V is expanded to the residual voxels. This process can be formulated in the following formulation and X is the distribution of full shape 3D model,

$$V = W_f x \odot (W_\omega x)^T = A \odot I, \quad (1)$$

$$X - \widetilde{f_{ae}} = D_{attention} \cdot V. \quad (2)$$

Attention Network is voxel complemental branch of AAN and shares the same input image with AE branch. It establishes the mapping relation between the local feature of image and the missing marginal voxels, and utilize the relationship to endow higher weight to more crucial region of image. Through the same decoding architecture as AE branch, Attention Network produces detailed volumetric occupancy ultimately.

3.3 Semantic Comparison

Inspired by the fact that the volumetric occupancy generated should get the same semantic information as the input image, we construct an additional subnet. We design a structure similar to GAN to confirm the semantic consistency of voxel and image.

Firstly, the image semantic releaser takes the image as input, and obtains the semantic information of the image through the image semantic releaser respectively as shown in Fig.

7(a). The residual block with depth=d, number=n, stride=s is shown in Fig. 6(a). The basic residual unit with depth=d stride=s is shown in Fig. 5(a).

Secondly, the voxel semantic releaser gets the result of AAN as another input, and obtains the semantic information of voxel as shown in Fig. 7(b). The 3D residual block with depth=d, number=n, stride=s is shown in Fig. 6(b). The basic 3D residual unit with depth=d stride=s is shown in Fig. 5(b).

Finally, feedback is obtained by comparing the similarities of semantic information between the two states.

Just like GAN's discriminator network, we compare the semantic information of generated voxel with the semantic information of the original image, and the semantic comparison phase is opposite to the generation phase. At this stage, by comparing the semantic information of the voxel and the semantic information of the input images, the loss function and the back propagation generate the information missing in generated voxel. Thereby the semantic information of the generated model is closer to the semantic information of the original image we expect to achieve. Of course, our semantic comparison module is not the same as GAN. At this stage, our semantic extraction modules are all completed by pre-trained modules. Compared with GAN, we don't have a cross-training phase of GAN, just fine-tuning the network on a pre-trained network. In this sense, we only borrowed the idea of GAN. We regard this stage as an independent semantic comparison module based on GAN.

3.4 The Proposed Network Architecture

The network architecture pipeline is shown in Fig. 8 and Fig. 9. In Fig. 8, we feed a single image which is sampled from arbitrary view into the two branches of AAN. In the upper half branch of Fig. 8, 3DAE network first uses several 2D convolutional layers which encode the 127×127 image into an eigenvector. Then, the decoder which is composed by 3D convolutional and deconvolutional layers extends the 343-dimensional eigenvector into $32 \times 32 \times 32$ voxel occupancy of 3D rough shape. In the lower half branch of Figure 8, the Attention Network endows high weights to unreconstructed regions by attention mechanism which is a concurrent convolutional network and uses the same decoder architecture to get the $32 \times 32 \times 32$ voxel occupancy of 3D detailed shape. In attention mechanism module, the upper sub-branch extracts the 2D image feature and the lower sub-branch learns the attention weight eigenvector.

For clear illustration, here we take a sample of airplane category as an example. The results of these two components of attention are shown in Fig. 4. The blue voxels describe a rough 3D shape and the pink ones add the details. We sum the results of these two components up for getting more precise 3D shape of object model.

As shown in Fig. 9, we then extract the image semantic information with 2D CNN model and the generated voxel semantic information with 3D CNN model. We compare the consistency of the information and get feedback to the AAN.

In general, we believe that for neural networks, the closer it is to the underlying layer, the closer it is to the true semantic information. Therefore, we first train a Res101 network on the ShapeNet image dataset, and intercept the parameters of the second-to-last layer of Res101 as the semantic information of the image. Similarly, we extend Res101 to 3D, train a 3D-Res101 model on the ShapeNet voxel data set, and intercept the 3D-Res101 second-to-last-order layer parameters as the voxel semantic information. We then compare the semantic information of two state and get feedback to the whole network.

3.5 Loss Function

Loss function is a critical factor for the convergence of neural network. A combined loss function would be introduced in this task.

As the representation of model is voxel grid which is a $32\times32\times32$ binary matrix, the reconstruction task can be split into 32768 logit tasks of whether this grid occupies or not. Due to the sparsity of 3D data, in which the empty voxels take significant larger percentage than occupied voxels, the reconstruction task could be regarded as an imbalanced distribution classification problem. This is a common problem in the task of medical image segmentation. The effectiveness of loss function weight Sigmoid Cross Entropy with Logits (ωSCE) has been proven in tackling imbalanced distribution classification problem. And the expression is written as follows:

$$\omega SCE = -\omega t \log(\sigma(o)) - (1-t)\log(1-\sigma(o)), \qquad (3)$$

where $\sigma(o) = \dfrac{1}{1+e^{-o}}$, $o = w_{ij}x_{ij} + b_{ij}$, t is the representation of occupancy which is either 0 or 1, o is the output of the network, ω is hyper parameter.

For our reconstruction task, we combine ωSCE with the commonly used MSE loss function to correct small errors, and the geometric loss function could be expressed as:

$$Loss = -\omega t \log(\sigma(o)) - (1-t)\log(1-\sigma(o)) + \lambda(t-\sigma(o))^2, \qquad (4)$$

where λ is a hyper parameter.

As $\dfrac{\delta\sigma(o)}{\delta o} = \dfrac{e^{-o}}{(1+e^{-o})^2} = \sigma(o)(1-\sigma(o))$, the divergence is close to 0 when $\sigma(o)$ is close to 0 or 1, we make another modification to improve training. We change the range of the output to $\{0.25, 0.75\}$. It means that we replace $\sigma(o)$ with $\dfrac{1+\sigma(o)}{2}$. This change increases the magnitude of the loss gradient throughout the domain of output, reducing the probability of vanishing gradients. The modified geometric loss function could be expressed as:

$$Loss = -\omega t \log\left(\frac{1}{4} + \frac{\sigma(o)}{2}\right) - (1-t)\log\left(\frac{3}{4} - \frac{\sigma(o)}{2}\right) + \lambda\left(t - \frac{\sigma(o)}{2} - \frac{1}{2}\right)^2, \quad (5)$$

Given the data (X_i, Y_i), $i \in C$, where X_i is the input image, Y_i is the target voxel, C is the index set. $Y_i = (y_i^1, y_i^2, \cdots, y_i^k)$, $k = 32 \times 32 \times 32$, $y_i^j \in \{0,1\}$, $H_0 = \{(X_i, y_i^j) | y_i^j = 0\}$, $H_1 = \{(X_i, y_i^j) | y_i^j = 1\}$.

As $\frac{\delta \omega SCE}{\delta b_{ij}} = \begin{cases} \sigma, & t=1 \\ \omega(\sigma-1), & t=0 \end{cases}$, $|H_0| \gg |H_1|$, ($|\cdot|$ is the size of the set) during training, we want to strongly penalizing false zeros while reducing the penalty for false ones. That is to say, $\omega > \frac{|H_0|}{|H_1|} = 92/8 = 11.5$. After some experiment, we find that the network with ωSCE reach the best result at $\omega = 20$. We also let $\omega = 20$ in training process with the combined loss function and then find the best hyper parameter $l = 1$. From the GAN perspective, the semantic comparison module is equivalent to a pre-trained discriminator. We take the similar loss function as the GAN discriminator. Suppose the semantic information of image is S_i, the semantic information of voxel is S_v, and the loss function we finally adopt in the semantic comparison module is as follows:

$$Loss_{semantic} = \log(s_i) + \log(1-s_v) - t\log(\sigma(o)) - (1-t)\log(1-\sigma(o)). \quad (6)$$

Through the gradient back propagation of the loss function, the whole network will gradually correct the error of the voxel semantic information compared to the image semantic information.

4. Experimental Results and Analysis

In this section, a set of experiments were performed to test the effectiveness and generalization of our proposed network.

4.1 Datasets and Implementation Details

Dataset The ShapeNet dataset is a richly-annotated, large-scale dataset of 3D shapes. It is collected by Princeton, Stanford and TTIC. We use a subset of the ShapeNet dataset as [13] which contains about 40,000 3D models over 13 common categories.

Implementation details We use the ADAM solver for stochastic optimization in all the experiments. During the training time, the learning rate is 10^{-4} (whole network) or 10^{-3} (pretrain) for the neural networks. The representation of accuracy is Intersection over Union (IoU). We initialize the 3DAE model with a pretrained AE model. We pretrain a 2D-res101 model and a 3D-res101 model on the ShapeNet, and use part of these pretrained parameters to initialize the image semantic information extractor and the voxel semantic information extractor, then jointly train the two networks to ensure that both get the same semantic information.

4.2 Quantitative Results on ShapeNet

We compare our results with the state-of-the-art deep learning 3D reconstruction methods 3D-R2N2, PSGN, ONet on ShapeNet dataset. The IoU accuracy of 13 categories for our AE, AAN and SAAN, and compared result are reported in Table I. As there is no traditional IoU score in the reference [22], we get the pretrained ONet model from **https:// github.com/autonomousvision/occupancy_networks** and test the IoU score.

According to Table I, our original AE model has achieved an IoU accuracy rate of 62.7%. Added with attention module, the accuracy rate of 1.3% has been improved, and the semantic comparison module has improved the accuracy by 1.2% rate. Our SAAN finally get the best IoU score. Compared to AE, AAN has been enhanced in categories with tiny components such as lamp and table. Compared to AAN, SAAN enhances results such as cars that can easily obtain semantic information from input images. However, the semantic module reduces the impact of the attention module in some way. Through the training process, we have achieved a balance between the attention module and the semantic module.

To verify the attention mechanism and the semantic comparison on this task, we also show the voxel grid visualization of our experimental results in Fig. 10. We compare our reconstruction results with 3D-R2N2 and 3D-AE for the qualitative analysis. The first column is the ground truth in the ShapeNet Dataset. The others are the network reconstruction results from the single 2D image. The forth column shows that AAN makes up the lacking details, especially in the thin structure such as the lower floor of a coffee table in the 4th row, so that we can get more precise results. The last column shows the effectiveness of SAAN. In particular, the output of SAAN in the first row has a clearer semantic information of 'table' than the result of AAN.

We save the model every certain number of epochs in the training process of SAAN. We visualize the results of these models in Fig. 11. Fig. 11 shows how the network gradually recovers the original semantic information after adding the semantic comparison module. There are still some limitations of our method. Fig. 12 shows two typical failure examples. In the first row, AAN did not complete the details due to the complexity of the hollow structure. In the second row, the semantic comparison module does not work, which may be caused by the acquisition of the wrong semantic information. We hope to solve these problems through more experiments in the future.

Table I 3D reconstruction IoU on the ShapeNet dataset test.

	3D-R2N2			PSGN	ONet	AE	AAN	SAAN
viewpoint	1	3	5	1	1	1	1	1
Plane	0.513	0.549	0.561	**0.601**	0.468	0.565	0.594	0.597
Bench	0.421	0.502	0.527	0.55	0.522	0.499	**0.74**	0.514
Cabinet	0.716	0.763	**0.772**	0.771	0.747	0.755	0.597	0.758
Car	0.798	0.829	**0.836**	0.831	0.784	0.828	0.709	0.835
Chair	0.466	0.533	0.55	0.544	0.553	0.605	0.562	**0.627**
monitor	0.468	0.545	0.565	0.552	**0.594**	0.591	0.59	0.59
Lamp	0.381	0.415	0.421	0.462	0.38	0.574	**0.771**	0.618
speaker	0.662	0.708	0.717	0.737	0.712	0.715	0.566	**0.774**
Firearm	0.544	0.593	0.6	**0.604**	0.516	0.508	0.547	0.541
Couch	0.628	0.69	0.706	0.708	**0.717**	0.693	0.588	0.709
Table	0.513	0.564	0.58	0.606	0.544	0.598	**0.716**	0.62
cellphone	0.661	0.732	0.754	0.749	0.742	0.697	**0.83**	0.714
watercraft	0.513	0.596	0.61	**0.611**	0.575	0.535	0.513	0.572
mean	0.56	0.617	0.631	0.64	0.604	0.627	0.64	**0.652**

Table II The IoU comparison between SAAN and 3DensiNet.

method	3DensiNet	SAAN (ours)
watercraft	0.554	**0.572**
cellphone	0.674	**0.714**
table	0.545	**0.62**
couch	0.668	**0.709**
monitor	0.487	**0.59**
chair	0.465	**0.627**
car	0.813	**0.835**
cabinet	0.748	**0.758**

4.3 Quantitative Result on Parts of ShapeNet

In recent work on 3D reconstruction, some articles have also been experimented on ShapeNet. But for some reason, these articles have only trained and tested under several categories of ShapeNet. Based on the consideration of experimental generalization, we do not compare with these articles in detail. Although we've considered more categories, we still achieved a full-scale lead compared to the experimental results in the reference [20] (as shown in Table II).

5. Conclusion

In this paper, we design a 3D reconstruction network SAAN. The proposed method decomposes the prediction into two parts. The first part is made of two parallel branches. The 3DAE branch produces rough 3D shape by a standard AE, and the attention branch establishes the correspondence between missing details in volumetric occupancy and regions in image to add the details for completing 3D model shape. In the other part, we then compare the semantic information between the input images and the generated voxel to generate more semantical representations. By comparing with state-of-art methods and analyzing the structure effectiveness, SAAN is verified to produce more precise 3D object models in qualitatively and quantitatively.

Declaration of Competing Interest

We declare that we have no financial and personal relationships with other 365 people or organizations that can inappropriately influence our work, there is no professional or other personal interest of any nature or kind in any product, service and/or company that could be construed as influencing the position presented in, or the review of, the manuscript entitled 'Semantic Based Autoencoder-Attention 3D Reconstruction Network'.

References

[1] HORN B K P. Shape from shading: a method for obtaining the shape of a smooth opaque object from one view[R]. Cambridge, MA, United States: Massachusetts Institute of Technology, 1970.

[2] ZHU Y F, ZHANG Y J. Multi-view stereo reconstruction via voxel clustering and optimization of parallel volumetric graph cuts[C]//Parallel Processing for Imaging Applications. SPIE, January 23-27, 2011, San Francisco Airport Hyatt Regency. SPIE, c2011, 7872: 261-271.

[3] REDDY D, SANKARANARAYANAN A C, CEVHER V, et al. Compressed sensing for multi-view tracking and 3-D voxel reconstruction[C]//The 15th IEEE International Conference on Image Processing, October 12-15, 2008, San Diego, CA, USA. IEEE, c2008: 221-224.

[4] FUENTES-PACHECO J, RUIZ-ASCENCIO J, RENDON-MANCHA J M. Visual simultaneous

localization and mapping: a survey[J]. Artificial intelligence review, 2015, 43: 55-81.

[5] HAMING K, PETERS G. The structure-from-motion reconstruction pipeline-a survey with focus on short image sequences[J]. Kybernetika, 2010, 46(5): 926-937.

[6] HORRY Y, ANJYO K I, ARAI K. Tour into the picture: using a spidery mesh interface to make animation from a single image[C]//The 24th Annual Conference on Computer Graphics and Interactive Techniques, August 3-8, 1997, Los Angeles, California, United States of America: ACM, c1997: 225-232.

[7] HOIEM D, EFROS A A, HEBERT M. Automatic photo pop-up[J]. ACM transactions on graphics, 2005, 24(3): 577-584.

[8] STURM P, MAYBANK S. A method for interactive 3d reconstruction of piecewise planar objects from single images[C]//The 10th British Machine Vision Conference, September 13-16, 1999, The University of Nottingham, United Kingdom. The British Machine Vision Association (BMVA), c1999: 265-274.

[9] GUILLOU E, MENEVEAUX D, MAISEL E, et al. Using vanishing points for camera calibration and coarse 3D reconstruction from a single image[J]. The Visual Computer, 2000, 16: 396-410.

[10] WU Z, SONG S, KHOSLA A, et al. 3d shapenets: a deep representation for volumetric shapes[C]//IEEE Conference on Computer Vision and Pattern Recognition, June 7-12, 2015, Boston, MA, USA. IEEE, c2015: 1912-1920.

[11] TATARCHENKO M, DOSOVITSKIY A, BROX T. Multi-view 3d models from single images with a convolutional network[C]//The 14th European Conference on Computer Vision, October 11–14, 2016, Amsterdam, the Netherlands. Springer International Publishing, c2016: 322-337.

[12] YAN X, YANG J, YUMER E, et al. Perspective Transformer Nets: Learning Single-View 3D Object Reconstruction Without 3D Supervision[C]//Proceedings of the 30th International Conference on Neural Information Processing Systems. Curran Associates Inc., Red Hook, NY, USA, 2016: 1704-1712.

[13] CHOY C B, XU D, GWAK J Y, et al. 3d-r2n2: a unified approach for single and multi-view 3d object reconstruction[C]//The 14th European Conference on Computer Vision, October 11–14, 2016, Amsterdam, The Netherlands. Springer International Publishing, c2016: 628-644.

[14] FAN H, SU H, GUIBAS L. A point set generation network for 3D object reconstruction from a single image[C]//IEEE Conference on Computer Vision and Pattern Recognition, July 21-26, 2017, Honolulu, Hawaii. IEEE Computer Society, c2017: 2463-2471.

[15] YI L, SHAO L, SAVVA M, et al. Large-scale 3d shape reconstruction and segmentation from shapenet core55[J]. arxiv preprint arxiv:1710.06104, 2017.

[16] GIRDHARI R, FOUHEY D F, RODRIGUEZ M, et al. Learning a predictable and generative vector representation for objects[C]//The 14th European Conference on Computer Vision, October 11–14, 2016, Amsterdam, The Netherlands. Springer International Publishing, c2016: 484-499.

[17] WU J, ZHANG C, XUE T, et al. Learning a Probabilistic Latent Space of Object Shapes via 3D Generative-Adversarial Modeling[C]//Proceedings of the 30th International Conference on Neural Information Processing Systems. Curran Associates Inc., Red Hook, NY, USA, 2016: 82-90.

[18] SUN X, WU J, ZHANG X, et al. Pix3d: dataset and methods for single-image 3d shape modeling[C]//IEEE Conference on Computer Vision and Pattern Recognition, June 18-23, 2018, Salt Lake City, UT, USA. IEEE, c2018: 2974-2983.

[19] KURENKOV A, JI J, GARG A, et al. Deformnet: free-form deformation network for 3d shape reconstruction from a single image[C]//IEEE Winter Conference on Applications of Computer Vision, March 12-15, 2018, Lake Tahoe, NV, USA. IEEE, c2018: 858-866.

[20] WANG M, WANG L, FANG Y. 3densinet: a robust neural network architecture towards 3d volumetric object prediction from 2d image[C]//The 25th ACM International Conference on Multimedia, October 23-27, 2017, Mountain View, CA USA. ACM: 2017: 961-969.

[21] WANG H, YANG J, LIANG W, et al. Deep single-view 3D object reconstruction with visual hull embedding[C]//The AAAI Conference on Artificial Intelligence, January 27-February 1, 2019, Honolulu, Hawaii, USA. AAAI, 33(01): 8941-8948.

[22] MESCHEDER L, OECHSLE M, NIEMEYER M, et al. Occupancy networks: learning 3d reconstruction in function space[C]//IEEE/CVF Conference on Computer Vision and Pattern Recognition, June 15-20, Long Beach, CA, USA. IEEE, c2019: 4460-4470.

[23] CHEN Z, ZHANG H. Learning implicit fields for generative shape modeling[C]//IEEE/CVF Conference on Computer Vision and Pattern Recognition, June 15-20, Long Beach, CA, USA. IEEE, c2019: 5939-5948.

[24] MNIH V, HEESS N, GRAVES A. Recurrent Models of Visual Attention[C]//Proceedings of the 27th International Conference on Neural Information Processing Systems. MIT Press, Cambridge, MA, USA, 2014, 2: 2204-2212.

[25] XU K, BA J, KIROS R, et al. Show, attend and tell: neural image caption generation with visual attention[C]//International Conference on Machine Learning, July 6-11, 2015, Lille, France. PMLR, c2015: 2048-2057.

[26] VASWANI A, SHAZEER N, PARMARN, et al. Attention is All You Need[C]//Proceedings of the 31st International Conference on Neural Information Processing Systems. Curran Associates Inc., Red Hook, NY, USA, 2017: 6000-6010.

[27] GENG Y, ZHANG G, LI W, et al. A novel image tag completion method based on convolutional neural transformation[C]//Artificial Neural Networks and Machine Learning, September 11-14, 2017, Alghero, Italy. Springer International Publishing, c2017: 539-546.

[28] ZHANG G, LIANG G, LI W, et al. Learning convolutional ranking-score function by query preference regularization[C]//Intelligent Data Engineering and Automated Learning, October 30–November 1, 2017, Guilin, China. Springer International Publishing, c2017: 1-8.

[29] ZHANG G, LIANG G, SU F, et al. Cross-domain attribute representation based on convolutional neural network[C]//Intelligent Computing Methodologies, August 15-18, 2018, Wuhan, China. Springer International Publishing, c2018: 134-142.

[30] YOSINSKI J, CLUNE J, NGUYEN A, et al. Understanding neural networks through deep visualization[J]. arxiv preprint arxiv:1506.06579, 2015.

[31] ZEILER M D, FERGUS R. Visualizing and understanding convolutional networks[C]//The 13th European Conference of Computer Vision, September 6-12, 2014, Zurich, Switzerland. Springer International Publishing, c2014: 818-833.

[32] KINGMA D P, BA J. Adam: a method for stochastic optimization[J]. arxiv preprint arxiv:1412.6980, 2014.

Flexible Light Field Angular Superresolution via a Deep Coarse-to-Fine Framework*

1. Introduction

The light field (LF) encodes the distribution of light into a high-dimensional function, contains rich scene visual information and has a wide range of applications in various fields, such as image refocusing, 3D scene reconstruction, depth inference, and virtual augmented reality. In order to obtain high-quality views without ghosting effects, many studies have focused on dense sampling of LF.

Dense sampling of LF means great acquisition difficulties. Early light field cameras, such as multicamera arrays and light field racks etc., are bulky and expensive in hardware. In recent years, the introduction of commercial and industrial light field cameras such as Lytro and RayTrix has brought light field imaging into a new era. Unfortunately, due to the limited resolution of the sensor, a trade-off must be made between spatial resolution and angular resolution.

One possible solution to this problem is view synthesis, which synthesizes novel views from a sparse set of input views. One type of existing previous work is to estimate the scene depth as auxiliary information, but it relies heavily on the depth estimation, which tends to fail in occluded regions, as well as in glossy or specular ones. The other is based on sampling and consecutive reconstruction of the plenoptic function. They do not use the depth information as an auxiliary mapping but suffer from either aliasing or blurring problem when the input LF is extremely undersampled.

Some learning-based methods have recently appeared, and they can be roughly classified into two categories: nondepth based and depth based. But most of them have average reconstruction quality under large parallax conditions. Besides, retraining is required for different scale factors, which increases the difficulty of actual acquisition.

---

* The paper was originally published in *Wireless Communications and Mobile Computing*, 2022, and has since been revised with new information. It was co–authored by Qian Wang, Li Fang, Long Ye, Wei Zhong, Fei Hu, and Qin Zhang.

In our previous work, we proposed a learning-based model for reconstructing densely-sampled LFs via angular superresolution, which is achieved by using an image superresolution network on epipolar plane images (EPIs). However, EPIs have very clear structure, which is very different from natural images. The performance is degraded by large-baseline sampling. In this paper, we provide a few distinguishable improvements and enable flexible and accurate reconstruction of a densely-sampled LF from very sparse sampling. We inherit the coarse-to-fine framework in the reference [20], that is, the proposed model consists of a coarse EPI angular superresolution module and an efficient EPI stack refinement module. Specifically, the coarse EPI angular superresolution module magnifies each EPI individually using a specially designed EPI superresolution network, where independent coefficients are used for row and column EPIs, respectively. We further refine the coarse results using 3DCNNs on stacked EPI based on photoconsistency between subaperture images. The overall frame explores pseudo 4DCNN, which is capable of making full use of LF data. In addition, we introduced perceptual loss to better fit the network training. Experimental results demonstrate the superiority of our method in reconstruction with higher numerical quality and better visual effect.

The rest of this paper is organized as follows: Latest developments of view synthesis and LF reconstruction are introduced in Section 2. Our presented approach is described in Section 3. The performance evaluation is given in Section 4. Finally, Section 5 summarizes this paper.

2. Related Works

The problem of reconstructing a complete densely-sampled LF from a set of sparsely sampled images has been extensively studied. These algorithms can be divided into depth dependent view synthesis that depends on depth information and depth-independent LF reconstruction that does not depend on depth information.

2.1 Depth-Dependent View Synthesis

Depth-dependent view synthesis approaches typically consist of two steps to synthesize the novel view of the scene, i.e., first estimating depth map at the novel view or the input view, and then using it to synthesize the novel, Kalantari et al. proposed the first deep learning system for view synthesis with two sequential networks that perform depth estimation and color prediction successively. Srinivasan et al. proposed to synthesize a 4D RGBD LF from a single 2D RGB image based on estimated 4D ray depth. Flynn et al. mapped input views to a set of depth planes of the same perspective through homography transform and then fused them together through two parallel CNNs for learning weights to average the color of each plane. Zhou et al. and Mildenhall and Ben trained a network that inferred alpha and multiplane images. Although utilizing depth information makes it easier to handle inputs with large disparities, most methods cannot achieve acceptable performance for large-baseline sampling, Jin et al. focus on the angular superresolution of light field images with a large-baseline and propose an end-to-end trainable method, by making full use of the intrinsic geometry information of light fields. However, unfortunately inaccurate depth estimation usually happens in challenging scenes that contain significant depth variations, complex lighting conditions, occlusions, non-Lambertian surfaces, etc.

2.2 Depth-Independent LF Reconstruction

Depth-independent LF reconstruction approaches can be considered as an angular dimension upsampling without any geometry information of the scene. Zhouchen and Heung-Yeung proved that for an LF which disparity between neighboring views is less than one pixel, and novel views can be produced using linear interpolation. Some methods have investigated LF reconstruction with specific sampling patterns. Levin and Durand exploited dimensionality gap priors to synthesize novel views from a set of images sampled in a circular pattern. Lixin et al. sampled only the boundary viewpoints or diagonal viewpoints to recover the full LF using sparsity analysis in the Fourier domain. These methods are far from practical application due to the difficulty in capturing input views in specific mode.

Recently, deep learning has been applied in many fields, such as 3D object detection and object recognition. Some learning-based approaches were also proposed for depth-independent reconstruction. Yoon et al. proposed a deep learning framework, in which two adjacent views are employed to generate the interview, while it can only generate novel views at 2x upsampling factor. Wu et al. proposed a blur-restoration-deblur framework as learning-based angular detail restoration on 2D EPIs. Wu et al. further discussed the trade-off either aliasing or blurring problem and designed a Laplacian pyramid EPI (LapEPI) structure that contains both low spatial scale EPI (for aliasing) and high-frequency residuals (for blurring) to solve the trade-off problem. However, the potential of the full LF data is under used in both works. Wang et al. and their subsequent work introduced an end-to-end learning-based pseudo 4DCNN framework using 3D LF volumes with a fixed interpolating rate. On the basis of the pseudo 4DCNN network, Ran et al. added an EPI repair module that magnifies the EPI with arbitrary scale factor.

3. The Proposed Approach

3.1 4D LF Representation

A 4D LF is usually denoted as $L(u, v, s, t)$, which uses the intersections of light rays with two parallel planes to record light rays, called the two plane representation (see Fig.

1). Each light ray travels from the spatial coordinates (u, v) on the focal plane then to the angular coordinates (s, t) on the camera plane. Therefore, a 4D LF is regarded as a 2D image array with $Cols \times Rows$ images sampled on a 2D angular grid, of which each image is of spatial resolution $W \times H$.

As shown in Fig. 1, by fixing u in spatial domain and S in angular domain (or v and t), we can get an EPI map denoted as $E_{u_0, s_0}(v, t)$ (or $E_{v_0, t_0}(u, s)$), which is a 2D slice of the 4D LF. A 3D volume $V_{t_0}(u, v, s)$ (or $V_{s_0}(u, v, t)$) can be produced if we stack EPIs from a column (or a row) views by fixing $t = t_0$ (or $S = S_0$). In the EPI, relative motion between the camera and object points manifests as lines with depth depending slopes. Thus, EPIs can be regarded as an implicit representation of the scene geometry. EPI has a highly clear structure compared with conventional photo images.

The goal of LF angular superresolution is to recover a densely-sampled LF with a resolution of $W \times H \times Cols \times Rows$ from a sparse one with a resolution of $W \times H \times cols \times rows$. The presented framework will be comprehensively introduced in the next section.

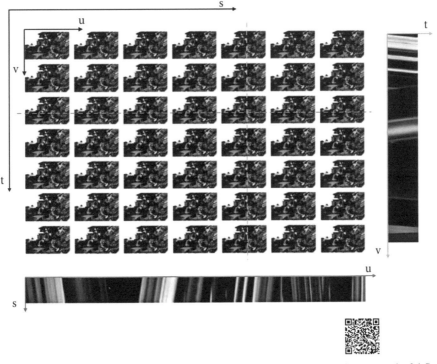

Scan the QR code to see colorful figures

Fig. 1 A 4D light field $L(u, v, s, t)$ visualization. The horizontal EPI is a 2D (u, s) slice $L(u, v_0, s, t_0)$ by positioning $v = v_0$ and $t = t_0$ (highlighted in red) and the vertical EPI (v, t) by positioning $u = u_0$ and $s = s_0$ (highlighted in blue).

3.2 Overview of the Proposed Method

In order to make full use of 4D LF data while circumventing high computational burden, we propose to first synthesize novel views in each row and then in each column, forming a pseudo 4DCNN. Specifically, as shown in Fig. 2, for each row/column of views, our network consists of two parts: coarse EPI angular superresolution module based on 2D metalearning and EPI stack refinement module using 3DCNNs. First, we perform these two parts on the 3D row EPI volumes, after by converting the angular from row to column. Then, perform the same operation on the 3D column EPI volumes.

Given an input sparse LF $LF_{in}(u, v, s, t)$, with the size of $W \times H \times cols \times rows$. First we fix the angular coordinate $s = s_0$, $s_0 \in \{1, 2, \cdots, cols\}$ to get the 3D row EPI volumes with the size of $W \times H \times rows$. Then, we fix the spacial axis $u = u_0$, $u_0 \in \{1, 2, \cdots, W\}$ and perform the angular superresolution metalearning based network $F(\cdot)$ on each EPI $E_{u_0, s_0}(v, t)$ to get $E^*_{u_0, s_0}(v, t)$ with the size of $H \times Rows$:

$$E^*_{u_0, s_0}(v, t) = F\left(E_{u_0, s_0}(v, t), r\right), \tag{1}$$

where $r = (Rows-1)/(rows-1)$ is the magnification scale factor. The network $N_c(\cdot)$ is followed to recover the high-frequency details of $E^*_{u_0, s_0}(v, t)$ and obtain the $LF_{inter}(u, v, s_0, t)$ with the size of $(W, H, cols, Rows)$:

$$LF_{inter}(u, v, s_0, t) = N_c\left(E^*_{s_0}(u, v, t)\right). \tag{2}$$

The angular conversion is performed on $LF_{inter}(u, v, s_0, t)$, that is, the angular dimension is changed from t to s. First, by fixing the angular coordinate $t = t_0$, $t_0 \in \{1, 2, \cdots, Rows\}$, we get the 3D column EPI volumes with the size of $W \times H \times cols$; then, we extract column EPI by fixing the spacial axis $v = v_0$, $v_0 \in \{1, 2, \cdots, H\}$ and perform the angular superresolution network $F(\cdot)$ on to obtain superresolved $E^*_{v_0, t_0}(u, s)$, with the size of $W \times Cols$:

$$E^*_{v_0, t_0}(u, s) = F\left(E_{v_0, t_0}(u, s), r\right). \tag{3}$$

Finally, we use the network $N_r(\cdot)$ to recover the high-frequency details of $E^*_{v_0, t_0}(u, s)$ and obtain the $LF_{out}(u, v, s, t)$ with the size of $W \times H \times Cols \times Rows$:

$$LF_{out}(u, v, s, t) = N_r\left(E^*_{t_0}(u, v, s)\right). \tag{4}$$

We use a learnable architecture for both EPI angular superresolution module and EPI stack refinement module, so that the proposed framework can be trained in an end-to-end strategy.

3.3 Coarse EPI Angular Superresolution Module

As shown in Fig. 3, for the angular superresolution of EPI, taking $E_{u_0, s_0}(v, t)$ as an example, the acquisition of $E^*_{u_0, s_0}(v, t)$ can be regarded as image superresolution on angular dimension t. We use a feature extraction module to learn the structure of the EPI and then upsample it to the desired resolution. In order to deal with arbitrary scale factor, we use the metalearning strategy in [36]. Since each pixel on the upsampled EPI is predicted via a local kernel. For different scale factors, both the number of the kernels and the weights of the kernels are different. We take advantage of the metalearning to predict the number of the kernels and the weights of the kernels for each scale factor.

Specifically, a feature learning module is settled to extract features from the low-resolution (LR) EPI; then, the upsample module dynamically predicts the weights of the upsampling filters by taking the magnification scale factor as input and using these weights to generate the high-resolution (HR) EPI of arbitrary size. We modify the network to perform magnification only on the angular dimension in EPI.

For the feature learning module, we use the residual dense network (RDN), which is composed of residual learning and dense connections. It consists of 3 layers of 2D CNN and 8 residual dense blocks (RDBs). Each RDB is composed of 8 layers of 2D CNN and ReLU with dense connections.

Taking $E_{u_0,s_0}(v,t)$ as an example, the resolution of the HR EPI is $(H, Rows)$, and the feature is denoted as feature $(H, rows)$. The upsample module can be regarded as the mapping function between EPI $E^*_{u_0,s_0}(v,t)$ with the size of $H \times Rows$ and feature $(H, rows)$. This part is mainly composed of three steps, namely, position projection, weight prediction, and feature mapping. First, we determine the corresponding pixels across the spatial resolution through the position projection operation, and the specific implementation is as follows:

$$(H, rows) = T(H, Rows) = \left(H, \left\lfloor \frac{Rows}{r} \right\rfloor\right), \tag{5}$$

where T refers to the conversion function, r is the magnification scale factor, and $\lfloor\ \rfloor$ refers to the floor operation. For different magnification scale factors r, we use the weight prediction module to predict the corresponding prediction filter weight:

$$\text{Weight}(H, Rows) = \varphi(V_{H, Rows}; \theta), \tag{6}$$

$$V_{H, Rows} = \left(H, \frac{Rows}{r} - \left\lfloor \frac{Rows}{r} \right\rfloor\right). \tag{7}$$

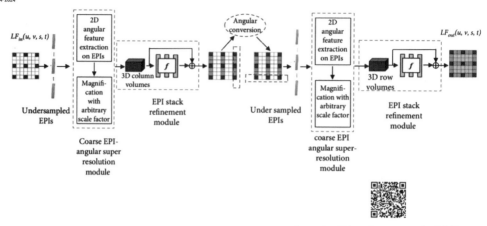

Fig. 2 The flowchart of the proposed method for reconstructing a densely-sampled LF from an undersampled LF. Our proposed model consists of two phases, i.e., coarse EPI angular superresolution module and EPI stack refinement module.

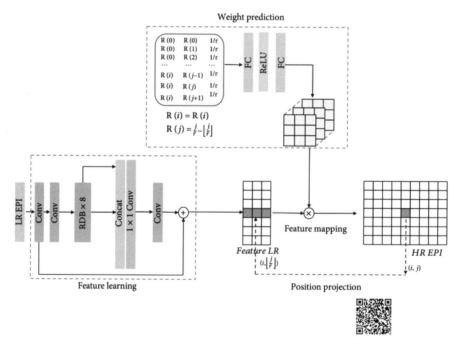

Fig. 3 Coarse EPI angular superresolution module structure.

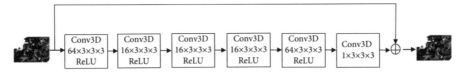

Fig. 4 Structure of the network for recovering details of 3D volumes. The first 5 layers are followed by a rectified linear unit (ReLU). The final detail restored volume is the sum of the predicted residual and the input.

Among them, $\varphi(\cdot)$ represents the weight prediction network, θ is the network related parameters, and $V_{H,Rows}$ is the vector of the pixel on the output EPI. Finally, the feature mapping step maps feature $(H, rows)$ with weight $(H, Rows)$ to get the final HR EPI, and matrix multiplication was chosen as its function.

3.4 EPI Stack Refinement Module

The proposed networks $N_c(\cdot)$ and $N_r(\cdot)$ are shown in Fig. 4. Both networks consist of an encoder and a decoder, where both of them comprise 3 convolution layers and are exactly symmetric. The first 3D convolutional layer comprises 64 channels with the kernel 3×3×3, where each kernel operates on 3×3 spatial region across 3 adjacent EPIs. Similarly, the second layer comprises 16 channels with the kernel 3×3×3. The last layer also comprises 16 channels with the kernel 3×3×3. Each layer uses a stride of 1 followed by a rectified linear unit (ReLU), i.e., $\sigma(x) = \max(0, x)$, excepting for the last one. The ultimate output of the network is the sum of the predicted residual and the input 3D EPI volume. To avoid border effects, we appropriately pad the input and feature maps before every convolution operation to maintain the input and output at the same size.

3.5 Perceptual Loss Function

As observed in prior work on image synthesis. Simply comparing the pixel colors of the synthesized image and the reference image could severely penalize perfectly realistic outputs. Instead, we adopt the perceptual loss. The basic idea is to match activations in a visual perception network that is applied to the synthesized image and separately to the reference image.

Let ϕ be a trained visual perception network (we use VGG-16). Layers in the network represent an image at increasing levels of abstraction: from edges and colors to objects and categories. Matching both lower-layer and higher-layer activations in the perception network guides the synthesis network to learn both fine-grained details and more global part arrangement. Here, g is the image synthesis network being trained and $\grave{E}$ is the set of parameters of this network.

Let ϕ_1 be a collection of layers in the network ϕ, such that ϕ_1 denotes the input image. Each layer is a three dimensional tensor. For a training pair $(I, L) \in D$, our loss is

$$l_{I,L}(\theta) = \sum_l \left\| \phi_l(I) - \phi_l(g(L;\theta)) \right\|_1. \tag{8}$$

Here, g is the image synthesis network being trained, and θ is the set of parameters of this network. For layers $\phi_l\,(l \geq 1)$, we use *conv*1_2, *conv*2_2, and *conv*3_2 in VGG-16.

4. Experiments and Results

4.1 Datasets and Training Details

We took real-world LF images captured with a Lytro Illume camera provided by Stanford Lytro LF Archive and Kalantari et al. as well as synthetic LF images from the 4D light field benchmark to train and test the proposed framework. Specifically, 20 synthetic images and 100 real-world images were used for training. 70 LF images captured by Lytro Illum camera were used for real-world scenes test, including 30 test scenes provided by Kalantari et al., 15 LF images from reflective dataset, and 25 LF images from occlusions dataset. 4 LF images from the HCI dataset and 5 LF images from the old HCI dataset were used for synthetic scenes test. We removed the border views and cropped the original LF data to 7×7 views as ground truth and then downsampled randomly to 2×2 and 3×3 views as the input. For each LF data, small patches in the same position of each view were extracted to formulate the training LF data. The spatial patch size was 32 × 32 and the stride was 20.

Similar to other methods, we only processed the luminance Y channel in the YCbCr color space. The framework was implemented with PyTorch. The optimization of end-to-end training was ADAM optimizer with $\beta_1 = 0.9$ and $\beta_2 = 0.999$, and the batch size was set to 1. The learning rate was initially set to 10^{-4} and then decreased by a factor of 0.1 every 10 epochs until the validation loss converges. The filters of 3DCNNs were initialized from a zero-mean Gaussian distribution with standard deviation 0.01, and all the bias were initialized to zero.

4.2 Angular Superresolution Evaluation

We used the average value of peak signal-to-noise ratio (PSNR) and structural similarity index (SSIM) over all synthesized novel views in each scene to quantitatively measure the quality of reconstructed densely-sampled LFs. We designed two experiments to test the angular superresolution quality at different angular resolutions, respectively, reconstructing a 7×7 densely-sampled light field from 2×2 and 3×3 sparse views. Fig. 5 demonstrates the sampling patterns.

Our experiments are compared with six state-of-the-art deep learning-based methods designed for densely-sampled light field angular superresolution, namely, Kalantari et al., Wu et al., Wu et al., Wang et al., Jin et al., and Ran et al. Table I shows the properties of the above methods and our proposed method. For the compared algorithms, although some of them do support different scale factors, they need to train different models for each scale factor separately, while our method is able to perform different scale factors with a single model.

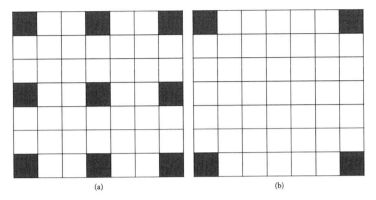

Fig. 5 Illustration of sampling patterns: (a) 3×3; (b) 2×2.

Table I Comparison of properties of different densely-sampled light field angular superresolution algorithm.

Methods	Based on deep learning	Based on geometry	Arbitrary scale factor	One model for various scale factor
Kalantari et al.	√	√	√	-
Wu et al.	√	-	-	-
Wu et al.	√	√	√	-
Wang et al.	√	-	-	-
Jin et al.	√	√	√	-
Li et al.	√	-	√	√
Ours	√	-	√	√

Table II Quantitative comparisons (PSNR/SSIM) of the proposed approach with the state-of-the-art ones under task 2×2 to 7×7 on synthetic scenes.

Methods	Kalantari et al.	Wu et al.	Wu et al.	Li et al.	Jin et al.	Ours
HCI	32.85/0.909	26.64/0.744	31.84/0.898	33.14/0.910	34.60/0.937	33.92/0.916
HCI old	38.58/0.944	31.43/0.850	37.61/0.942	38.54/0.944	40.84/0.960	41.50/0.975
Average	36.03/0.928	29.30/0.803	35.05/0.922	36.14/0.928	38.07/0.949	38.13/0.949

For the task 2×2 to 7×7, we used the 9 synthetic LF images with angular resolution of

9×9, including 4 LF images from the HCI dataset (bicycle bedroom, herbs, and dishes) and 5 LF images from the old HCI dataset (Buddha Buddha2, StillLife, Papillon, and Mona). The central 7×7 views were extracted as ground truth, and 2×2 corner images were taken as input. We carried out comparison with the methods by Wu et al., Kalantari et al., Ran et al., and Wanner and Goldluecke. Table II shows the quantitative evaluation of the proposed approach on the synthetic dataset compared with the above methods.

On the two datasets, our proposed method provides an average angular superresolution advantage of 0.06 dB in terms of PSNR. On the old HCI dataset, our proposed method provides an average angular superresolution advantage of 0.66 dB in terms of PSNR and an advantage of 0.015 in terms of SSIM. On the HCI dataset, our method is inferior to Wanner and Goldluecke since they used depth information to connect correspondence in views, but it is still better than other depth-dependent methods such as Kalantari et al. and Wu et al. This is the cost for depth-free reconstruction. In general, our method has an absolute advantage in non-Lambertian scenes with smaller sampling baseline (the old HCI dataset), while it is also competitive on scenes with large sampling baseline (the HCI dataset). Therefore, we can say that our network has an advantage under task 2×2 to 7×7. We also provided visual comparisons of different methods, as shown in Fig. 6. In contrast, our method produces high-quality images which are closer to the ground truth ones.

For the task 3×3 to 7×7, we used the real-world scene LF images with angular resolution of 7×7. We carried out comparison with the method by Wu et al., Wang et al., and Ran et al.

As shown in Table III, our proposed method performs better for all datasets than comparing methods: with 2.63 dB angular superresolution advantage over Wu et al., 0.53 dB over Wang et al., and 0.41 dB over Ran et al. in terms of PSNR. Experimental results have further proven the advantages of our method.

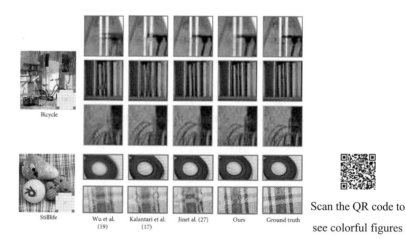

Fig. 6 Visual comparison between Wu et al., Kalantari et al., Jin et al., and our approach in synthetic scenes.

Table III Quantitative comparisons (PSNR/SSIM) of the proposed approach with the state-of-the-art ones under task 3×3 to 7×7 on real-world scenes.

Methods	Wu et al.	Wang et al. [35]	Li et al.	Ours
30scenes	41.02/0.988	43.82/0.993	43.83/0.993	44.07/0.988
Occlusions	39.80/0.981	41.23/0.983	41.89/0.984	42.42/0.988
Reflective	40.53/0.984	42.33/0.985	42.34/0.982	42.36/0.986
Average	40.48/0.985	42.58/0.988	42.82/0.987	43.11/0.988

4.3 Ablation Study

We conducted ablation experiments in terms of network structure. First, in order to verify the effectiveness of the two modules from our coarse-to-fine framework, we removed the EPI stack refinement module and tested only the coarse EPI angular superresolution module. Then, in order to maximize the reconstruction quality and find the optimal structure of the EPI stack refinement module, we evaluated three different structure solutions and selected the best one as the one used in this paper. The specific structure is shown in Fig. 7. Scheme 1 is composed of 3 3D convolutional layers, the filter sizes are 5×5×3, 1×1×3, and 9×9×3, and the numbers of output channels are 64, 32, and 1. Each convolutional layer is followed by a ReLU, excepting for the last layer. The filter bank learns the residuals, and the output of the last layer is added to the input as the final result. In general, scheme 1 is used as a residual block, and two residual blocks are connected to form a detail recovery module. The second scheme is an extension of scheme 1 and also refers to the idea of residual network. Scheme 2 is composed of 6 3D convolutional layers and the filter size is set to 3×3×3. The third one is the method used in this paper. Based on the three schemes, the angular superresolution performance was tested. The specific experimental design was to combine the coarse EPI angular superresolution module with the three schemes.

The experiment was carried out on the HCI dataset, and the task is 2×2 to 7×7. Table IV shows the experimental results. It can be seen that the angular superresolution quality of the EPI coarse angular superresolution network itself is on the high side. And the addition of the EPI stack refinement module of any scheme can increase the angular superresolution performance on the HCI datasets. This shows that both the coarse EPI angular superresolution module and the EPI stack refinement module play a positive role in improving the quality of the reconstructed light field, and the design of the entire frame structure is reasonable and effective. Between these two modules, the main function of the coarse EPI angular superresolution module is to complete the magnification with any scale factor. The EPI stack refinement module plays an important role in the quality of the reconstructed view, which is responsible for the accurate generation of image textures and complex regions. For the EPI stack refinement module, among the three different structure solutions, scheme 3 has obvious advantages by 1.96 dB.

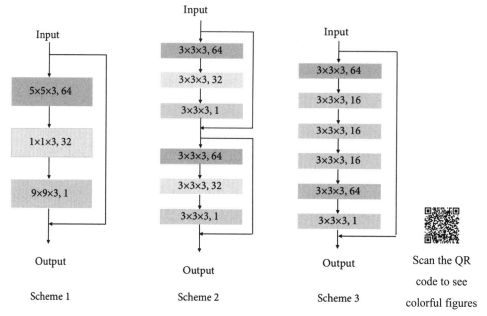

Fig. 7 Three different schemes of EPI stack refinement module.

Table IV Quantitative comparisons (PSNR/SSIM) on synthetic scenes with different EPI stack refinement module under task 2×2 to 7×7.

Methods	HCI
Only 2D EPI angular superresolution module	31.15/0.871
Coarse EPI angular superresolution module + solution 1	31.64/0.902
Coarse EPI angular superresolution module + solution 2	31.96/0.911
Coarse EPI angular superresolution module + solution 3 (ours)	33.92/0.916

Table V Quantitative comparisons (PSNR/SSIM) on synthetic scenes with different loss functions under task 2×2 to 7×7.

Methods	Edge enhancement loss function	VGG–16 perception loss
HCI	33.31/0.910	33.92/0.916
HCI old	41.14/0.971	41.50/0.975
Average	37.66/0.944	38.13/0.949

With the same training environment and training parameters, we used two different loss functions for training and observed the final training results. The first loss function is the edge enhancement loss function proposed in our previous work, and the second is the perceptual loss described in Section 3.5. The experiment was carried out on the HCI dataset and the old HCI dataset under the task 2×2 to 7×7. Table V shows the experimental results.

It can be seen that training with perceptual loss has an advantage of 0.47 dB in terms of PSNR and 0.005 in terms of SSIM, compared with training with edge enhancement loss function. We believe that this is because the perceptual loss uses the features extracted by CNN as the object of the loss function, making the global structure and high-level semantics of the synthetic image and the ground truth closer. Compared with edge enhancement loss function, which pays more attention to texture information, perceptual loss performs better on angular superresolution tasks.

5. Conclusion

In this paper, an end-to-end coarse-to-fine framework is proposed to directly synthesize novel views of 4D densely-sampled LF from sparse input views. We combine magnification with arbitrary scale factor network for coarse EPI angular superresolution and 3DCNNs for EPI stack refinement, working in a coarse-to-fine manner and forming a pseudo 4DCNN which can make full use of 4D LF data while circumventing high computational burden. Our framework first reconstructs an intermediate LF by recovering EPI row volumes and then works on EPI column volumes to synthesize the final densely-sampled LF. In addition, metalearning is utilized to upsample EPIs in the coarse EPI angular superresolution module, which enables magnification with arbitrary scale factor with one model. We conducted ablation experiments on the network structure and loss function, which proved the feasibility of our proposed network framework and the superiority of the perceived loss. Experimental results show that the proposed framework outperforms other state-of-the-art methods.

Data Availability

The light field image data supporting the findings of this study are from previously reported studies and datasets, which have been cited. The processed data are available from the corresponding author upon request.

Conflicts of Interest

The author(s) declare(s) that they have no conflicts of interest.

Acknowledgments

This work is supported by the National Natural Science Foundation of China (Grant Nos. 62001432 and 61971383), National Key R&D Program of China (Grant No. SQ2020YFF0426386), and the Fundamental Research Funds for the Central Universities (Grant Nos. CUC19ZD006 and CUC21GZ007).

References

[1] MARC L, PAT H. Light field rendering[C]//The 23rd Annual Conference on Computer Graphics and Interactive Techniques, August, 1996. ACM, c1996: 31-42.

[2] IHRKE I, RESTREPO J, MIGNARD-DEBISE L. Principles of light field imaging: briefly revisiting 25 years of research[J]. IEEE signal processing magazine, 2016, 33(5): 59-69.

[3] FISS J, CURLESS B, SZELISKI R. Refocusing plenoptic images using depth-adaptive splatting[C]//EEE International Conference on Computational Photography, May 02-04, 2014, Santa Clara, CA, USA. IEEE, c2014: 1-9.

[4] KIM C, ZIMMER H, PRITCH Y, et al. Scene reconstruction from high spatio-angular resolution light fields[J]. ACM transactions on graphics, 2013, 32(4): 73:1-73:12.

[5] CHEN J, HOU J, NI Y, et al. Accurate light field depth estimation with superpixel regularization over partially occluded regions[J]. IEEE transactions on image processing, 2018, 27(10): 4889-4900.

[6] HUANG F C, CHEN K, WETZSTEIN G. The light field stereoscope[J]. ACM transactions on graphics, 2015, 34(4): 1-12.

[7] CHAI J X, TONG X, CHAN S C, et al. Plenoptic sampling[C]//The 27th Annual Conference on Computer Graphics and Interactive Techniques, July, 2000. ACM, c2000: 307-318.

[8] RAYTRIXA. 3D light field camera technology [EB/OL]. [2021-10-01]. https://raytrix.de.

[9] BISHOP T E, FAVARO P. The light field camera: Extended depth of field, aliasing, and superresolution[J]. IEEE transactions on pattern analysis and machine intelligence, 2011, 34(5): 972-986.

[10] BOOMINATHAN V, MITRA K, VEERARAGHAVAN A. Improving resolution and depth-of-field of light field cameras using a hybrid imaging system[C]//IEEE International Conference on Computational Photography, May 02-04, 2014, Santa Clara, CA, USA. IEEE, c2014: 1-10.

[11] PENNER E, ZHANG L. Soft 3d reconstruction for view synthesis[J]. ACM transactions on graphics, 2017, 36(6): 1-11.

[12] FLYNN J, NEULANDER I, PHILBIN J, et al. Deepstereo: learning to predict new views from the world's imagery[C]//The IEEE Conference on Computer Vision and Pattern Recognition, June 27-30, 2016, Las Vegas, NV, USA. IEEE, c2016: 5515-5524.

[13] MARWAH K, WETZSTEIN G, BANDO Y, et al. Compressive light field photography using overcomplete dictionaries and optimized projections[J]. ACM transactions on graphics, 2013, 32(4): 1-12.

[14] SHI L, HASSANIEH H, DAVIS A, et al. Light field reconstruction using sparsity in the continuous fourier domain[J]. ACM transactions on graphics, 2014, 34(1): 1-13.

[15] VAGHARSHAKYAN S, BREGOVIC R, GOTCHEV A. Light field reconstruction using shearlet transform[J]. IEEE transactions on pattern analysis and machine intelligence, 2018, 40(1): 133-147.

[16] VAGHARSHAKYAN S, BREGOVIC R, GOTCHEV A. Accelerated shearlet-domain light field reconstruction[J]. IEEE journal of selected topics in signal processing, 2017, 11(7): 1082-1091.

[17] KALANTARI N K, WANG T C, RAMAMOORTHI R. Learning-based view synthesis for light field cameras[J]. ACM transactions on graphics, 2016, 35(6): 1-10.

[18] WU G, ZHAO M, WANG L, et al. Light field reconstruction using deep convolutional network on EPI[C]//The IEEE Conference on Computer Vision and Pattern Recognition, July 21-26, 2017, Honolulu, HI, USA. IEEE, c2017: 6319-6327.

[19] WU G, LIU Y, DAI Q, et al. Learning sheared EPI structure for light field reconstruction[J]. IEEE transactions on image processing, 2019, 28(7): 3261-3273.

[20] LI R. Light field reconstruction with arbitrary angular resolution using a deep coarse-To-fine framework[C]//Digital TV and Wireless Multimedia Communication: 17th International Forum, December 2, 2020, Shanghai, China. Springer Nature, c2020: 402–414.

[21] SIMONYAN K, ZISSERMAN A. Very deep convolutional networks for large-scale image recognition[C]//Proceedings of the 3rd International Conference on Learning Representations, 2015: 1-14.

[22] CHAURASIA G, DUCHENE S, SORKINE-HORNUNG O, et al. Depth synthesis and local warps for plausible image-based navigation[J]. ACM transactions on graphics, 2013, 32(3): 1-12.

[23] WANNER S, GOLDLUECKE B. Variational light field analysis for disparity estimation and super-resolution[J]. IEEE transactions on pattern analysis and machine intelligence, 2014, 36(3): 606-619.

[24] SRINIVASAN P P, WANG T, SREELAL A, et al. Learning to synthesize a 4D RGBD light field from a single image[C]//The IEEE International Conference on Computer Vision, July 21-26, 2017, Honolulu, HI, USA. IEEE, 2017: 2243-2251.

[25] ZHOU T, TUCKER R, FlLYNN J, et al. Stereo magnification: learning view synthesis using multiplane images[J]. arXiv preprint arXiv:1805.09817, 2018.

[26] MILDENHALL B, SRINIVASAN P P, ORTIZ-CAYON R, et al. Local light field fusion: practical view synthesis with prescriptive sampling guidelines[J]. ACM transactions on graphics, 2019, 38(4): 1-14.

[27] JIN J, HOU J, YUAN H, et al. Learning light field angular super-resolution via a geometry-

aware network[C]//The AAAI Conference on Artificial Intelligence, February 7-12, 2020, Hilton New York Midtown, New York, USA. AAAI, c2020: 11141-11148.

[28] LIN Z, SHUM H Y. A geometric analysis of light field rendering[J]. International journal of computer vision, 2004, 58: 121-138.

[29] LEVIN A, DURAND F. Linear view synthesis using a dimensionality gap light field prior[C]//IEEE Computer Society Conference on Computer Vision and Pattern Recognition, June 13-18, 2010, San Francisco, CA, USA. IEEE, c2010: 1831-1838.

[30] YAN M, LI Z, YU X, et al. An end-to-end deep learning network for 3D object detection from RGB-D data based on Hough voting[J]. IEEE access, 2020, 8: 138810-138822.

[31] YANG S, WANG J, ARIF S, et al. SAL-net: self-supervised attribute learning for object recognition and segmentation[J]. Wireless communications and mobile computing, 2021: 1-13.

[32] YOON Y, JEON H G, YOO D, et al. Light-field image super-resolution using convolutional neural network[J]. IEEE signal processing letters, 2017, 24(6): 848-852.

[33] WU G, LIU Y, FANG L, et al. Lapepi-net: A laplacian pyramid EPI structure for learning-based dense light field reconstruction[J]. arXiv preprint arXiv:1902.06221, 2019.

[34] WANG Y, LIU F, WANG Z, et al. End-to-end view synthesis for light field imaging with pseudo 4DCNN[C]//The European Conference on Computer Vision, September 8-14, 2018, Munich, Germany. Springer, c2018: 333-348.

[35] WANG Y, LIU F, ZHANG K, et al. High-fidelity view synthesis for light field imaging with extended pseudo 4DCNN[J]. IEEE transactions on computational imaging, 2020, 6: 830-842.

[36] HU X, MU H, ZHANG X, et al. Meta-SR: A magnification-arbitrary network for super-resolution[C]//IEEE/CVF Conference on Computer Vision and Pattern Recognition, June 15-20, 2019, Long Beach, CA, USA. IEEE, c2019: 1575-1584.

[37] ZHANG Y, TIAN Y, KONG Y, et al. Residual dense network for image super-resolution[C]//IEEE/CVF Conference on Computer Vision and Pattern Recognition, June 18-22, 2018, Salt Lake City, Utah, USA. IEEE, c2018: 2472-2481.

[38] CHEN Q, KOLTUN V. Photographic image synthesis with cascaded refinement networks[C]//The IEEE International Conference on Computer Vision, October 22-29, 2017, Venice, Italy. IEEE, c2017: 1511-1520.

[39] HONAUER K, JOHANNSEN O, KONDERMANN D, et al. A dataset and evaluation methodology for depth estimation on 4D light fields[C]//The 13th Asian Conference on Computer Vision, November 20-24, 2016, Taipei, Taiwan. Springer, c2016: 19-34.

[40] WANNER S, MEISTER S, GOLDLUECKE B. Datasets and benchmarks for densely sampled 4D light fields[C]//International Symposium on Vision, Modeling, and Visualization, September 11-13, 2013, Lugano, Switzerland. c2013: 225-226.

Cross-Domain Feature Similarity Guided Blind Image Quality Assessment*

1. Introduction

Objective image quality assessment (IQA) aims to enable computer programs to predict the perceptual quality of images in a manner that is consistent with human observers, which has become a fundamental aspect of modern multimedia systems. Based on how much information the computer program could access from the pristine (or reference) image, objective IQA could be categorized into full-reference IQA (FR-IQA), reduced-reference IQA (RR-IQA) and no-reference (or blind) IQA (NR-IQA/BIQA). The absence of reference information in most real-world multimedia systems calls for BIQA methods, which are more applicable but also more difficult.

Deep neural network (DNN) has significantly facilitated various image processing tasks in recent years due to its powerful capacity in feature abstraction and representation. It is also worth noting that the success of deep-learning techniques is derived from large amounts of training data, which is often leveraged to adjust the parameters in the DNN architecture to guarantee that both the accuracy and generalization ability are satisfying. Unfortunately, image quality assessment is typically a small-sample problem since the annotation of the ground-truth quality labels calls for time-consuming subjective image quality experiments. Inadequate quality annotations severely restrict the performance of DNN-based BIQA models in terms of both accuracy and generalization ability.

In order to address the problem caused by limited subjective labels, data augmentation is firstly employed to increase the training labels, [e.g., Kang et al. (2014)] proposed to split the image with quality labels into multiple patches, and each of the patches is assigned with a quality score which is the same with the whole image. However, some distortion types are inhomogeneous, i.e., the perceptual quality of local patches might differ from the overall quality of the whole image. Therefore, transfer learning has gained more attention

---

* The paper was originally published in *Frontiers in Neuroscience*, 2021, 15, and has since been revised with new information. It was co-authored by Chenxi Feng, Long Ye, and Qin Zhang.

to relieve the small-sample problem (Li et al., 2016). Specifically, the BIQA framework is comprised of two stages: which are pre-training and fine-tuning. In the pre-training stage, the parameters in the DNN architecture are trained by other image processing tasks such as object recognition, whilst in the fine-tuning stage, images with subjective labels are employed as training samples. Such a transfer-learning scheme is feasible since the low-level feature extraction procedure across different image processing tasks are shared.

More recently, various sources of external knowledge are incorporated to learn a better feature representation for the BIQA issue. For example, hallucinated reference is generated via a generative network and employed to guide the quality-aware feature extraction. The distortion identification is incorporated as the auxiliary sub-task in MEON model, by which the distortion type information is transparent to the primary quality prediction task for better quality prediction. Visual saliency is employed in Yang et al. (2019) to weight the quality-aware features more reasonably. Semantic information is also employed for better understanding of the intrinsic mechanism of quality prediction, e.g., multi-layer semantic features are extracted and aggregated through several statistical structures in Casser et al. (2019). An effective hyper network is employed in Su et al. (2020) to generate customized weights from the semantic feature for quality prediction, i.e., the quality perception rule differs as the image content changes.

Unlike other studies, this paper employs the cross-domain feature similarity as an extra restraint for better quality-aware feature representation. Specifically, the transfer-learning based BIQA approach is pre-trained in one domain (say, object recognition in the semantic domain) and is fine-tuned in the perceptual quality domain with similar DNN architectures, we have observed that the cross-domain (Semantic vs. Quality) feature similarity would, in turn, contribute to the quality prediction task (as shown in **Fig. 1**).

By thoroughly analyzing the intrinsic interaction between object recognition task and quality prediction task, we think the phenomenon represented in **Fig. 1** is sensible. As shown in **Fig. 2**, previous works have revealed that human observers would take different strategies to assess the perceptual quality when viewing images with different amounts of degradation: when judging the quality of a distorted image containing near-threshold distortions, one tends to rely primarily on visual detection of any visible local differences, in such a scenario, semantic information is instructive for quality perception since distortion in the semantic-sensitive area would contribute more in the quality decision and vice versa. On the other hand, when judging the quality of a distorted image with clearly visible distortions, one would rely much less on visual detection and much more on the overall image appearance, in such a scenario, the quality decision procedure is much more independent with semantic information.

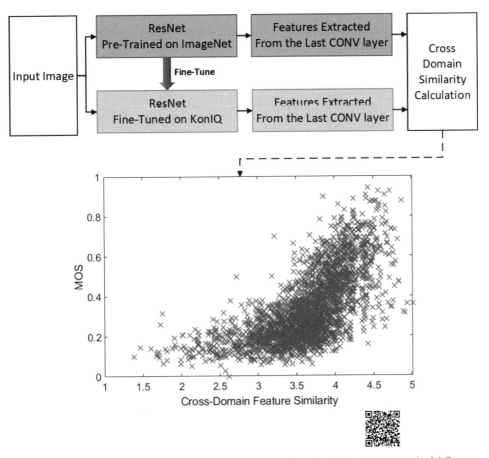

Scan the QR code to see colorful figures

Fig. 1 The overall framework of our proposed CDFS guided BIQA approach. As shown in the lower part, the cross-domain feature similarity is highly correlated with the perceptual quality. The 'cross-domain similarity calculation' is obtained by: (1) Extracted the features from the last convolutional layer of pre-trained ResNet (denoted as R_s) and fine-tuned ResNet (denoted as R_q); (2) Calculate the similarity matrix W according to Equation 1; (3) Obtaining the eigen values of W by $\vec{v} = eig(W)$; (4) The similarity Sim is calculated by $Sim = \dfrac{1}{std(v)}$, in which $std(\cdot)$ denotes the standard deviation operator.

Fig. 2 Illustration of different strategies that the human visual system would take to assess the perceptual quality when viewing images with different amounts of degradation. Specifically, when judging the quality of a distorted image containing near-threshold distortions (Left), one tends to rely primarily on visual detection of any visible local differences, e.g., the distortions in red boxed are slighter than that in the green box even though the noise intensity is the same. On the other hand, when judging the quality of a distorted image with clearly visible distortions (Right), one would rely much less on visual detection and much more on overall image appearance e.g., the distortions in each image area are roughly the same.

Considering the effectiveness of cross-domain feature similarity (CDFS), this work leverages CDFS as an extra restraint to improve the prediction accuracy of BIQA models. As shown in **Fig. 3**, the parameters in our CDFSNet are updated according to both the basic loss and the extra loss, which would restrain the network yielding quality predictions as similar as the ground-truth label whilst maintaining that the CDFS also correlates well with the perceptual quality, in such a manner that, the accuracy of the DNN architecture would get improved according to the experimental results presented in section 3.

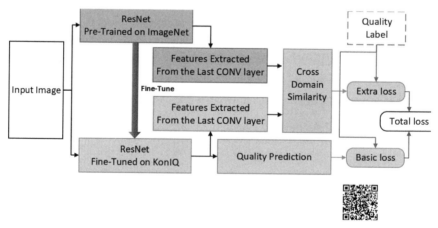

Fig. 3 The overall pipeline of our proposed CDFS-based IQA approach.

Compared to the aforementioned works, the superiority of the cross-domain feature similarity guided BIQA framework is embodied in the following aspects:

(1) The proposed cross-domain feature similarity is self-contained for transfer-learning based BIQA models since the transfer-learning procedure itself is comprised of the training in two different domains (i.e., object recognition and quality prediction). Therefore, no extra annotation procedure (such as distortion identification in Ma et al., 2017b and visual saliency in Yang et al., 2019) is needed.

(2) The proposed cross-domain feature similarity is more explicable since it is derived from the intrinsic characteristic of interactions between semantic recognition and quality perception.

(3) In addition to general-purpose IQA, the performance of our proposed CDFS guided BIQA framework is also evaluated on other specific scenarios such as screen content and dehazing oriented IQA. The experimental results indicate that CDFS guided BIQA has significant potential towards diverse types of BIQA tasks.

The rest part of the paper is organized as follows: Section 2 illustrates the details of our CDFS-based BIQA framework and Section 3 shows the experimental results; Section 4 is the conclusion.

2. Materials and Methods

2.1 Problem Formulation

Let x denote the input image, conventional DNN based BIQA works usually leverage an pre-trained DNN architecture $f(\cdot;\theta)$ (with learnable parameters θ) to predict the perceptual quality of x via $\hat{q} = f(x;\theta)$, where $\hat{q}$ denotes the prediction of perceptual quality q.

Our work advocates employing the cross-domain feature similarity to supervise the update of parameters in a quality prediction network. Specifically, let $f(\cdot;\theta_{Smtc})$ denotes the DNN with fixed and pre-trained parameters oriented towards semantic recognition, and $f(\cdot;\theta_{Qlty})$ denotes the DNN with learnable parameters oriented towards quality prediction. It should be noticed that $f(\cdot;\theta_{Smtc})$ and $f(\cdot;\theta_{Qlty})$ share the same architectures whilst having own different parameters. This work attempts to further improve the quality prediction accuracy by analyzing the similarity between the features extracted for different tasks, i.e., features extracted for semantic recognition $ft_s = f(x;\theta_{Smtc})$, and features extracted for quality regression $ft_q = f(x;\theta_{Qlty})$.

Given three-dimensional features ft_s and ft_q with size $[C, H, W]$, where C, H, W denotes the channel size, height, and width of the features respectively, ft_s and ft_q are firstly reshaped into R_q and R_s with size $[C, H \times W]$. The similarity Sim between R_q and R_s is obtained via the following steps.

Step 1, employ linear regression to express R_q via R_s, i.e., $R_q = W \times R_s + e$, where W denotes the weighting matrix and e denotes the prediction error of linear regression. Therefore, W could be obtained by

$$W = \left(R_s^T \times R_s\right)^{-1} \times R_s^T \times R_q \tag{1}$$

Step 2, a learnable DNN architecture $g(\cdot; \gamma)$ is employed to yield the similarity between ft_s and ft_q given W, i.e., $Sim = g(W; \gamma)$.

2.2 Network Design

The architecture of our proposed network is shown in **Fig. 4**, which mainly consists of a semantically oriented feature extractor, perceptual-quality oriented feature extractor, and cross-domain feature similarity predictor. More details are described as follows.

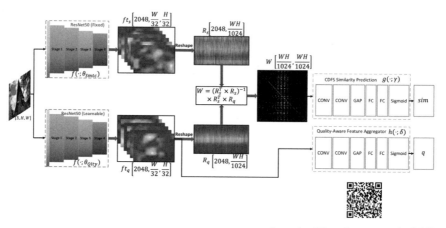

Scan the QR code to see colorful figures

Fig. 4 The detailed architecture of our proposed Cross-Domain Feature Similarity Guided Network. The 'CONV' denotes convolutional layers followed by batch normalization and ReLU layer, the 'FC' denotes the fully-connected layer, and the 'GAP' denotes the global average pooling layer.

2.2.1 Semantic Oriented Feature Extractor

The DNN pre-trained in large-scale object recognition datasets (e.g., ImageNet Deng et al., 2009) are leveraged as the semantic oriented feature extractor.

Specifically, this work employs the activations of the last convolutional layers in ResNet50 to represent the semantic-aware features ft_s of a specific image, i.e., $ft_s = f(x; \theta_{Smtc})$.

It is worth noting that θ_{Smtc} is fixed during the training stage since the proposed DNN framework will be fine-tuned in IQA datasets in which the semantic label is unavailable.

2.2.2 Perceptual-Quality Oriented Feature Extractor

The architecture of perceptual-quality oriented feature extractor $f(\cdot;\theta_{Qlty})$ is quite similar with semantic oriented feature extractor. However, the parameters θ_{Qlty} in $f(\cdot;\theta_{Qlty})$ are learnable and independent with θ_{Smtc}.

The quality-aware features $ft_q = f(x;\theta_{Qlty})$ are further leveraged to aggregate the prediction of subjective quality score, i.e., $\hat{q} = h(ft_q;\delta)$, in which q denotes the subjective quality score (MOS), $\hat{q}$ is the prediction of q, and $h(\cdot;\delta)$ stands for the MOS prediction network given quality-aware features with learnable parameters δ.

2.2.3 Cross-Domain Feature Similarity Predictor

As illustrated in section 1, the cross-domain feature similarity would contribute to the prediction of perceptual quality. However, directly evaluating the similarity between ft_s and ft_q via Minkowski-Distance or Wang-Bovik metric is not as efficient, as shown in **Fig. 6**. We think the invalidation of the Wang-Bovik metric is mainly attributed to its pixel-wise sensitivity, i.e., any turbulence during the parameter initializing and updating of the DNN framework would result in a significant difference between ft_s and ft_q.

To this end, this work proposes to depict the cross-domain feature similarity through a global perspective. Specifically, the similarity is derived from the weighting matrix W which is employed to reconstruct ft_q given ft_s via linear regression. Since the W is derived from the features amongst all channels, it is less likely to suffer from the instability of the DNN during initializing and updating. The experiments reported in section 3.3 also demonstrate the superiority of our proposed similarity measurement for cross-domain features. In our CDFS-guided BIQA framework, the CDFS is incorporated as follows:

Linear regression is employed for the reconstruction and the weighting matrix W could be obtained according to equation 1 and Step 1 in section 2.1

A stack of convolutional layers (denoted as $g(\cdot;\gamma)$) is followed to learn the cross-domain feature similarity given W.

During the training stage, the cross-domain similarity is employed as a regularization item to supervise the quality prediction network.

2.2.4 Loss Function

The loss function L of our proposed network is designed as

$$L_1 = \mathrm{argmin}_{[\theta_{Qlty},\delta]} \| q - h(f(x;\theta_{Qlty});\delta) \| \qquad (2)$$

$$L_2 = \mathrm{argmin}_{[\theta_{Qlty},\gamma]} \| q - g(W;\gamma) \| \qquad (3)$$

and

$$L = L_1 + \lambda L_2 \qquad (4)$$

where $\|\cdot\|$ denotes the $L1$ norm operator, W is calculated according to equation 1, and λ is a hyper parameter controlling the weights of L_1 and L_2.

2.3 Implementation Details

We use ResNet50 as the backbone model for both the semantically oriented feature extractor and the perceptual-quality oriented feature extractor. As aforementioned, the pre-trained model on ImageNet is used for network initialization. During the training stage, the θ_{Smtc} is fixed whilst θ_{Qlty} is learnable. In our network, the last two layers of the origin ResNet50, i.e., an average pooling layer and a fully connected layer, are removed to output features ft_s and ft_q.

For quality regression, a global average pooling (GAP) layer is used to pool the features ft_q into one-dimensional vectors, then three fully-connected (FC) layers are followed with size 2048-1024-512-1 and activated by ReLu, except for the last layer (activated by sigmoid).

The $g(\cdot;\gamma)$ in cross-domain feature similarity predictor is implemented by 3 stacked convolutional layers, a GAP layer, and three FC layers. The architectures of convolutional layers are $in(1)-out(32)-k(1)-p(0)$, $in(32)-out(64)-k(3)-p(1)$, and $in(64)-out(128)-k(3)-p(1)$, respectively, where $in(\alpha)-out(\beta)-k(x)-p(y)$ denotes the input channel size and output channel size is α and β, the kernel size is x, and the padding size is y. Each of the convolutional layer is followed by a batch normalization layer and a ReLu layer. The GAP layer and the FC layers are the same with quality regression except that the size of FC layers is 128-512-512-1.

The experiment is conducted on Tesla V100P GPUs, while the DNN modules are implemented by Pytorch. The size of minibatch is 24. Adam is adopted to optimize the loss function with weight decay 5×10^{-4} and learning rate 1×10^{-5} for parameters in baseline (ResNet) and 1×10^{-4} for other learnable parameters. As mentioned, the parameters in semantic oriented feature extractor are fixed, i.e., the learning rate is 0 for θ_{Smtc}.

3. Experimental Results

3.1 Datasets and Evaluation Metrics

Three image databases including KonIQ-10k, LIVE Challenges (LIVEC) (Ghadiyaram and Bovik, 2015), and TID2013 are employed to validate the performance of our proposed network. The KonIQ-10k and LIVEC are authentically distorted image databases containing 10073 and 1162 distorted images, respectively, and the TID2013 is a synthetic image database containing 3000 distorted images.

Two commonly used criteria, Spearman's rank order correlation coefficient (SRCC) and Pearson's linear correlation coefficient (PLCC), are adopted to measure the prediction monotonicity and the prediction accuracy. For each database, 80% images are used for training, and the others are used for testing. The synthetic image database is split according to reference images. All the experiments are under five times random train-test splitting operation, and the median SRCC and PLCC values are reported as final statistics.

3.2 Comparison with the State-of-the-Art Methods

Ten BIQA methods are selected for performance comparison, including five hand-crafted based and five DNN-based approaches. The experimental results are shown as in **Table I**.

Table I Performance comparison in terms of PLCC and SRCC on KonIQ, LIVEC, and TID2013, respectively.

SRCC	KonIQ	LIVEC	TID2013
BRISQUE	0.665	0.608	0.572
ILNIQE	0.507	0.432	0.521
HOSA	0.671	0.640	0.688
BPRI	-	-	0.899
BMPRI	-	-	**0.929**
SFA	0.856	0.812	-
DBCNN	0.875	0.851	-
HyperIQA	0.906	0.859	-
SGDNet	0.903	0.851	0.843
DeepFL	0.877	0.734	0.858
ours	**0.918**	**0.865**	0.899
PLCC	**KonIQ**	**LIVEC**	**TID2013**
BRISQUE	0.681	0.645	0.651
ILNIQE	0.523	0.508	0.648
HOSA	0.694	0.678	0.764
BPRI	-	-	0.892
BMPRI	-	-	**0.947**
SFA	0.872	0.833	-
DBCNN	0.884	0.869	-
HyperIQA	0.917	**0.882**	-
SGDNet	0.920	0.872	0.861
DeepFL	0.887	0.769	0.876
ours	**0.928**	0.875	0.880

Values in bold represents the highest value.

As shown in **Table I**, our method outperforms all the SOTA methods on the two authentic image databases in terms of SRCC. As for PLCC measurement, our method achieves the best performance on KonIQ and competing (the second) performance on LIVEC. This suggests that calculating cross-domain feature similarity for quality prediction refinement is effective. Though we do not especially modify the networks for synthetic image feature extraction, the proposed network has achieved competing performance in TID2013. Specifically, the proposed approach achieves the second-highest performance in terms of SRCC and the third-highest performance in terms of PLCC on TID2013.

3.3 Cross-Domain Feature Similarity Visualization

In order to further illustrate the superiority of our proposed CDFS, we firstly present the scatter plot of CDFS vs. MOS on KonIQ in **Fig. 5**, indicating the CDFS is well correlated with perceptual quality.

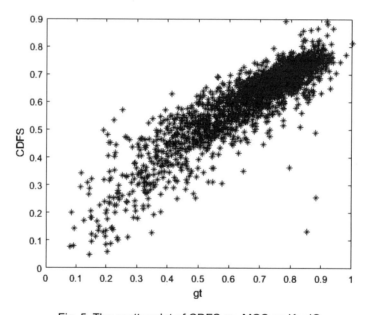

Fig. 5 The scatter plot of CDFS vs. MOS on KonIQ.

In addition, we also investigate several non-learnable approaches for calculating CDFS: (1) $Sim_1 = \text{mean}\left(\dfrac{2 \times ft_s \times ft_q + C}{ft_s^2 + ft_q^2 + C}\right)$, where C denotes the constant to avoid numerical singularity; and (2) $Sim_2 = std(eig(W))$; (3) $Sim_3 = \text{mean}\left(\dfrac{2 \times \vec{v} \times \vec{1} + C}{\vec{v}^2 + \vec{1}^2 + C}\right)$, and $\vec{v} = eig(W)$, in which $\vec{1}$ denotes the vectors with the same size as $\vec{v}$ whilst whose elements are all 1.

Therefore, the calculation of Sim_1 is directly comparing the difference between

ft_s and ft_q, and the calculation of Sim_2 and Sim_3 is based on the W derived according to equation 1. As shown in **Fig. 6**, Sim_2 and Sim_3 is more correlated with the subjective score, demonstrating that measuring the cross-domain feature similarity based on W is more effective.

3.4 Ablation Study

Ablation study is conducted on KonIQ-10k to validate the efficiency of our proposed components, including the ResNet50 backbone (BaseLine), the similarity predictor (SP) obtained by Wang-Bovik metric (SP_wang, similar as Sim_1 in section 3.3), and the similarity predictor derived from the weighting metric W (SP_W). The results are shown in **Table II**, indicating that incorporating a cross-domain similarity predictor could significantly improve the accuracy of quality prediction. Our proposed similarity measurement has achieved a great PLCC improvement (1.8%) compared to SP_wang and a more significant SRCC improvement (2.7%).

Table II Ablation results in terms of SRCC and PLCC on KonIQ.

Modules	BaseLine	+SP_wang	+SP_W
SRCC	0.842	0.895	0.918
Gain (%)	-	6.3	9.0
PLCC	0.849	0.913	0.928
Gain (%)	-	7.5	9.3

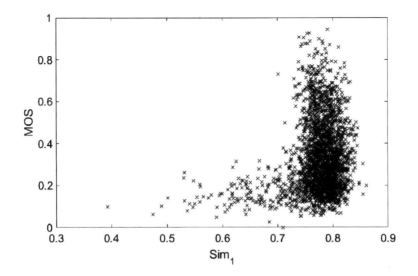

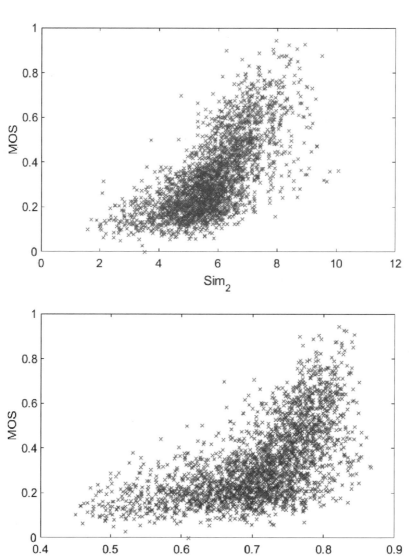

Fig. 6 The scatter plot of Sim_1, Sim_2, and Sim_3 vs. MOS on KonIQ.

The impact of λ in equation 4 is also investigated, i.e., we set $\lambda = [0.2, 0.4, 0.6, 0.8, 1.0]$, respectively and observe the corresponding performance as shown in **Fig. 7**. Therefore, we select $\lambda = 0.4$ for performance comparison and the following experiments.

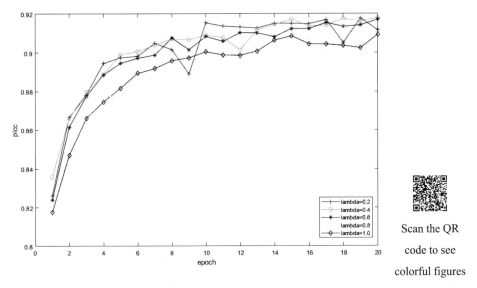

Fig. 7 Impact on selections of different λ. The experimental result is conducted on KonIQ, and a total of 20 epochs are involved.

Scan the QR code to see colorful figures

3.5 Cross-Database Validation

In order to test the generalization ability of our network, we train the model on the entire KonIQ-10k and test on the entire LIVEC. The four most competing IQA models in terms of generalization ability are involved in the comparison, which are PQR, DBCNN, HyperIQA, and DeepFL. The validation results are shown in **Table III**, indicating the generalization ability of our approach is higher than existing SOTA methods for assessing authentically distorted images.

Table III Cross data base validation (Trained on KonIQ-10k and Tested on LIVEC).

Modules	DeepFL	DBCNN	HyperIQA	PQR	Ours
SRCC	0.704	0.755	0.770	0.785	0.817
Gain (%)	-	7.2	9.4	11.5	16.1

However, if the network is trained on KonIQ-10k and directly applied for a synthetic image database, its generalization ability is not satisfactory, and the SRCC on TID2013 is only 0.577. That is mainly because the distortion mechanisms between synthetic and authentically distorted image databases are widely different. Training the network solely on authentically-distorted image databases could not learn the specific synthetic distortion patterns such as JPEG compression, transmission errors, or degradation caused by denoising, etc.

3.6 Further Validation on Other Specific IQA Tasks

In order to further validate the robustness of our BIQA framework towards other specific IQA tasks, the performance of CDFS guided BIQA network is evaluated on CCT, DHQ, and SHRQ. The CCT contains 1320 distorted images with various types of images including natural scene images (NSI), computer graphic images (CGI), and screen content images (SCI); The DHQ contains 1750 dehazed images generated from 250 real hazy images; The SHRQ database consists of two subsets, namely: regular and aerial image subsets, which include 360 and 240 dehazed images created from 45 and 30 synthetic hazy images using 8 image dehazing algorithms, respectively.

The training pipeline is similar with section 3.1, i.e., 80% of the CCT, DHQ or SHRQ are involved as the training set and the other 20% is the testing set. Considering that the scale of the subset is not adequate for the training of DNN, we merge the subsets in each datasets. For example, the NSI, CGI, and SCI are merged as the training set of CCT.

As shown in **Table IV**, the predictions of our CDFS guided BIQA framework show significant consistency with subjective scores, indicating that our proposed BIQA approach is feasible to be generalized into other types of IQA tasks.

Table IV SRCC and PLCC performance on CCT, DHQ, and SHRQ.

		SRCC	PLCC
CCT	20-%Test	0.9655	0.9672
	100-%Test	0.5758	0.6193
DHQ	20-%Test	0.9533	0.9223
	100-%Test	0.6819	0.6678
SHRQ	20-%Test	0.8875	0.9082
	100-%Test	0.4233	0.4761

Furthermore, if the network is trained on KonIQ-10k and directly applied on CCT, DHQ, and SHRQ, the accuracy is not satisfactory, as shown in **Table IV**. Such phenomenon is similar to the cross-database validation results discussed in section 3.5, indicating that training the network solely on authentically-distorted natural image databases could not sufficiently learn the quality-aware features for CGI, SCI, etc.

4. Conclusion

This work aims to evaluate the perceptual quality based on cross-domain feature similarity. The experimental results on KonIQ, LIVEC, and TID2013 demonstrate the superiority of our proposed methods.

We would further investigate such CDFS-incorporated BIQA framework in the following aspects: (1) investigating more efficient approaches of CDFS measurement; (2) investigating more types of DNN baselines in addition to ResNet.

Data Availability Statement

The original contributions presented in the study are included in the article/supplementary material, further inquiries can be directed to the corresponding author.

Author Contributions

CF established the BIQA framework and adjusted the architecture for better performance. LY and CF conducted the experiments and wrote the manuscripts. QZ designed the original method, and provided resource support (e.g., GPUs) for this manuscript. All authors contributed to the article and approved the submitted version.

Funding

This work is supported by the National Key R&D Program of China under Grant No. 2021YFF0900503, the National Natural Science Foundation of China under Grant Nos. 61971383 and 61631016, and the Fundamental Research Funds for the Central Universities.

Acknowledgments

We would like to thank Li Fang, Wei Zhong, and Fei Hu for some swell ideas.

Conflict of Interest

The authors declare that the research was conducted in the absence of any commercial or financial relationships that could be construed as a potential conflict of interest.

Publisher's Note

All claims expressed in this article are solely those of the authors and do not necessarily represent those of their affiliated organizations, or those of the publisher, the editors and the reviewers. Any product that may be evaluated in this article, or claim that may be made by its manufacturer, is not guaranteed or endorsed by the publisher.

Copyright © 2022 Feng, Ye and Zhang. This is an open-access article distributed under the terms of the Creative Commons Attribution License (CC BY). The use, distribution or reproduction in other forums is permitted, provided the original author(s) and the copyright owner(s) are credited and that the original publication in this journal is cited, in accordance with accepted academic practice. No use, distribution or reproduction is permitted which does not comply with these terms.

References

[1] CASSER V, PIRK S, MAHJOURIAN R, et al. Depth prediction without the sensors: leveraging structure for unsupervised learning from monocular videos[C]//AAAI Conference on Artificial Intelligence, January 27-February, 2019, Honolulu, Hawaii, USA. AAAI, c2019: 8001-8008.

[2] CHANG H W, YANG H, GAN Y, et al. Sparse feature fidelity for perceptual image quality assessment[J]. IEEE transactions on image processing, 2013, 22: 4007-4018.

[3] DENG J, DONG W, SOCHER R, et al. ImageNet: a large-scale hierarchical image database [C]//IEEE Conference on Computer Vision and Pattern Recognition, June 20-25, 2009, Miami, FL, USA. IEEE, c2009: 248-255.

[4] FANG H S, XIE S, TAI Y W, et al. RMPE: regional multi-person pose estimation[C]//IEEE International Conference on Computer Vision, October 22-29, 2017, Venice, Italy. IEEE, c2017: 2334-2343.

[5] GHADIYARAM D, BOVIK A C. Massive online crowdsourced study of subjective and objective picture quality[J]. IEEE transactions on image processing, 2015, 25: 372-387.

[6] GHOSAL D, MAJUMDER N, PORIA S, et al. Dialoguegcn: a graph convolutional neural network for emotion recognition in conversation[J]. arXiv preprint arXiv:1908.11540.

[7] HE K, ZHANG X, REN S, et al. Deep residual learning for image recognition[C]//IEEE Conference on Computer Vision and Pattern Recognition, June 27-30, 2016, Las Vegas, NV, USA. IEEE, c2016: 770-778.

[8] HOSU V, LIN H, SZIRANYI T, et al. Koniq-10k: an ecologically valid database for deep learning of blind image quality assessment[J]. IEEE transactions on image processing, 2020, 29: 4041-4056.

[9] KANG L, YE P, LI Y, et al. Convolutional neural networks for no-reference image quality assessment[C]//IEEE Conference on Computer Vision and Pattern Recognition, June 23-28, 2014, Columbus, OH, USA. IEEE, c2014: 1733-1740.

[10] KIM J, LEE S. Fully deep blind image quality predictor[J]. IEEE Journal of Selected Topics in Signal Processing, 2016, 11: 206-220.

[11] KINGMA D P, BA J. Adam: a method for stochastic optimization[J]. arXiv preprint arXiv: 2014. 1412.6980.

[12] LARSON E C, CHANDLER D M. Most apparent distortion: fullreference image quality assessment and the role of strategy[J]. Journal of electronic imaging, 2010, 19(1): 011006.

[13] LI D, JIANG T, LIN W, et al. Which has better visual quality: the clear blue sky or a blurry animal?[J]. IEEE transactions on. multimedia, 2018, 21: 1221-1234.

[14] LI S, ZHANG F, MA L, et al. Image quality assessment by separately evaluating detail

losses and additive impairments[J]. IEEE transactions on multimedia, 2011, 13: 935-949.

[15] LI Y, PO L M, FENG L, et al. No-reference image quality assessment with deep convolutional neural networks[C]//IEEE International Conference on Digital Signal Processing, October 16-18, 2016, Beijing, China. IEEE, c2016: 685-689.

[16] LIN K Y, WANG G. Hallucinated-iqa: No-reference image quality assessment via adversarial learning[C]//IEEE Conference on Computer Vision and Pattern Recognition, June 18-23, 2018, Salt Lake City, UT, USA. IEEE, c2018: 732-741.

[17] LIU T J, LIN W, KOU C C J. Image quality assessment using multi-method fusion[J]. IEEE transactions on image processing, 2012, 22: 1793-1807.

[18] LIU X, VAN DE WEIJER J, BAGDANOV A D. Rankiqa: Learning from rankings for no-reference image quality assessment[C]//IEEE International Conference on Computer Vision, October 22-29, 2017, Venice, Italy. IEEE, c2017: 1040-1049.

[19] MA K, LIU W, LIU T, et al. Dipiq: blind image quality assessment by learning-to-rank discriminable image pairs[J]. IEEE transactions on image processing, 2017, 26: 3951-3964.

[20] MA K, LIU W, ZHANG K, et al. End-to-end blind image quality assessment using deep neural networks[J]. IEEE transactions on image processing, 2017, 27: 1202-1213.

[21] MIN X, GU K, ZHAI G, et al. Blind quality assessment based on pseudo-reference image[J]. IEEE transactions on multimedia, 2017, 20: 2049-2062.

[22] MIN X, MA K, GU K, et al. Unified blind quality assessment of compressed natural, graphic, and screen content images[J]. IEEE transactions on image processing, 2017, 26: 5462-5474.

[23] MIN X, ZHAI G, GU K, et al. Blind image quality estimation via distortion aggravation[J]. IEEE transactions on Broadcast, 2018, 64: 508-517.

[24] MIN X, ZHAI G, GU K, et al. Objective quality evaluation of dehazed images[J]. IEEE transactions on intelligent transportation systems, 2018, 20: 2879-2892.

[25] MIN X, ZHAI G, GU K, et al. Quality evaluation of image dehazing methods using synthetic hazy images[J]. IEEE transactions on multimedia, 2019, 21: 2319-2333.

[26] MIN X, ZHAI G, ZHOU J, et al. Study of subjective and objective quality assessment of audio-visual signals[J]. IEEE transactions on image processing, 2020, 29: 6054-6068.

[27] MIN X, ZHOU J, ZHAI G, et al. A metric for light field reconstruction, compression, and display quality evaluation[J]. IEEE transactions on image processing, 2020, 29:3790-3804.

[28] MITTAL A, MOORTHY A K, BOVIK A C. No-reference image quality assessment in the spatial domain[J]. IEEE transactions on image processing, 2012, 21: 4695-4708.

[29] PAN D, SHI P, HOU M, et al. Blind predicting similar quality map for image quality assessment[C]//IEEE Conference on Computer Vision and Pattern Recognition, June 18-23, 2018, Salt Lake City, UT, USA. IEEE, c2018: 6373-6382.

[30] PARK S J, SON H, CHO S, et al. Srfeat: single image super-resolution with feature discrimination[C]//European Conference on Computer Vision), September 8-14, 2018, Munich, Germany. Springer, c2018: 439-455.

[31] PONOMARENKO N, JIN L, IEREMEIEV O, et al. Image database tid2013: peculiarities,

results and perspectives[J]. Signal processing: image communication, 2015, 30: 57-77.

[32] REHMAN A, WANG Z. Reduced-reference image quality assessment by structural similarity estimation[J]. IEEE transactions on image processing, 2012, 21: 3378-3389.

[33] SHEIKH H R, BOVIK A C. Image information and visual quality[J]. IEEE transactions on image processing, 2006, 15: 430-444.

[34] SU S, YAN Q, ZHU Y, et al. Blindly assess image quality in the wild guided by a self-adaptive hyper network[C]//IEEE/CVF Conference on Computer Vision and Pattern Recognition, June 13-19, 2020, Seattle, WA, USA. IEEE, c2020: 3667-3676.

[35] SUN W, MIN X, ZHAI G, et al. Blind quality assessment for inthe-wild images via hierarchical feature fusion and iterative mixed database training[J]. arXiv preprint arXiv: 2021. 2105.14550.

[36] TALEBI H, MILANFAR P. Nima: neural image assessment[J]. IEEE transactions on image processing, 2018, 27: 3998-4011.

[37] TAN C, SUN F, KONG T, et al. A survey on deep transfer learning[C]//International Conference on Artificial Neural Networks, 2018, Rhodes. Springer, c2018: 270-279.

[38] WANG Z, BOVIK A C. Reduced-and no-reference image quality assessment[J]. IEEE signal processing magazine, 2011, 28: 29-40.

[39] WANG Z, BOVIK A C, SHEIKH H R, et al. Image quality assessment: from error visibility to structural similarity[J]. IEEE transactions on image processing, 2004, 13: 600-612.

[40] WANG Z, SIMONCELLI E P. Reduced-reference image quality assessment using a wavelet-domain natural image statistic model[J]. IS&T/SPIE electronic imaging, 2005: 149-159.

[41] WANG Z, SIMONCELLI E P, BOVIK A C. Multiscale structural similarity for image quality assessment[C]//The 37th Asilomar Conference on Signals, Systems & Computers, 2003, Pacific Grove, CA. IEEE, 2003: 1398-1402.

[42] MIN X K, GU K, ZHAI G T, et al. Screen content quality assessment: overview, benchmark, and beyond[J]. ACM computing surveys, 2021, 5(9): 1-36.

[43] XU J, YE P, LI Q, et al. Blind image quality assessment based on high order statistics aggregation[J]. IEEE transactions on image processing, 2016, 25: 4444-4457.

[44] XUE W, ZHANG L, MOU X, et al. Gradient magnitude similarity deviation: a highly efficient perceptual image quality index[J]. IEEE transactions on image processing, 2013, 23: 684-695.

[45] YANG S, JIANG Q, LIN W, et al. Sgdnet: an end-to-end saliency guided deep neural network for no-reference image quality assessment[C]//The 27th ACM International Conference on Multimedia, October 21-25, 2019, Nice, France. ACM, c2019: 1383-1391.

[46] ZENG H, ZHANG L, BOVIC A C. A probabilistic quality representation approach to deep blind image quality prediction[J]. arXivV preprint: 2017. 1708.08190.

[47] ZHAI G, MIN X. Perceptual image quality assessment: a survey[J]. Science china information sciences, 2020, 63: 211301.

[48] ZHANG L, SHEN Y, LI H. Vsi: a visual saliency-induced index for perceptual image quality assessment[J]. IEEE transactions on image processing, 2014, 23: 4270-4281.

[49] ZHANG L, ZHANG L, BOVIK A C. A feature-enriched completely blind image quality evaluator[J]. IEEE transactions on image processing, 2015, 24: 2579-2591.

[50] ZHANG L, ZHANG L, MOU X, et al. Fsim: a feature similarity index for image quality assessment[J]. IEEE transactions on image processing, 2011, 20: 2378-2386.

[51] ZHANG R, ISOLA P, EFROS A A, et al. The unreasonable effectiveness of deep features as a perceptual metric[C]//IEEE Conference on Computer Vision and Pattern Recognition, June 18-23, 2018, Salt Lake City, UT, USA. IEEE, c2018: 586-595.

[52] ZHANG W, MA K, YAN J, et al. Blind image quality assessment using a deep bilinear convolutional neural network[J]. IEEE transactions on circuits and systems for video technology, 2018, 30(1): 36-47.

A Dataset and Benchmark for 3D Scene Plausibility Assessment*

1. Introduction

In recent years, 3D scene synthesis has gained considerable attention. The intricate nature of synthesizing functional and plausible 3D scenes, requiring extensive knowledge to select appropriate object categories and arrange them logically, poses inherent challenges. The absence of robust scene plausibility assessment impedes the generation of high-quality scenes using neural networks. Consequently, there is a growing emphasis on 3D scene plausibility assessment due to its significant impact on scene quality.

The assessment of 3D scene plausibility involves determining the realism and believability of a given scene. This task has become increasingly vital as 3D scenes find applications in diverse fields such as virtual reality (VR), augmented reality (AR), video games, simulation and planning, robotics, architecture and design, and movie and TV production.

In VR and AR applications, the credibility of 3D scenes is crucial for delivering a realistic and immersive user experience. If scenes lack plausibility, users may experience disorientation or discomfort. In video games, plausible 3D scenes contribute to creating more convincing game worlds, enhancing player engagement. For simulation and planning applications, such as architectural design and robotics, the use of accurate and representative 3D scenes is imperative. Otherwise, simulation and planning outcomes may be unreliable, impacting real-world applications such as robotic navigation planning.

Differentiating from general 3D model quality assessment, 3D scene plausibility assessment presents two key distinctions. Firstly, it must consider interactions between objects within the scene, ensuring logical arrangements (e.g., chairs not within walls).

---

* The paper was originally published in *IEEE Transactions on Multimedia*, 2024, and has since been revised with new information. It was co-authored by Fei Hu, Wei Zhong, Long Ye, Xinyan Yang, Li Fang, and Qin Zhang. The first two authors contributed equally to this work and should be considered co-first authors.

Secondly, it must account for the contextual aspect of the scene, recognizing that a scene with numerous objects in a confined space may be implausible despite high-quality individual models. Existing studies either rely on subjective experiments or fixed criteria, both of which have limitations in handling complex scenarios. In response, this paper draws inspiration from image quality assessment and employs neural networks to model objective scores accurately. Image quality assessment (IQA) has witnessed significant advancements, with researchers introducing various methods and datasets to objectively evaluate image quality. Learning-based IQA approaches, particularly those leveraging convolutional neural networks (CNNs), have achieved state-of-the-art performance. Despite these advances, there is a noticeable gap in the study of 3D scene plausibility. This paper addresses this gap by presenting a plausibility assessment method for 3D scenes, highlighting key differences from image quality assessment:

- The absence of a dedicated dataset for 3D scene plausibility assessment.
- The impracticality of using the entire scene representation as direct input.
- The crucial importance of scene layout and object attributes for scene assessment tasks.

To tackle these challenges, we introduce the 3D-SPAD dataset and propose a method for 3D scene plausibility assessment, as illustrated in Fig. 1. The dataset comprises 1500 manually labeled scenes from the SUNCG dataset and 1500 scenes generated through data augmentation with automatic labels. The proposed method involves data preprocessing and the evaluation of the 3D-SPAN networks. The input scene undergoes processing to generate a scene graph containing information about the scene layout and object attributes. This scene graph is then fed into the network to obtain plausibility scores for both individual objects and the entire scene.

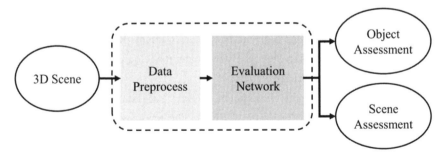

Fig. 1 The overall framework of our method.

In summary, the main contributions of this paper include:

- The construction of a dataset containing 3000 3D scenes, representing the first dedicated dataset for scene plausibility assessment.

- The introduction of a method for edge relation extraction based on a Gaussian mixture model, offering simplicity and efficiency.
- The proposal of a framework based on neural networks for assessing the plausibility of 3D scenes, encompassing individual objects and the scene as a whole.
- Demonstration of the method's robust generalization to real-world scenes, validated on the ScanNet dataset without the need for fine-tuning or additional features.

2. Related Work

2.1 3D Scene Graph

Scene graphs are widely used for 3D scene synthesis and retrieval. Typically, a scene graph represents objects in a 3D scene as vertices and relationships between object pairs as edges. Graphs can efficiently represent rich information in a scene. Current research on scene graphs mainly focuses on enhancing semantic representation and improving efficiency. Luo et al. and Ma et al. represent objects and relationships in 3D scenes as vertices to enhance semantic representation, while edges only represent connection states. Armeni et al. represent the 3D scene graph as a multi-layer form to contain more diverse information. These improvements not only make the 3D scene graph contain richer scene semantic information but also make the 3D scene graph more complex. Wald et al. employ deep learning methods to efficiently construct vertices and edges. The learning method improves the construction efficiency of 3D scene graphs but requires a large amount of labeled data to train the network. In our work, we build scene graphs via Gaussian mixture models to avoid collecting annotation data and simplify the process.

2.2 Image Quality Assessment

Image Quality Assessment (IQA) has garnered considerable attention in recent years due to the escalating demand for high-quality visual content. Researchers have devised numerous methods and datasets to objectively evaluate image quality, employing a range of image features and machine learning techniques.

Reference-based image quality assessment involves evaluating the quality of an image by comparing it to a reference or pristine image. The primary objective is to measure the extent of distortion or differences between the test image and a known high-quality reference. Nevertheless, a notable limitation of this method is the challenge of acquiring well-aligned or partially aligned pristine-quality images with content similar to the query image. Consequently, no reference/blind image quality methods, which assesses image quality without referencing any pristine-quality image, has emerged as a predominant research direction. Blind quality assessment is a promising approach for evaluating the quality of experience (QoE) of various visual content, including images, videos, and light field images.

CG-DIQA centers on character gradient-based assessment for document images. Alaei et al. utilize Hast derivations for assessing document image quality. Shahkolaei et al. contribute a blind quality assessment metric tailored for degraded document images, addressing degradation classification. MetaIQA stands out for its incorporation of deep meta-learning into the realm of no-reference image quality assessment. Another notable advancement is the work of Su et al. on blind image quality assessment in real world scenarios, leveraging a self-adaptive hyper network. Additionally, the Transformer-based approach exemplifies the application of transformer architectures in image quality assessment. Min et al. unify blind quality assessment across compressed natural, graphic, and screen content images. The exacerbation of distortions for blind image quality estimation presents another innovative avenue.

Pertinent databases are significant for evaluation. The CID2013 database has played a pivotal role in evaluating no-reference IQA algorithms. Ghadiyaram et al.'s extensive online crowd-sourced study provides valuable insights into both subjective and objective picture quality. The KonIQ-10k database is notable for its ecological validity in deep learning-based blind image quality assessment.

The recent advancements in image quality assessment also serve as a foundation for endeavors to enhance learning in the context of 3D scene plausibility assessment, employing neural networks. The incorporation of innovative architectures, datasets, and evaluation methodologies, inspired by the progress in image quality assessment, is instrumental in developing effective neural network models for addressing the nuanced intricacies of 3D scene plausibility assessment.

2.3 3D Indoor Scene Synthesis Assessment

Research on the assessment of 3D indoor scene synthesis can be summarized into the following categories: speed assessment, richness assessment, accuracy assessment, and plausibility assessment.

Speed assessment directly calculates the time complexity of the synthesis algorithm. Richness assessment evaluates the diversity of synthetic scenes. Luo et al. define the variety of scene layouts as the standard deviation of boxes and angles. Zhang et al. adopt the percentage of style combinations recognized by users to compare the diversity. Li et al. measure the scene similarity by graph kernel to show the diversity of scenes. Scene accuracy assessment is to assess whether the synthesized scene meets the specified objectives. Scene accuracy assessment is to assess whether the synthesized scene meets the specified objectives. Luo et al. adopt a L1 bounding box loss to evaluate the accuracy of generated scene layouts. Wald et al. evaluate the accuracy with scene graph prediction task and scene retrieval tasks. Qi et al. evaluate the scene by calculating the total variation distance and Hellinger distance between the energy maps of the synthetic and real samples. Zhang et al. measure the accuracy by 1NN accuracy and maximum mean discrepancy.

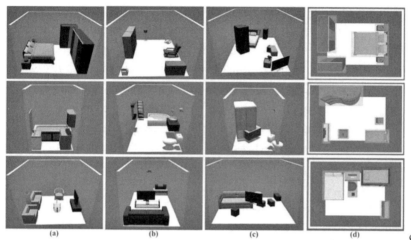

Scan the QR code to see colorful figures

Fig. 2 Examples of the 3D scenes in 3D-SPAD. (a) The sample of "good" scenes. (b) The sample of "bad" scenes. (c) The sample of scenes generated by data augmentation. (d) The sample of objects. The green mask denotes a "good" object, and the red mask denotes a "bad" object.

Plausibility assessment is to assess whether the synthesized scenes are plausible and natural. Plausibility assessment can directly judge whether the synthetic scene can satisfy the subjective feelings of human beings, which is significant for the scene synthesis task. Ma et al. adopt the average plausibility-naturalness scores by subjects. Zhang et al. use the single-stimulus continuous mass grading method and the co-occurrence sort method. Zhang et al. adopt the two-stimuli continuous mass grading method. Yu et al. evaluate the realism and functionality of the furniture arrangements by perceptual study. Fisher et al. collect human subjective evaluation using the single stimulus continuous mass grading method. Li et al. adopt perceptual studies to compare plausible scenes. It is worth noting that the diversity of subjective evaluation makes the relevant research without a unified and reliable comparison scale. In the study of objective methods, the verification part of 3D-SLN is limited in accuracy and efficiency, and 3D-SLN only considers the physical and spatial attributes of the 3D scene. Moreover, the HCMs are calculated according to ergonomic priors and probability and cannot fully reflect human subjective perception. Our study aims to explore an objective 3D scene plausibility assessment method to promote the assessment research of 3D indoor scene synthesis.

3. Dataset

To the best of our knowledge, there is a paucity of established datasets for 3D scene plausibility assessment. Inspired by previous studies in image quality assessment and 3D datasets, we construct the 3D scene plausibility assessment dataset (3D-SPAD), which is utilized with the deep learning method.

3.1 Data Collection

We selected 1500 3D indoor scenes from the SUNCG dataset, including three types of scenes: bedroom, bathroom, and living room. To save labor costs, we prioritize data availability over abundance. Therefore, the type and number of scene objects are appropriately limited. In addition, in order to avoid calculation errors caused by irregular objects, objects that may cause ambiguity, such as doors and windows, are deleted.

3.2 Human Subjective Evaluation

Next, we construct a basic dataset by subjectively evaluating the collected 3D scenes through Single Stimulus Continuous Quality Evaluation (SSCQE).

The evaluation proceeds with the following steps:

- **Environment and configuration:** We've developed a scene annotation application based on the Unity 3D engine. With this application, participants could freely change their perspectives to observe 3D indoor scenes and evaluate them.
- **Evaluation process:** 26 participants participated in the human subjective evaluation. The participants ranged in age from 22 to 24 years, with a male-to-female ratio of 9:17. Each participant needed to evaluate the object and the entire scene independently. We first gave each participant a brief introduction to the purpose of the assessment, and then the participant watched a demonstration video of the assessment process. To collect more authentic subjective evaluation data, participants only need to score according to their personal subjective feelings, rather than strict scoring standards.
- **Evaluation statistics:** The scene data and annotation software were disseminated to 36 evaluators using manual distribution. To guarantee each scene's assessment by three distinct evaluators, we organized evaluators into three groups. Each group thoroughly annotated all scenes to ensure evaluation accuracy. Employing the single-stimulus continuous quality grading method, evaluators were presented with one scene at a time for assessment using a predefined scale. Participants rated options based on their subjective perceptions of both individual objects and the overall indoor three-dimensional scene. Evaluation options were categorized as either

"good" or "poor." When evaluating scene objects, walls and floors were defaulted to an "empty" evaluation since they also contribute to rationality calculations. Subsequently, we synthesized the scene annotation data and employed a voting method to derive the final evaluation results.

3.3 Data Augmentation

There are problems of insufficient data volume and insufficient annotation quality in the basic dataset. We employ data augmentation to address these problems. The "bad" scene is expanded by randomly disrupting the object placement of the scene in the basic dataset, and the label is verified by three independent participants. We call the expanded dataset 3D-SPAD. As shown in Table I, data augmentation expands the scale of 3D-SPAD but also introduces data imbalance. We explore the impact of data augmentation in Section V-D4.

The statistical information of 3D-SPAD is shown in Table I. Fig. 2 shows examples of the scenes and objects in 3D-SPAD.

Table I The statistics of 3D-SPAD. R1 is defined as the ratio of good scenes to bad scenes. R2 is defined as the ratio of good objects to bad objects.

Scene	Total	Object Class	R1	R2
Bedroom	1000	23	0.22	0.52
Bathroom	1000	12	0.13	0.51
Living room	1000	27	0.21	0.49
All	3000	46	0.19	0.51

4. Method

In this section, we will describe the 3D scene plausibility assessment method. As shown in Fig. 1, our method can be divided into two parts: the data preprocess and the assessment network (3D-SPAN).

4.1 Data Preprocess

In order to express sufficient semantic information and make the input data regular, this paper transforms the 3D scene into a scene graph.

1) Gaussian Mixture Model: The Gaussian mixture model has been widely used in 3D scene synthesis, particularly for 3D scene optimization. Ma et al. propose that the spatial distribution of object pairs in a scene follows a Gaussian mixture distribution.

Consequently, 3D scene optimization is reformulated as an optimization problem involving probability distributions. We have further developed this method and successfully applied it to edge relation extraction.

By assuming that the spatial distribution between object pairs follows a Gaussian mixture distribution, we can extract edge relationships based on the prior probability of the object spatial distribution. Our method offers a simpler and more efficient approach compared to the methods presented in [3], [8].

2) 3D Scene Graph Construction: In our work, a 3D scene graph G is a pair of sets $G = (V, E)$, where V and E denote the set of vertices and edges, respectively. Fig. 3 shows the process of 3D scene graph construction.

On the one hand, vertices represent objects as feature vectors. Each vertex $v \in V$ encapsulates information about the corresponding object. $v \in \mathbb{R}^{33} = (v_{sm}, v_{sp}, v_{py})$ is as follows:

- **Semantic attributes** $v_{sm} \in \mathbb{R}^{18}$: Word embedding of object category information.
- **Spatial attributes** $v_{sp} \in \mathbb{R}^{6}$: Position relative to the center of the scene and the rotation relative to the initial orientation of the object.
- **Physical attributes** $v_{py} \in \mathbb{R}^{9}$: Physical characteristics, such as the boundary box and the scale of the object.

On the other hand, edges represent the relationships between object pairs as an adjacency matrix. In this study, we adopt an object-centric reference frame to extract relationships between objects. We categorize the extraction of edge relationships into three types: object-supported, wall-supported, and unsupported. The extraction process is as follows:

- **Object-supported:** An object-supported relationship exists when the bounding box of one object is in contact with the top of another object.
- **Wall-supported:** Every object in the scene has a default relationship with the wall and floor, defined as the wall-supported relationship.
- **Unsupported:** A spatial distribution where the probability of the Gaussian mixture distribution between object pairs exceeds a threshold of 0.05 is defined as an unsupported relationship.

If the relationship between a pair of objects falls into one of the three types mentioned above, the corresponding value in the matrix E is set to '1'; otherwise, it is set to '0'.

4.2 3D Scene Plausibility Assessment Network (3D-SPAN)

As depicted in Fig. 4, the 3D Scene Plausibility Assessment Network (3D-SPAN) comprises two modules: 3D-SPAN-Object (3D-SPAN-O) and 3D-SPAN-Scene (3D-SPAN-S). 3D-SPAN-O and 3D-SPAN-S, trained independently, assess the plausibility of individual objects and the overall scene, respectively.

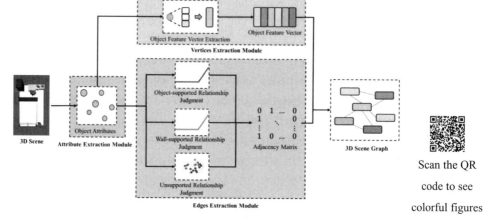

Scan the QR code to see colorful figures

Fig. 3 3D scene graph construction consists of three steps: attribute extraction, vertex extraction and edge extraction. The attribute extraction module uses scene data as input and outputs semantic (category information), physical (boundary box and scale), and spatial attributes (position and rotation) of objects to the vertex extraction module and edge extraction module. In the vertex extraction module, semantic attributes, spatial attributes, and physical attributes are processed into tensors, and vertices are the concatenation of these tensors. The edge extraction module represents edge relations as an adjacency matrix. Finally, the 3D scene graph is composed of vertices and edges.

1) Graph Attention Mechanism in 3D-SPAN: Introduced by Veličković et al., the graph attention network (GAT) integrates the attention mechanism into the graph neural network, leveraging the spatial domain. It can dynamically learn asymmetric attention weights for neighbor vertices and then update vertex features through weighted aggregation.

Human perception of a scene arises not only from object properties but also from analyzing the relationships between objects. Given GAT's ability to effectively capture node associations, we employ it to calculate attention coefficients for different object pairs under varying spatial distributions.

The vertex feature update process can be defined as:

$$\tilde{v}_i = \sum_{i \neq j} \sigma\left(A(v_i, v_j) \cdot E_{ij} \cdot W \cdot v_j\right), v_i, v_j \in V, \tilde{v}_i \in \tilde{V}, \tag{1}$$

where A denotes the calculation of graph attention coefficients, E_{ij} denotes the adjacency between v_i and v_j, W denotes the weight matrix, V denotes the vertex features, $\tilde{V}$ denotes the updated vertex features, and σ denotes activation function.

The vertex feature update process of the multi-head graph attention layer (M-GAT) can be defined as:

$$\tilde{v}_i = \|_k \left[\sum_{i \neq j} \sigma\left(A(v_i, v_j) \cdot E_{ij} \cdot W \cdot v_j\right)\right] \tag{2}$$

where $\|$ denotes the vector concatenation operation, and k denotes the number of heads. Other symbols are the same as Equation (1).

2) 3D-SPAN-O: 3D-SPAN-O takes as input a 3D scene graph represented by $\mathbf{G} = (V, E)$, where $V \in \mathbb{R}^{N \times M}$, $N = 14$, $M = 33$ represents the set of vertices and E represents the set of edges. The model embeds the set of vertices V into a set of features $\mathbf{F}$ using a vertices embedding module VE with trainable parameters θ_F:

$$\mathbf{F} = VE(\mathbf{G}, \theta_F). \tag{3}$$

The updated feature set $\mathbf{F}$ is then processed by the graph attention module GAM with trainable parameters θ_A, resulting in the plausibility evaluation of objects $\mathbf{O} \in \mathbb{R}^{N \times C}$ ($C=3$, where C represents the number of vertex categories).

$$\mathbf{O} = GAM(\mathbf{F}, \theta_A). \tag{4}$$

In more detail, GAM consists of a multi-head graph attention layer ($k=4$) and two graph attention layers. It first aggregates the graph representation vector via the multi-head graph attention layer and then processed by two graph attention layers that generate the $\mathbf{O} = \{o_1, o_2, \cdots, o_N\}$ as output. Note that $\mathbf{O}$ fully encodes a specific 3D scene by the corresponding set of vertices, hence it can be regarded as a local representation of the scene.

Given a scene with objects evaluation $L = \{l_1, l_2, \cdots, l_N\}$, we train all parameters end-to-end via a cross entropy loss on subjective evaluation of objects:

$$\arg\min \left\{ -\sum_{i \in N} l_i \cdot \log[softmax(o_i)] \right\}, l_i \in L, o_i \in \mathbf{O}. \tag{5}$$

3) 3D-SPAN-S: 3D-SPAN-S integrates information about the scene at a global scale by fusing local representation and input graphs. It starts by concatenating along the channel dimension original vertices V and the local representation $\mathbf{O}$, obtained from the enhanced 3D scene graph $\mathbf{G}' \in \mathbb{R}^{N \times (M+C)}$. A vertices embedding module VE' with trainable parameters $\theta_{F'}$ embeds original features and local features into a set of features $\mathbf{F}'$:

$$\mathbf{F}' = VE'(\mathbf{G}', \theta_{F'}) \tag{6}$$

The $\mathbf{F}'$ is then processed by a graph classification module GCM with trainable parameters θ_B that adopts two graph attention layers to aggregate graph representation and two full connection layers to obtain the plausibility evaluation of the overall 3D scene $\mathbf{S} \in \mathbb{R}^{C'}$ ($C' = 2$, where C' represents the number of graph categories). GCM in 3D-SPAN-S focuses on the correlation between objects and scenes, while GAM in 3D-SPAN-O only focuses on the correlation between objects:

$$\mathbf{S} = GCM(\mathbf{F}', \theta_B). \tag{7}$$

3D-SPAN-S effectively captures the global context of the scene by integrating local representation and input graphs. This fusion of information allows 3D-SPAN-S to assess scene plausibility at a broader scale, considering the overall arrangement and relationships between objects. In contrast, 3D-SPAN-O focuses solely on object-level correlations, analyzing the plausibility of individual objects and their immediate surroundings.

To train the model, we utilize subjective scene evaluations represented by Y. We employ an end-to-end training approach, optimizing the model's parameters using a cross-entropy loss function as shown in Equation (8) based on these subjective evaluations. This ensures that the model learns to effectively assess scene plausibility in accordance with human perception.

$$\arg\min_{\theta_{F'},\theta_B}\{-Y\cdot\log[softmax(\mathbf{S})]\}. \tag{8}$$

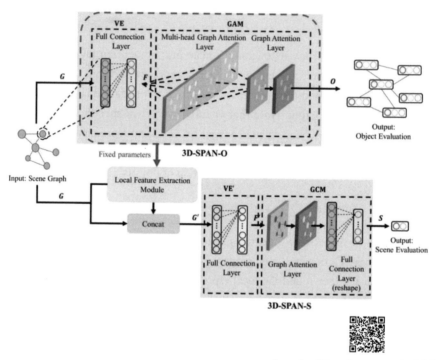

Scan the QR code to see colorful figures

Fig. 4 The 3D Scene Plausibility Assessment Network (3D-SPAN) comprises two primary modules: 3D-SPAN-O and 3D-SPAN-S. First, the 3D-SPAN-O module processes the input scene graph G and generates plausibility scores for the individual objects within the 3D scene. Subsequently, the local feature extraction module utilizes a fixed-parameter version of 3D-SPAN-O. Using the local feature extraction module, we enhance the features extracted from the scene graph G, resulting in the enhanced scene graph G'. Finally, the 3D-SPAN-S module employs the enhanced scene graph G' to assess the overall plausibility of the 3D scene.

5. Experiments and Analysis

Despite there being feasible methods for 3D scene plausibility assessment, the generalization of deep learning-based 3D scene plausibility assessment models may fall behind the conventional state-of-the-art methods (subjective evaluation) due to the lack of effective training data and well-designed network architectures. With the 3D-SPAD, we propose a Network framework based on GAT for 3D scene plausibility assessment, called 3D-SPAN. The purpose of the proposed 3D-SPAN as a baseline is to call for the development of deep learning-based 3D scene plausibility assessment and demonstrate the generalization of the 3D-SPAD for training 3D-SPAN. Note that the proposed 3D-SPAN is only a baseline model which can be further improved by well-designed network architectures, task-related loss functions, and the like.

5.1 Experimental Details

In this study, we evaluate the performance of our model using average recall, accuracy, and F1 score. Due to the class imbalance in the 3D-SPAD dataset, we prioritize average recall over other metrics. We split the dataset into three sets: training (80%), validation (10%), and test (10%). In all experiments, we use the Adam optimizer with a learning rate of 0.0001 to minimize the loss function. We selected the Adam optimizer for its efficiency and stability in training deep learning models. We experimentally determined that a learning rate of 0.0001 was optimal for our model's convergence and performance.

5.2 Baselines

To evaluate the performance of 3D-SPAN, we consider four categories of baseline methods:

- **Measure-based methods:** Following Luo et al., we define two baselines based on the task: 2D box measure and 3D box measure. We regard the training and validation sets as ground truth and measure the L1 bounding box loss from the test set to the ground truth. The label of the ground truth with the minimum loss is used as the prediction.
- **Encoder-based methods:** Encoder models such as multilayer perceptrons (MLPs) and convolutional neural networks (CNNs) are designed to learn effective graph representations and obtain graph classifications via readout functions. We set the number of layers in the MLP and CNN baselines to 5.
- **Graph network methods:** Graph convolution network (GCN) and graph attention networks (GATs) are two graph neural network (GNN) models designed to learn vertex representations in graphs. Vertex representations are aggregated into a graph

representation via a readout function, which is then used for graph classification. We set the number of layers in the GCN and GAT baselines to 5.

- **Graph classification methods:** This group of methods combines GNNs with pooling mechanisms to better learn graph-level representations and classify graphs. We compare with three algorithms: SAGPool, HGP-SL and GraphSSL-Reg.

5.3 Results

We conduct quantitative and qualitative analyses to compare the performance of our method and the baseline methods on the 3D-SPAD dataset.

1) Quantitative Analysis: Table II shows the 3D scene plausibility assessment results. 3D-SPAN outperforms all baseline methods on all evaluation metrics. 3D-SPAN achieves an average recall of 75.27%, an F1 score of 80.86%, and an accuracy of 87.33%.

Graph classification methods based on graph pooling perform poorly on this task for two reasons. First, these methods are trained on biological datasets, which have significantly different graph structures than indoor scene graphs. Second, the network architectures of graph classification methods are designed for large-scale graphs, which are not suitable for the small-scale graphs of indoor scenes.

Encoder-based methods, graph network methods, and graph classification methods all perform worse than measure-based methods. This indicates that it is challenging to capture sufficient information from the original graph and vertex features alone. We will validate this point in our ablation experiments.

2) Qualitative Analysis: We select several representative baselines to compare with our method and conduct qualitative analysis through visual experiments.

As shown in Fig. 5, 3D-SPAN is more stable in its evaluation performance than the other methods. In the second column, 2D box measure incorrectly evaluates a very bad scene as "good." In the third column, GAT and GraphSSL-Reg also make incorrect judgments. Our method correctly evaluates almost all scenes, except for the ambiguous case in the fourth column.

Table II Performance (%) on the 3D-SPAD dataset W.R.T. average recall, accuracy, and F1. the best result among the methods is highlighted.

Method	Recall↑	Accuracy↑	F1↑
2D box measure	71.30%	82.33%	76.82%
3D box measure	63.77%	78.00%	70.68%
MLP	53.15%	83.00%	64.81%
CNN	52.51%	85.67%	65.11%

Table II Continued

Method	Recall↑	Accuracy↑	F1↑
GCN	57.44%	86.67%	69.09%
GAT	54.55%	87.33%	67.16%
SAGPool	50.00%	87.00%	63.50%
HGP-SL	52.18%	87.00%	65.24%
GraphSSL-Reg	50.00%	87.00%	63.50%
3D-SPAN(our)	**75.27%**	**87.33%**	**80.86%**

5.4 Ablation Study and Analysis

In this section, we conduct ablation experiments to verify the effectiveness and design principles of 3D-SPAN. We also explore the impact of data augmentation, loss function, and training method on network performance.

Table III Ablation experiment of the local feature extraction module. ('W/O LF' denotes model without local feature extraction module while 'W/ LF' denotes model with local feature extraction module.)

Model	w/o LF			w/ LF		
	Recall↑	Accuracy↑	F1↑	Recall↑	Accuracy↑	F1↑
3D-SPAN-S-MLP	53.15%	83.00%	64.81%	53.40%	85.33%	65.69%
3D-SPAN-S-GCN	57.44%	86.67%	69.09%	71.81%	87.00%	78.68%
3D-SPAN-S-CNN	52.51%	85.67%	65.11%	55.26%	86.67%	67.49%
3D-SPAN-S-(M-GAT+GAT)	61.48%	**88.00%**	72.39%	63.60%	86.00%	73.12%
3D-SPAN-S-(GCN+CNN)	57.19%	84.33%	68.16%	50.00%	87.00%	63.50%
3D-SPAN-S-(GAT+CNN)	56.03%	**88.00%**	68.46%	50.00%	87.00%	63.50%
3D-SPAN-S-(our)	54.55%	87.33%	67.16%	**75.27%**	87.33%	**80.86%**

Input Model						
2D box measure	Bad	Good	Good	Bad	Good	Bad
GAT	Good	Bad	Bad	Bad	Bad	Bad
GraphSSL-Reg	Bad	Bad	Bad	Bad	Bad	Bad
3D-SPAN(our)	Good	Bad	Good	Good	Good	Good
Ground-Truth	Good	Bad	Good	Bad	Good	Good

Scan the QR code to see colorful figures

Fig. 5 Visual experiment on the 3D scene plausibility assessment. Green indicates that the prediction is consistent with ground-truth, and red indicates that the prediction is inconsistent with ground-truth.

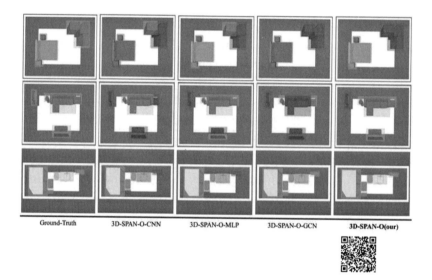

Scan the QR code to see colorful figures

Fig. 6 Qualitative comparison on the plausibility assessment of the object. From top to bottom are bedroom, living room and bathroom. The objects with a green mask indicate that they are evaluated as "good", and those with a red mask indicate that they are evaluated as "bad". All the models adopt the vertices embedding module. 3D-SPAN-O is equal to 3D-SPAN-O-(M-GAT+GAT). Since 3D-SPAN-O, as an extension of 3D-SPAN-O-GAT, outperforms 3D-SPAN-O-GAT on all metrics, we only list 3D-SPAN-O in this figure.

Table IV Results for object plausibility assessment. ("W/O VE" denotes model without vertices embedding module while "W/ VE" denotes model with vertices embedding module.)

Model	w/o VE			w/ VE		
	Recall↑	Accuracy↑	F1↑	Recall↑	Accuracy↑	F1↑
3D-SPAN-S-MLP	66.67%	83.31%	74.06%	78.33%	87.14%	82.50%
3D-SPAN-S-CNN	36.06%	38.05%	37.02%	78.06%	87.14%	82.35%
3D-SPAN-S-GCN	78.95%	87.26%	82.90%	83.65%	89.43%	86.44%
3D-SPAN-S-GAT	84.33%	89.81%	86.99%	85.33%	90.05%	87.63%
3D-SPAN-S-(our)	87.38%	91.31%	89.30%	**87.86%**	**91.45%**	**89.62%**

We evaluate 3D-SPAN and the less powerful 3D-SPAN variants:

- 3D-SPAN-O-N: This variant replaces the graph attention module (GAM) of 3D-SPAN-O with a network N such as MLP, CNN, GCN, or GAT.
- 3D-SPAN-S-N: This variant replaces the graph attention layers of 3D-SPAN-S with a network N such as MLP, CNN, GCN, or M-GAT. The "+" symbol denotes a combination of different networks. For example, "GCN+CNN" denotes a combination of GCN and CNN.

1) *Local Feature Extraction*: In this section, we conduct ablation experiments to illustrate the role of the local feature extraction module in scene plausibility assessment.

To test the effectiveness of local features, we design the experiments by removing the local feature extraction module. As shown in Table III, the performance of most models improved after adding the local feature extraction module. Notably, in 3D-SPAN-S, the average recall increased by 20.72%, and the F1 score increased by 13.7%.

Additionally, we observe that local features significantly improve the performance of models using graph attention networks (GATs) or graph convolutional networks (GCNs), while local features have less impact on the performance of models without GATs or GCNs. Therefore, local features need to be combined with the relationships between objects to play a better role in this task. This conclusion also corroborates our design idea that global features can be described according to local features in the scene.

2) *Impact of the Vertex Embedding and Graph Attention Module on Object Plausibility Assessment*: In this section, we conduct quantitative and qualitative analyses to analyze the design principles of the vertex embedding (VE) and graph attention module (GAM). We focus on the impact of the vertex embedding module and graph attention module on the model.

The experimental results on object plausibility assessment are shown in Table IV. From

this systematic evaluation, we observe the following:
- Models with VE perform better than models without VE, indicating that embedding the original features is crucial for object plausibility assessment.
- Models using GAT or GCN perform better than other models, confirming the importance of graph representation. Both GAT and GCN aggregate the features of adjacent vertices to the central vertex. This phenomenon verifies that the plausibility of objects in a scene is affected by other objects, and the interaction between objects plays a key role in plausibility evaluation.
- 3D-SPAN-O-GAT and 3D-SPAN-O perform slightly better than 3D-SPAN-O-GCN. This is why we chose GAT over GCN when designing 3D-SPAN-O. This is likely because the graph attention network (GAT) successfully processes the interaction between object pairs.
- Our method achieves the best performance on all evaluation metrics, proving that 3D-SPAN-O can effectively evaluate objects in a scene.

Fig. 6 shows the visual qualitative results on object plausibility assessment. From these results, we can draw the following conclusions:
- Compared to 3D-SPAN-O-CNN and 3D-SPAN-O-GCN, our model does not tend to give the same evaluation for clustered objects. For example, in the first row of Fig. 6, 3D-SPAN-O-CNN and 3D-SPAN-O-GCN both evaluate the double bed and the stands on both sides of the double bed as "bad". Our model can consider the interaction between object pairs and calculate asymmetric weights to avoid errors.
- The graph attention mechanism enables our model to evaluate objects in complex scenes. For example, in the second row of Fig. 6, 3D-SPAN-O-MLP evaluates the TV stand under the television as "bad" because it only considers the features of the object itself, making its evaluation ability for complex examples weak. Our model considers the interaction between object pairs, giving it better evaluation ability for complex examples.

3) Impact of the Vertex Embedding and Graph Classification Module on Scene Plausibility Assessment: In this section, we conduct quantitative and qualitative analyses to analyze the design principles of the vertex embedding (VE') and graph convolution module (GCM). We use 3D-SPAN-O as the common local feature extraction module and compare the performance of different networks.

Scan the QR code to see colorful figures

Fig. 7 Qualitative comparison on the plausibility assessment of the scene. Green indicates that the evaluation is consistent with the ground-truth, and red indicates inconsistency. All the models adopt the vertices embedding module. 3D-SPAN-S is equal to 3D-SPAN-S-GAT. Since 3D-SPAN-S-(M-GAT+GAT), as an extension of 3D-SPAN-S, underperforms 3D-SPAN-S on all metrics, we only list 3D-SPAN-S in this figure.

Table V Results for scene plausibility assessment. ("W/O VE' " denotes model without vertices embedding module while "W/ VE' '" denotes model with vertices embedding module.)

Model	w/o VE'			w/ VE'		
	Recall↑	Accuracy↑	F1↑	Recall↑	Accuracy↑	F1↑
3D-SPAN-S-MLP	52.51%	85.67%	65.11%	53.40%	85.33%	65.59%
3D-SPAN-S-CNN	-	-	-	71.81%	87.00%	78.68%
3D-SPAN-S-GCN	63.47%	**87.67%**	73.63%	71.81%	87.00%	78.68%
3D-SPAN-S-(M-GAT+GAT)	70.47%	84.67%	76.92%	63.60%	86.00%	73.12%
3D-SPAN-S-(our)	60.33%	86.00%	70.91%	**75.27%**	87.33%	**80.86%**

Table V shows the results of the scene plausibility assessment. We can observe the following:

- Models with VE' perform better than models without VE' in most cases, confirming the importance of the vertex embedding module for scene plausibility assessment.
- Graph-based methods significantly outperform other methods. Even 3D-SPAN-S-MLP with VE', which performs the best among MLP models, is 5.07% (average recall) lower than 3D-SPAN-S without VE', which performs the worst among GAT models. This echoes the conclusion of Section V-D2 and verifies the importance of

the interaction between objects.

- Our model outperforms 3D-SPAN-S-GCN with VE' on all metrics, implying that GAT is better at exploiting the correlation between objects than GCN to evaluate the overall scene. This verifies the effectiveness of the graph attention mechanism.
- Fig. 7 shows the visual qualitative results on scene plausibility assessment. We can conclude the following:
- Our model 3D-SPAN-S, with its graph attention mechanism, can evaluate complex scenes. In the second column, the clock on the wall is too large in proportion, which means that the scene should be evaluated as "bad". Graph-based models consider the impact of individual objects on the whole, while MLP and CNN only consider most objects. Therefore, only 3D-SPAN-S and 3D-SPAN-S-GCN can make a correct evaluation.
- Our model is better at capturing local features and more reliably combines local and global features than 3D-SPAN-S-GCN. For example, there are some small defects in the layout of the third and last scenes. In our subjective evaluation, we classify these two scenes as spurious. 3D-SPAN-S-GCN has a weak ability to infer global correlations, so it cannot distinguish these situations.

4) *More Details:* Moreover, we also perform ablation experiments on the data augmentation, loss function and training method to explore the impact of more details on network performance.

Table VI The impact of data augmentation on the performance of 3D-SPAN-S. "Basic" denotes the dataset before augmentation. "Balance" denotes the dataset after balanced.

Dataset	R1	Total	Recall↑	Accuracy↑	F1↑
Basic	0.45	1500	64.92%	71.33%	67.98%
Balance	1	936	65.17%	63.44%	64.29%
3D-SPAD	0.19	3000	**75.27%**	**87.33%**	**80.86%**

Table VII The impact of loss function on the performance of 3D-SPAN-S.

Loss	Recall↑	Accuracy↑	F1↑
Binary cross entropy loss	70.27%	86.33%	77.52%
Cross entropy loss(our)	75.27%	87.33%	**80.86%**

Data Augmentation Table VI shows ablation experiments on data augmentation. It can be observed that data augmentation effectively improves the performance of the network and solve the problems of insufficient data volume and annotation quality. The results prove that the advantages of data increase far outweigh the disadvantages of data imbalance.

Loss Function Table VII presents the performance of 3D-SPAN-S using different loss functions. Our results indicate that cross entropy loss outperforms binary cross entropy loss. This difference can be attributed to the use of softmax function and sigmoid function. The softmax function enlarges the distance between input vector elements and normalizes them to a probability distribution, resulting in significantly different individual probabilities in the output. The maximum value probability is closer to 1, which brings the output distribution closer to the true distribution.

Training Method Table VIII shows an unexpected result that joint training leads to a lower performance than separation training. This may be due to the local conflict brought by the cascading structure.

5.5 Real-World Application

In this section, we aim to test the domain generalization of our method in the real world using the ScanNet dataset. To construct the scene graph in the ScanNet dataset, we first prune nodes based on the object category in the 3D scene. We then retain nodes from the 3D-SPAD dataset that support the object category, and extract the 3D bounding box, spatial coordinates, and semantic information of the object using the object segmentation annotation provided by the ScanNet dataset to construct the node feature vector. Finally, we fill in any missing rotation and direction features with zeros.

To enhance the performance of our 3D-SPAN model in processing multi-size scene graphs, we replace the penultimate fully connected layer with a global max pooling layer during pre-training on the 3D-SPAD dataset. Using the training parameters from the 3D-SPAD dataset, our model achieves an average recall rate of 60.72%, an accuracy rate of 80.56%, and an F1 score of 70.64% on the ScanNet dataset.

We also perform a visual analysis on a subset of samples from the ScanNet dataset to evaluate the plausibility of our method. This analysis reveals that our approach can successfully assess the plausibility of scenes with missing or too many walls. However, the lack of object direction information results in a decline in the performance of the 3D-SPAN model. As shown in Fig. 8, the chair in the second image of the first row has obvious direction confusion, but 3D-SPAN cannot obtain relevant information, resulting in an error in the assessment. Additionally, the quality of point clouds affects the performance of the 3D-SPAN model. Incomplete point clouds, such as the mirror in the second image of the second row, result in inaccuracies in the boundary box and coordinate information of objects in the 3D scene, which negatively impact the construction of the scene graph and the performance of the 3D-SPAN model.

Despite these challenges, our 3D-SPAN model based on 3D-SPAD pre-training maintains an acceptable performance on ScanNet samples. It can effectively generalize to different domains, even in the presence of adverse circumstances such as a lack of object direction information, inaccuracies in boundary box information, the lack of information about objects with unknown classifications, and irregular walls. Overall, our findings demonstrate that our method has good generalization and application ability in the real world.

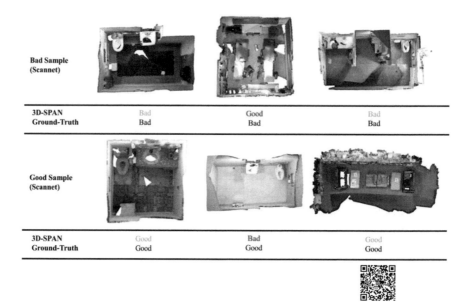

Scan the QR code to see colorful figures

Fig. 8 Visualization on the scenes of Scannet dataset. The top view of 3D point cloud is adopted. On the output of 3D-SPAN, green means the same as ground truth (human subjective evaluation), while red means different.

Table VIII The impact of training strategy on the performance of 3D-SPAN-S.

Method	Recall↑	Accuracy↑	F1↑
Joint training	73.42%	86.00%	79.21%
Separation training(our)	**75.27%**	**87.33%**	**80.86%**

6. Conclusion

In this paper, we present a novel plausibility assessment dataset, 3D-SPAD, which includes a variety of 3D scenes and reference plausibility scores. This dataset can be used to train neural networks for scene plausibility assessment. We also propose a graph attention network (GAT)-based model trained on this dataset, which outperforms existing baseline methods. Our method's ability to generalize to the ScanNet dataset indicates its potential for application in real-world scenarios. These results demonstrate the effectiveness and generalization ability of our approach, which has significant potential for further development in the field of deep learning-based scene plausibility assessment.

Acknowledgment

This work is supported by National Key R&D Program of China under Grant No. 2021YFF0900504, the National Natural Science Foundation of China under Grant Nos. 61971383 and 62001432, and the Fundamental Research Funds for the Central Universities under Grant No. CUC210C013, No. CUC21GZ007 and No. CUC18LG024.

References

[1] ZHANG S H, ZHANG S K, XIE W Y, et al. Fast 3D indoor scene synthesis by learning spatial relation priors of objects[J]. IEEE transactions on visualization and computer graphics, 2022, 28(9): 3082-3092.

[2] WU H, YAN W, LI P, et al. Deep texture exemplar extraction based on trimmed T-CNN[J]. IEEE transactions on multimedia, 2021, 23: 4502-4514.

[3] WALD J, DHAMO H, NAVAB N, et al. Learning 3D semantic scene graphs from 3D indoor reconstructions[C]//IEEE/CVF Conference on Computer Vision and Pattern Recognition, June 13-19, 2020, Seattle, WA, USA. IEEE Computer Society, c2020: 3960-3969.

[4] ZHANG S, HAN Z, LAI Y K, et al. Active arrangement of small objects in 3D indoor scenes[J]. IEEE transactions on visualization & computer graphics, 2021, 27(04): 2250-2264.

[5] ZHANG Z, SUN W, ZHOU Y, et al. EEP-3DQA: efficient and effective projection-based 3D model quality assessment[C]//IEEE International Conference on Multimedia and Expo, July 10-14, 2023, Brisbane, Australia. IEEE, c2023: 2483-2488.

[6] MA R, PATIL A G, FISHER M, et al. Language-driven synthesis of 3D scenes from scene databases[J]. ACM transactions on graphics, 2018, 37(6): 1-16.

[7] ZHANG S H, ZHANG S K, XIE W Y, et al. Fast 3D indoor scene synthesis with discrete and exact layout pattern extraction[J/OL]. arXiv, 2020. [2021-04-19]. https://arxiv.org/pdf/2002.00328.pdf. DOI:10.48550/arXiv.2002.00328.

[8] LUO A, ZHANG Z, WU J, et al. End-to-end optimization of scene layout[C]//IEEE/CVF Conference on Computer Vision and Pattern Recognition, June 13-19, 2020, Seattle, WA, USA. IEEE Computer Society, c2020: 3753-3762.

[9] FU Q, FU H, YAN H, et al. Human-centric metrics for indoor scene assessment and synthesis[J]. Graphical models, 2020, 110: 101073.

[10] GU K, WANG S, YANG H, et al. Saliency-guided quality assessment of screen content

images[J]. IEEE transactions on multimedia, 2016, 18(6): 1098-1110.

[11] CHEN W, GU K, ZHAO T, et al. Semi-reference sonar image quality assessment based on task and visual perception[J]. IEEE transactions on multimedia, 2020, 23: 1008-1020.

[12] CHEN H, CHAI X, SHAO F, et al. Perceptual quality assessment of cartoon images[J]. IEEE transactions on multimedia, 2021, 25: 140-153.

[13] MIN X, ZHAI G, GU K, et al. Quality evaluation of image dehazing methods using synthetic hazy images[J]. IEEE transactions on multimedia, 2019, 21(9): 2319-2333.

[14] ZHAI G, MIN X. Perceptual image quality assessment: a survey[J]. Science china information sciences, 2020, 63: 1-52.

[15] MIN X, GU K, ZHAI G, et al. Screen content quality assessment: overview, benchmark, and beyond[J]. ACM computing surveys (CSUR), 2021, 54(9): 1-36.

[16] SONG S, YU F, ZENG A, et al. Semantic scene completion from a single depth image[C]// IEEE Conference on Computer Vision and Pattern Recognition, July 21-26, 2017, Honolulu, Hawaii. IEEE Computer Society, c2017: 190-198.

[17] ARMENI I, HE Z Y, ZAMIR A, et al. 3D scene graph: a structure for unified semantics, 3D space, and camera[C]//IEEE/CVF International Conference on Computer Vision, Oct. 27-Nov, 2019, Seoul, Korea (South). IEEE, c2019: 5663-5672.

[18] MIN X, ZHAI G, GU K, et al. Objective quality evaluation of dehazed images[J]. IEEE transactions on intelligent transportation systems, 2019, 20(8): 2879-2892.

[19] MIN X, ZHAI G, GU K, et al. Blind image quality estimation via distortion aggravation[J]. IEEE transactions on broadcasting, 2018, 64(2): 508-517.

[20] SUN W, MIN X, LU W, et al. A deep learning based no-reference quality assessment model for ugc videos[C]//Proceedings of the 30th ACM International Conference on Multimedia, October 10-14, 2022, Lisboa Portugal. New York: Association for Computing Machinery, c2022: 856-865.

[21] SUN W, MIN X, TU D, et al. Blind quality assessment for in-the-wild images via hierarchical feature fusion and iterative mixed database training[J]. arxiv preprint arxiv:2105.14550, 2021.

[22] MIN X, ZHAI G, ZHOU J, et al. Study of subjective and objective quality assessment of audio-visual signals[J]. IEEE transactions on image processing, 2020, 29: 6054-6068.

[23] MIN X, ZHAI G, ZHOU J, et al. A multimodal saliency model for videos with high audio-visual correspondence[J]. IEEE transactions on image processing, 2020, 29: 3805-3819.

[24] MIN X, ZHAI G, HU C, et al. Fixation prediction through multimodal analysis[C]//Visual Communications and Image Processing, December 13-16, 2015, Singapore. IEEE, c2015: 1-4.

[25] MIN X, ZHOU J, ZHAI G, et al. A metric for light field reconstruction, compression, and display quality evaluation[J]. IEEE transactions on image processing, 2020, 29: 3790-3804.

[26] LI H, ZHU F, QIU J. CG-DIQA: No-reference document image quality assessment based on character gradient[C]//The 24th International Conference on Pattern Recognition, August 20-24, 2018, Beijing, China. IEEE, c2018: 3622-3626.

[27] ALAEI A. A new document image quality assessment method based on hast derivations[C]// International Conference on Document Analysis and Recognition, September 20-25, 2019,

Sydney, Australia. IEEE, c2019: 1244-1249.

[28] SHAHKOLAEI A, BEGHDADI A, CHERIET M. Blind quality assessment metric and degradation classification for degraded document images[J]. Signal processing: image communication, 2019, 76: 11-21.

[29] ZHU H, LI L, WU J, et al. MetaIQA: deep meta-learning for no-reference image quality assessment[C]//IEEE/CVF Conference on Computer Vision and Pattern Recognition, June 13-19, 2020, Seattle, WA, USA. IEEE, c2020: 14131-14140.

[30] SU S, YAN Q, ZHU Y, et al. Blindly assess image quality in the wild guided by a self-adaptive hyper network[C]//IEEE/CVF Conference on Computer Vision and Pattern Recognition, June 13-19, 2020, Seattle, WA, USA. IEEE, c2020: 3664-3673.

[31] YOU J, KORHONEN J. Transformer for image quality assessment[C]//IEEE International Conference on Image Processing, September 19-22, Anchorage, AK, USA. IEEE, c2021: 1389-1393.

[32] MIN X, MA K, GU K, et al. Unified blind quality assessment of compressed natural, graphic, and screen content images[J]. IEEE transactions on image processing, 2017, 26(11): 5462-5474.

[33] VIRTANEN T, NUUTINEN M, VAAHTERANOKSA M, et al. CID2013: a database for evaluating no-reference image quality assessment algorithms[J]. IEEE transactions on image processing, 2015, 24(1): 390-402.

[34] GHADIYARAM D, BOVIK A C. Massive online crowdsourced study of subjective and objective picture quality[J]. IEEE transactions on image processing, 2016, 1(25): 372-387.

[35] HOSU V, LIN H, SZIRANYI T, et al. KonIQ-10k: an ecologically valid database for deep learning of blind image quality assessment[J]. arxiv preprint arxiv:1910.06180, 2019.

[36] MERRELL P, SCHKUFZA E, LI Z, et al. Interactive furniture layout using interior design guidelines[J]. ACM transactions on graphics, 2011, 30(4): 1-10.

[37] QI S, ZHU Y, HUANG S, et al. Human-centric indoor scene synthesis using stochastic grammar[C]//IEEE Conference on Computer Vision and Pattern Recognition, June 18-23, 2018, Salt Lake City, UT, USA. IEEE, c2018: 5899-5908.

[38] ZHANG S K, XIE W Y, ZHANG S H. Geometry-based layout generation with hyper-relations among objects[J]. Graphical models, 2021, 116: 101104.

[39] ZHANG S, HAN Z, LAI Y K, et al. Stylistic scene enhancement GAN: mixed stylistic enhancement generation for 3D indoor scenes[J]. The visual computer, 2019, 35: 1157-1169.

[40] LI M, PATIL A G, XU K, et al. Grains: Generative recursive autoencoders for indoor scenes[J]. ACM transactions on graphics, 2019, 38(2): 1-16.

[41] YU L F, YEUNG S K, TANG C K, et al. Make it home: automatic optimization of furniture arrangement[J]. ACM transactions on graphics, 2011, 30(4): 1-12.

[42] FISHER M, RITCHIE D, SAVVA M, et al. Example-based synthesis of 3D object arrangements[J]. ACM transactions on graphics, 2012, 31(6): 1-11.

[43] HUA B S, PHAM Q H, NGUYEN D T, et al. Scenenn: a scene meshes dataset with annotations[C]//International Conference on 3D Vision (3DV), October 25-28, 2016, Stanford, CA, USA. IEEE, c2016: 92-101.

[44] LIU Y, XUE F, HUANG H. Urbanscene3d: a large scale urban scene dataset and simulator[J]. arxiv preprint arxiv: 2107.04286, 2021, 2(3).

[45] ADAMS W, ELDER J, GRAF E, et al. Perception of 3D structure and natural scene statistics: the southampton-york natural scenes (SYNS) dataset[J]. Journal of vision, 2015, 15(12): 726-726.

[46] GARCIA A, MARTINEZ-GONZALEZ P, OPREA S, et al. The robotrix: an extremely photorealistic and very-large-scale indoor dataset of sequences with robot trajectories and interactions[C]//IEEE/RSJ International Conference on Intelligent Robots and Systems. IEEE, 2018: 6790-6797.

[47] DAI A, CHANG A X, SAVVA M, et al. ScanNet: richly-annotated 3D reconstructions of indoor scenes[C]//IEEE Conference on Computer Vision and Pattern Recognition, July 21-26, 2017, Honolulu, HI, USA. 2017: 2432-2443.

[48] LI W, SAEEDI S, MCCORMAC J, et al. InteriorNet: mega-scale multi-sensor photo-realistic indoor scenes dataset[J]. arXiv:1809.00716, 2018.

[49] WALD J, AVETISYAN A, NAVAB N, et al. RIO: 3D object instance re-localization in changing indoor environments[C]//IEEE/CVF International Conference on Computer Vision, October 27-November 02, 2019, Seoul, Korea (South). IEEE, c2019: 7657-7666.

[50] VELICKOVIC P, CUCURULL G, CASANOVA A, et al. Graph attention networks[J]. arXiv:1710.10903, 2017.

[51] KIPF T N, WELLING M. Semi-supervised classification with graph convolutional networks[J]. arXiv:1609.02907, 2016.

[52] LEE J, LEE I, KANG J. Self-attention graph pooling[C]//International Conference on Machine Learning, June 9-15, 2019, Long Beach, California. PMLR, c2019: 3734-3743.

[53] ZHANG Z, BU J, ESTER M, et al. Hierarchical graph pooling with structure learning[J]. arXiv:1911.05954, 2019.

[54] ZENG J, XIE P. Contrastive self-supervised learning for graph classification[J]. arXiv:2009.05923, 2020.

[55] DAI A, CHANG A X, SAVVA M, et al. ScanNet: richly-annotated 3D reconstructions of indoor scenes[C]//IEEE Conference on Computer Vision and Pattern Recognition, July 21-26, 2017, Honolulu, HI, USA. IEEE Computer Society, c2017: 2432-2443.

第四部分

人工智能

Interpretability Diversity for Decision-Tree-Initialized Dendritic Neuron Model Ensemble*

1. Introduction

In recent decades, ensemble learning gained extensive attraction in machine learning because it can solve practical application problems well. Dasarathy and Sheela first proposed the idea of ensemble learning. It is a machine-learning paradigm where multiple learners are trained to solve the same problem. In contrast to ordinary machine learning approaches which try to learn one hypothesis from training data, ensemble methods try to construct a set of hypotheses and combine them to use. An ensemble contains a number of learners that are called base learners. In this article, we deal with a classification problem, a base learner is also known as a single classifier or weak learner. Ensemble learning indeed agrees with an old oriental proverb stating that "three humble shoe-makers' brainstorming makes a great leader." There are three fundamental reasons for making ensemble learning superior to single learning

1) Statistical: A training set cannot provide adequate information for selecting the best learner.
2) Computational: A search procedure of learning algorithms tends to be imperfect.
3) Representational: Any hypothesis space cannot represent the true target function being selecting.

It is generally believed that high sensitivity, great diversity, and efficient learning ability must be considered in the selection of base learners for an ensemble. The base learners can be a decision tree (DT), neural network, or other kinds of machine learners. Based on dendritic mechanisms from the Koch-Poggio-Torre model, Tang et al. proposed

---

* The paper was originally published in *IEEE Transactions on Neural Networks and Learning Systems*, 2023, and has since been revised with new information. It was co-authored by Xudong Luo, Long Ye, Xiaolan Liu, Xiaohao Wen, Mengchu Zhou, and Qin Zhang.

a dendritic neuron model (DNM) in which they introduced multiplications into a neuron model. It has a powerful ability to deal with nonlinear problems. It has an excellent effect on classification, approximation, and prediction demonstrated in [7], [8], [9], [10], [11], [12], [13], [14], [15], [16], [17], [18], [19], [20], [21], [22], [23], [24], and [25]. Luo et al. proposed a DT-initialized DNM (DDNM) with a convincing performance, which can reduce the number of dendrites in DNM while improving training efficiency without affecting accuracy. It is reasonable to believe that both DNM and DDNM are excellent base learners in ensemble learning. Diversity, which refers to the difference among base learners, plays a critical role in ensemble learning. Understanding diversity in ensemble learning can be achieved by decomposing the generalization error. There are two famous error decomposition schemes for ensemble methods, namely the error-ambiguity decomposition and the bias-variance-covariance one. There are some effective heuristic mechanisms for diversity generation in ensemble learning, such as manipulating data samples, input features, learning parameters, and output representations. Improving ensemble learning performance requires consideration of both base learners' performance and diversity. Ultimately, the success of ensemble learning usually lies in achieving an excellent tradeoff between base learners' individual performance and diversity.

Interpretability is indeed an essentially important concept in machine learning , i.e., the transparency of learned knowledge and the ability to explain its reasoning process. Murdoch et al. defined interpretable machine learning. To achieve high interpretability, Wu et al. proposed regional tree regularization that promoted a deep model to be well-approximated. Townsend et al. proposed that deep neural networks (DNNs) could be described in terms of long-established taxonomies and frameworks presented in early neural-symbolic literature. In the reference [41], some existing methods and potential research directions about extracting rules from multilayer perceptron and DNN are discussed. DT is a machine learning technique with high interpretability; while a neural network is more like a black-box technique because the learned knowledge is implicitly encoded in many connections. Because of multiplication at its dendrite layer and summation at its membrane layer, DNM can be substituted by a logic circuit. Then DNM is interpretable because the former uses logical rules. Because of a mapping between DT and DNM, DDNM uses DT to initiate DNM, making it highly interpretable.

According to the reference [43], a deficiency of existing diversity measures lies in the fact that they consider only the diversity of results, i.e., classifiers' results when making predictions, while neglecting that classifiers may be potentially different even when they make identical predictions. We offer the following perspectives: for interpretable machine learning algorithms, diversity is interpretability diversity in their interpretability is easy-to-understand and perhaps most useful in ensemble learning. Take rule-based machine learning as an example. Two base learners with the same rules have the same interpretability, and two base learners with different rules have different one. It is possible to measure this difference from the level of interpretability, thus making the difference and its measurement interpretable.

This work main contributes to the field of ensemble learning in the following aspects.

1) The concept of base learners' interpretability diversity (LID) is proposed as an important basis for constructing an ensemble, thus providing a novel perspective for the research of ensemble learning diversity.
2) We propose an LID measurement method that quantifies the difference between two DDNMs from an interpretability perspective. By selecting DDNMs with high interpretability diversity, we can achieve excellent ensemble learning effects.
3) Unlike traditional diversity measures that are based on training results, the proposed LID measure allows us to assess the interpretability diversity among DDNMs before training, which enables us to reduce the number of DDNMs that need to be trained.
4) We provide extensive experiments to compare the advantages and disadvantages of the LID-based ensemble selection method proposed in this article with one of the state-of-the-art methods, margin distance minimization (MDM), in terms of both the performance and the compression scale of the DDNM ensemble.

Section II reviews previous work on diversity measures and the selection of ensemble learning. Section III introduces the fundamental structures and functions of DNM and DDNM and defines interpretability diversity. Section IV presents random-forest-initialized DNM (RDNM). Section V shows experimental results for RDNM and its peers on seven different problems. Section VI concludes this article.

2. Related Work

In this section, we will review the existing research on ensemble methods, diversity measures, and ensemble selection methods, which are crucial for improving the performance and generalization of ensemble learning.

2.1 Ensemble Methods

There are various fashionable ensemble methods, but in this section, we will focus on introducing two representative methods.

1) The first method is Bagging (short for bootstrap aggregating), where there is no strong dependence between base learners. A series of base learners can be generated in parallel. It is worth mentioning that random forest (RF) is considered one of the most powerful ensemble methods to date, based on the DT classifier model and Bagging.
2) The second method is Boosting, where there is a strong dependency between base learners. A series of base learners need to be generated serially. Boosting is a family of algorithms with numerous variants, and the most famous algorithm is AdaBoost.

Therefore, we conduct our ensemble studies within these two methods to ensure a fair and valid comparison.

2.2 Diversity Measure

Plenty of diversity measures have been independently proposed, including four averaged pair-wise measures: Q statistic, correlation, disagreement, and double fault. Additionally, six non-pair-wise measures exist: entropy of votes, difficulty index, Kohavi-Wolpert variance, measurement of interrater agreement K, generalized diversity, and percentage correct diversity measure. Different from the diversity previously proposed, information-theoretic diversity might provide a promising direction, as noted in the reference [33]. Some researchers used negative correlation learning to improve the diversity among base learners. Yu et al. proposed a diversity regularized machine (DRM) in a mathematical programming framework, which efficiently generates an ensemble of diverse support vector machines (SVMs) and can significantly improve generalization ability. Sun and Zhou proposed structural diversity as a complement to output diversity for the difference of DTs according to the difference of structure in different DTs. A structure is a concrete form of interpretability. Motivated by the success of using structural diversity in ensemble learning, we believe that interpretability is an excellent way to consider diversity in ensemble learning.

2.3 Ensemble Selection Methods

Ensemble selection methods, also known as ensemble pruning methods, are designed to reduce the complexity of ensemble models. Zhou et al. reveal that many could be better than all. The aim of these methods is to identify a subset of ensemble members that exhibit performance levels equivalent to or surpassing that of the original ensemble. The primary objective is to enhance the efficiency and generalizability of the model while preserving its predictive performance. Over the past decade, many effective ensemble selection methods have been proposed, which can generally be classified into three categories: ordering-based, clustering-based, and optimization-based.

Ordering-based selection methods attempt to rank base learners according to some criterion and only include the top-ranked learners in the final ensemble. The main difference among these selection methods is the diversity measure used for ranking. For instance, Margineantu and Dietterich proposed Kappa selection, which sorts base learners in ascending order based on their Kappa measures, and aggregates those with smaller Kappa measures. Martínez-Muñoz and Suárez and Martínez-Muñoz et al. proposed ordering-based methods, including the complementarity measure and MDM, which found that the selected base learners are complementary. Guo and Boukir proposed Margin-based Ordered AGgregation for ensemble pruning, which selects base learners with larger margin-based criteria to constitute the final ensemble. Numerous studies have demonstrated that ordering-based selection outperforms the original ensemble. For outstanding representative ordering-based ensemble selection, MDM can also be combined with other heuristic algorithms to improve the ensemble effect. Zhu et al. proposed an improved discrete artificial fish swarm algorithm combined with MDM for ensemble pruning (IDAFMEP). First, MDM is used to preselect base learners in the constructed initial pool. Second, the proposed improved discrete artificial fish swarm algorithm (IDAFSA) is used to further reduce the size of the ensemble. We introduce an LID measure and propose a method that ranks base learners in descending order of LID. The approach then aggregates the base learners with high interpretability diversity. And it is proven that LID is a promising integration perspective by comparing it with MDM.

3. DNM and DDNM

3.1 Dendritic Neuron Model

According to the characteristics of biomimetic neurons, DNM is built to comprise four layers: synaptic, dendrite, membrane, and soma ones. The synaptic layer uses a sigmoid function on the received input and converts the external signal into a neuron signal. The interaction among synaptic signals occurs on each dendrite layer, and the dendrite layer performs a multiplication function on the outputs of the synaptic layer. The membrane layer adds all dendrite layer outputs. The soma layer calculates the output of the membrane via another sigmoid function and generates its final result or transmits it to the next neuron. In [8], [9], and [26], its detailed structure and algorithm are described.

3.2 DT-Initialized DNM

Comparing DT and DNM, it is evident that their advantages and drawbacks are complementary. For instance, people understand the knowledge representation of DT better than DNM. DT has some trouble dealing with noise in a training set, but this is not the case for DNM. DT learns quickly and DNM learns relatively slowly. DDNM combines the advantages of DT and DNM. First, we build a DT which is used to initialize DNM, then train it.

In general, $D = \{(x_1, y_1), (x_2, y_2), \ldots, (x_m, y_m)\}$ represents an m-sample training set, and each sample is described by d attributes. The attribute set is $A = \{a_1, a_2, \ldots, a_d\}$. Each sample $x_i = (x_{i1}; x_{i2}; \ldots, x_{id})$ is a vector in d sample space $X = (0, 1)^d$. $x_i \in X$, x_{ij} is the value of x_i on the j th attribute. We use (x_i, y_i) to represent the i th example, where $y_i \in Y$ is the label of sample x_i. For binary classification problems, $Y = \{0, 1\}$. DDNM includes the following steps.

1) Use a k–fold cross validation method [70] to generate a training set $S = \{(x_1, y_1), (x_2, y_2), \ldots, (x_m, y_m)\}$ from a dataset.
2) Use the min-max normalization method to generate a new training set $D = \{(x_i, y_i)_{i=0}^{m} \mid x_i \in X, y_i \in Y\}$ from S.
3) Generate the corresponding DT structure by a learning algorithm, e.g., CART which is commonly used for classification. It avoids overfitting by setting the minimum number of samples for each leaf node (MSL).
4) Transform all paths of DT into dendrite layers in DNM.

Algorithm 1 outlines the process of initializing DNM for a DT. Since DNM initialization is not random, θ_{ij} is generated according to the threshold μ_{ij} of the nonleaf node.

3.3 LID Measure

A DT comprises a set of nodes and branches. Each nonleaf node is associated with an attribute to split and each leaf node is associated with a class label. The rules of a DT are like DNM. It makes the combination of DT and DNM easier by establishing a mapping relationship between them. The corresponding relation between DT and DNM is shown in Table I. This correspondence is a critical link between DT and DNM. It provides powerful support for DNM research by using the idea of DT. The four connection states of DNM represent four logical rules. According to these four connection states, it is easy to give the difference in logic rules between two DNMs, and this difference is interpretable.

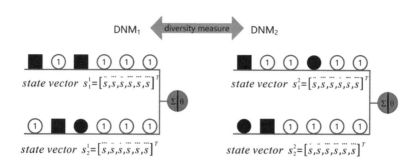

$$S^1 = \left[s_1^1, s_2^1\right] \qquad\qquad S^2 = \left[s_1^2, s_2^2\right]$$

$$\varphi(s^1, s^2) = \min\left\{\phi_L\left(s_1^1, s_1^2\right), \phi_L\left(s_1^1, s_1^2\right)\right\} + \min\left\{\phi_L\left(s_2^1, s_1^2\right), \phi_L\left(s_2^1, s_2^2\right)\right\}$$

$$= \min(2, 3) + \min(4, 2)$$

$$= 4$$

Fig. 1 Two DNMs distance calculation case details

Algorithm 1 DT to Neural Network Mapping

Input: DT paths: $\Gamma = \{\gamma_1, \gamma_2, \ldots, \gamma_j, \ldots, \gamma_N\}$;

$\gamma_j = \left(a_1 \langle \mu_{1j}, a_2 \rangle = \mu_{2j}, \ldots, a_i >= \mu_{ij}, \ldots, a_M <= \mu_{Mj}\right)$, $\mu_{ij} \in (0, 1)$

Initialize $w_{ij} = 1.0, i = 1, \ldots, M, j = 1, \ldots, N$

Initialize $\theta_{ij} = -0.5, i = 1, \ldots, M, j = 1, \ldots, N$

for $\gamma_j \in \Gamma$ **do**

  for $a_i \in \gamma_j$ **do**

    if $a_i >= \mu_{ij}$ **then**

      $w_{ij} = 1.0$ and $\theta_{ij} = \mu_{ij}$

    else if $a_i < \mu_{ij}$ **then**

      $w_{ij} = -1.0$ and $\theta_{ij} = -\mu_{ij}$

    end if

    $i = i + 1$

  end for

  $j = j + 1$

end for

Output: $w_{ij}; \theta_{ij}$

Table I Corresponding relation between DT and DNM

DT	DNM
Non-leaf Nodes of DT	Synaptic layers
Paths of DT	Dendrite layers
Leaf Nodes	Soma layer & Membrane layer

Therefore, we propose LID to measure the diversity of ensemble DNM. It is a pair-wise DNM diversity measure method and is defined as calculating the state similarity between their dendritic layers. Its detailed process is illustrated in Fig. 1. First, we use the 0/1 function to calculate the state difference in the two synapses, which can be defined as

$$\phi(p,q) = \begin{cases} 0: & if \quad p = q \\ 1: & p \neq q. \end{cases} \tag{1}$$

$S = [s_1, s_2, ..., s_j, ..., s_M]$ stands for DNMs state matrix. The state vector of the j th dendrite layer is characterized by $s_j = [s_{1,j}, s_{2,j}, ..., s_{i,j}, ..., s_{N,j}]^T$. $S_{i,j}$ represents the state of the i th synapse in the j th dendrite layer. DNM has four connection states and each is represented by four specific symbols: Direct connection ($\dot{s}$), Inverse connection ($\ddot{s}$), Constant 1 connection ($\dddot{s}$), and Constant 0 connection ($\ddddot{s}$), therefore, $s_{ij} \in \{\dot{s}, \ddot{s}, \dddot{s}, \ddddot{s}\}$. The difference between the two dendrite layers can be calculated by

$$\phi_L(s_{j1}, s_{j2}) = \sum_{i=1}^{N} \phi(s_{i,j1}, s_{i,j2}) \tag{2}$$

The difference between the two DNMs ϕ can be expressed as

$$\psi(s_{j1}^1, s^2) = \min \phi_L(s_{j1}^1, s_{j2}^2), \quad j2 \in \{1, 2, ..., M2\} \tag{3}$$

$$\varphi(s^1, s^2) = \sum_{j=1}^{M1} \psi(s_{j1}^1, s^2) \tag{4}$$

Note that this measure is symmetric for the two DNMs. The larger the value of φ, the more different their interpretability. The overall diversity of an ensemble with over two DNMs is defined as the average of all pair-wise measures. The measure could be normalized by the maximum pair-wise value among the ensemble. If the maximum pair-wise value is φ_{max}, then $\varphi_{1,2} = \varphi(s^1, s^2)/\varphi_{max}$. If there are H DNMs for an ensemble learning, the matrix of pair-wise DNMs diversity measures is shown in

$$\Gamma = \begin{bmatrix} \varphi_{1,1} & \varphi_{1,2} & \cdots & \varphi_{1,H-1} & \varphi_{1,H} \\ \varphi_{2,1} & \varphi_{2,2} & \cdots & \varphi_{2,H-1} & \varphi_{2,H} \\ \vdots & \vdots & \cdots & \cdots & \cdots \\ \varphi_{H-1,1} & \varphi_{H-1,2} & \cdots & \varphi_{H-1,H-1} & \varphi_{H-1,H} \\ \varphi_{H,1} & \varphi_{H,2} & \cdots & \varphi_{H,H-1} & \varphi_{H,H} \end{bmatrix} \tag{5}$$

The average distance between the v th DNM and others is denoted by ξ_v.

$$\xi_v = \frac{1}{H-1} \sum_{h=1 \, and \, h \neq v}^{H} \varphi_{v,h}. \tag{6}$$

The larger the value of ξ_v, the greater the interpretability difference from other DNMs. Assuming $\xi = [\xi_1 ... \xi_v ... \xi_H]$ represents distance vector of all DNMs, we can sort it by magnitude to obtain a new vector $\hat{\xi} = [\hat{\xi}_1 ... \hat{\xi}_v ... \hat{\xi}_H]$ where $\hat{\xi}$ contains the same elements as ξ but in a sorted order. We can select some DNMs with large ξ_v values for the ensemble. When we select H' DNMs from a group of H DNMs for the ensemble, we can extract the corresponding H' values from $\hat{\xi}$ and create the vector $\hat{\xi}' = [\hat{\xi}_1 \cdots \hat{\xi}_{h'} \cdots \hat{\xi}_{H'}]$. The scalar κ is used to represent the diversity of the ensemble, and is defined as follows:

$$\kappa = \sum_{h'=1}^{\hat{\xi}_{H'}} \hat{\xi}_{h'} \cdot \Omega D_h T_h \qquad (7)$$

Since the initial weights of the original DNM are randomly generated, it becomes difficult to measure the difference between them using LID. Because of the mapping between DT and DNM, DDNM has high interpretability. DT initializes the DNM so that LID can measure the difference between two DNMs before training.

4. RF-Initialized DNMS

DT is often considered weaker than other machine learning algorithms, such as BPNN, DNM, and DDNM. However, DTs are highly interpretable, and their different interpretations ensure an interpretable distance between them. Assembling many DTs can result in excellent results. DDNM uses DT to initiate DNM, making it highly interpretable. However, DDNM, like other single learners, has performance constraints. Therefore, using multiple DTs with bagging (DTs + bagging) to initialize multiple DNMs is expected to enhance the performance of the DNM ensemble. This approach is referred to as DDNM + bagging. RF is an excellent representative of DTs + bagging, and it often has higher prediction accuracy than traditional DTs + bagging methods. Therefore, using RDNM is likely to further enhance the performance of the DNM ensemble. Additionally, a batch of DDNMs with the largest difference was selected for training according to the interpretability diversity measure.

The flowchart illustrating the RDNM combined with LID (RDNM + LID) approach is presented in Fig. 2 and Algorithm 2, which consists of five stages.

1) Constructing the RF, where DTs generated within the RF exhibit significant diversity.
2) Mapping the RF to DNMs, which endows each DNM with unique logical rules and potentially places it near a local or global optimum solution.
3) Computing the distance vector ξ, which calculates the difference of logical rules between each DNM and the other DNMs using LID.
4) Sorting and selecting the ensemble DNMs, which involve selecting DNMs that are significantly different from other DNMs for the ensemble to improve the diversity.

5) Training the selected H' DNMs.

Algorithm 2 RDNMs Combined With LID

Input: Training set $D = \{(x_1,y_1),(x_2,y_2),\ldots,(x_m,y_m)\}$

Number of base learners H.

Number of attributes d.

for h=1 to H **do**

  D_h = Bootstrap(D); % Generate a bootstrap sample % from D

  T_h = BuildRandomTreeModel ($D_h, \tilde{d}$) %Select $\tilde{d}$ % attributes at random from the d attributes at each node

  Initialize a DNM D_h with T_h, according to Algorithm 1.

end for

Form a DNMs set $\Omega = \{D_1, D_2, \ldots, D_H\}$.

According to $\hat{\xi}$, a new DNMs ordered set $\bar{\Omega} = \{\tilde{D}_1, \tilde{D}_2, \ldots, \tilde{D}_H\}$ can be obtained by sorting Ω.

We select H' DNMs and form a new DNMs set $\tilde{\Omega} = \{\tilde{D}_1, \tilde{D}_2, \ldots, \tilde{D}_H\}$

for $h' = 1$ to H' **do**

  Use training set $D_{h'}$ to train $\tilde{D}_{h'}$

end for

Output: $F(x) = \text{undersety} \in Y \arg\max \sum_{h'=1}^{H'} \delta(y = \tilde{D}_{h'}(x))$ % The value of is 1 if a is true and 0 otherwise.

The computational complexity of RDNM + LID is analyzed as follows: The algorithmic complexity of Stage 1 can be obtained from [29]. By analyzing Algorithm 1, we can obtain the complexity of mapping DT to DNM, which allows us to derive the complexity of mapping RF to DNMs. Equations (1)-(6) can be used to determine the algorithmic complexity of Stage 3. In Stage 4, quicksort serves as the sorting algorithm, and its complexity has been analyzed in [72]. The results are summarized in Table II.

Table II Results of the analysis of the algorithmic complexity (including time and space complexity) of each stage of RDNM + LID

Stages	Time Complexity	Space Complexity
1)	$O(m*\log(m)*N*H)$	$O(p*H)$
2)	$O(N*M*H)$	$O(N*M)$
3)	$O(N*M^2*H^2)$	$O(N*M*H^2)$
4)	$O(H*\log(H))$	$O(H)$
5)	$H'*O(\bar{T})$	$O(\bar{S})$

1 m: The number of samples

2 N: The number of features

3 M: The number of dendritic layers

4 H: The total number of DDNMs

5 H': The number of DDNMs selected for ensemble

6 $O(\bar{T})$: The time complexity of training DDNM

7 $O(\bar{S})$: The space complexity of training DDNM

5. Experiments

5.1 Datasets

In this section, we perform experiments on a number of well-known benchmark datasets for pattern classification problems: **Glass, Vertebral Column, Blood Transfusion, Parkinsons, Ionosphere, Heart**, and **Wisconsin Breast-Cancer** from the University of California at Irvine Machine Learning Repository. Table III describes these datasets.

Table III Dataset Description

Dataset	Samples	Attributes	Classes
Glass	214	9	2
Vertebral Column	310	6	2
Blood Transfusion	748	4	2
Ionosphere	351	34	2
Heart	270	13	2
Parkinsons	195	22	2
Wisconsin Breast-Cancer	699	9	2

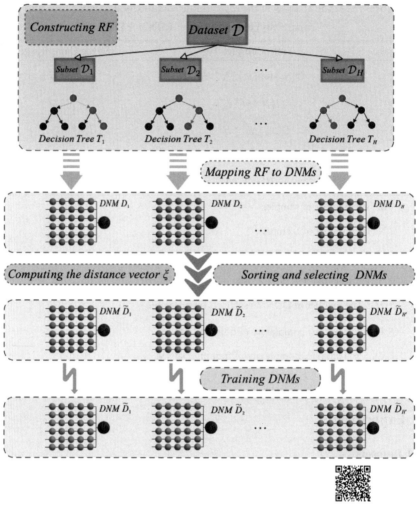

Scan the QR code to see colorful figures

Fig. 2 RDNMs combined with LID.

5.2 Experimental Condition and Evaluation Method

We evaluate the performance of the DDNM ensemble (including RDNM) by comparing it with other ensemble classification methods based on two ensemble learning frame-works (Bagging, Adaboost) and three base learners (DNM, BPNN, and DT). To assess the effectiveness of LID, we have incorporated ablation studies into the overall experiment and conducted two sets of ablation studies on seven datasets. Specifically, we have compared the performance of RDNM with and without LID, as well as DDNM + Bagging with and without LID. MDM is one of the best ensemble selection methods.

To further demonstrate the benefits of LID, we will also compare LID with MDM in the experiment. The number of base learners (H) in the full ensemble is set to 100. We traverse all possible values of the ensemble size H' (where $H' \leq H$) to obtain a more comprehensive observation of the advantages of LID.

In addition, for a fair comparison, the main parameters' settings and descriptions following the characteristics of each base learner are shown in Tables IV and V. There are three key parameters in both DNM and DDNM (i.e., the synaptic parameter (k_c) in the connection sigmoid function and two soma parameters (k_s and θ_s) in the output sigmoid function), and in DT (i.e., the MSL and $\tilde{d}$). The grid search method is used to acquire a reasonable (likely the best) selection of five parameters. These parameters ensure the methods achieve their relatively optimal experimental results. In evaluating the performance of a method, testing is a significant step. A K-fold cross-validation is a classical approach to evaluate it. In such validation, the original samples are randomly and averagely partitioned into K subsamples. Each subsample becomes a testing set once, and the rest of $K-1$ subsamples are used as a training set. Generally, $K = 5$. To ensure the independence and effectiveness of the experiments, we randomly repeat different partitions six times. Hence, the experiments are performed for $6K$ times with these subsamples. The average performance of the experiments is considered as their final performance. The advantage of this method is that all observations are used for both training and testing so that overfitting and lack of learning are avoided to generate persuasive results.

Table IV Main parameters used in base classification mehods

Method	Parameter	Value
DT	Algorithm	CART
DNM	Learning rate Activation function Learning algorithm	0.01 Sigmoid Adam
DDNM	Learning rate Activation function DT algorithm Learning algorithm	0.01 Sigmoid CART Adam
BPNN	Learning rate Activation function Learning algorithm	0.01 Sigmoid Adam

Table V Reasonable selection of parameters of various algorithms for seven tested problems, respectively

	DNM and DDNM			DT	
Problem	k_c	K_s	θ_s	MSL	MF
Glass	6	12	0.7	1	1
Vertebral Column	5	15	0.7	13	1
Blood Transfusion	16	10	0.6	11	1
Ionosphere	5	15	0.8	1	1
Heart	10	17	0.2	1	1
Parkinsons	4	6	0.6	3	1
Wisconsin Breast-Cancer	9	10	0.5	3	1

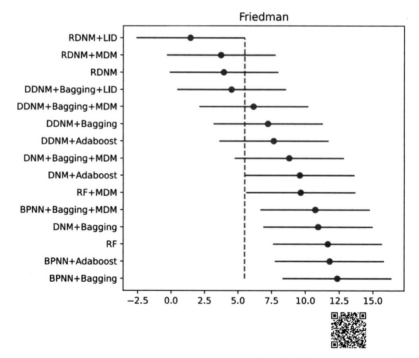

Scan the QR code to see colorful figures

Fig. 3 Comparison of all ensemble classification methods against each other with the Nemenyi test.

Interpretability Diversity for Decision-Tree-Initialized Dendritic Neuron Model Ensemble

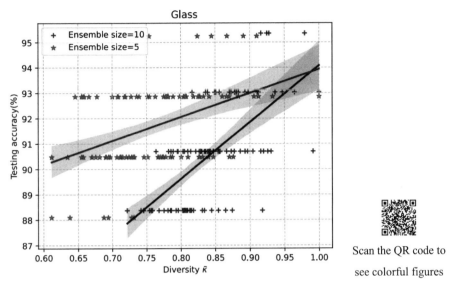

Fig. 4 Relationship graph between ensemble accuracy and diversity. Each point represents an ensemble. A linear regression is performed on the 100 points obtained from each set of experiments. The green and blue points correspond to the two sets of experiments with ensemble sizes of 5 and 10, respectively. Both sets exhibit a positive correlation between diversity and accuracy. Here, $\bar{\kappa}$ represents the normalized value of $\mathcal{K}$.

Statistical test is implemented for measuring and assessing whether there are significant differences between RDNM + LID and the other methods. Wilcoxon signed ranks test is used to detect if there are significant differences between two classification methods. In Tables VI and XII, $\bar{W}$ indicates the average number of adjusted weights in a base learner. A $p-value$ smaller than 0.05 signifies that RDNM + LID performs significantly better than others. Average accuracy ($\bar{A}$) and $F1-score$ ($F1-score$ is called F-measure in other studies) are adopted as the crucial evaluation metrics. $\bar{H}$ represents the ensemble size (H') with the highest test accuracy. Observing the four metrics ($\bar{H}$, train accuracy, test accuracy, and $F1-score$) in Tables V and XII, it can be found that in 21 out of 28 cases, RDNM + LID performs better than others, suggesting that the proposed LID is a promising diversity measurement method. Furthermore, the results can be summarized as follows.

1) DDNM ensemble has a better performance than others, e.g., BPNN and DNM. It also often requires a smaller number of adjusted weights and fewer iterations.
2) When the number of attributes is large, the DNM ensemble may experience difficulty in convergence as shown in Tables IX and X regarding **Parkinsons** and **Ionosphere**. However, the DDNM ensemble is able to fix this issue.
3) LID can improve the performance and significantly reduce the ensemble size

of the DDNM ensemble.

4) Compared to MDM, LID reduces test errors and the ensemble size of the DDNM ensemble even more.

5) RDNM + LID is an excellent representative of the DDNM ensemble.

Considering the fact that our experiments examine quintessential problems of comparing multiple ensemble classification methods over multiple datasets, we decide to use the Friedman aligned-rank test together with its post-hoc test (i.e., Nemenyi test) as our statistical test methods. We compare $k_d = 15$ classification methods on $N_d = 7$ datasets. Exact critical value $F_{\alpha=0.05} = 1.811$ is computed. According to Table XIII, we calculate the Friedman statistics τ_F, which is $8.345 > F_{\alpha=0.05}$. Hence, the null-hypothesis (15 classification methods perform alike) is rejected.

Table VI Experimental results for glass

Method	$\bar{H}$	$\bar{W}$	Iterations	Train $\bar{A}(\%)$	Test $\bar{A}(\%)$	F1-score	p-Value
RF	56	-	-	97.5	94.1	0.867	0.000002
BPNN + Bagging	25	145	1000	96.0	93.0	0.840	0.000002
BPNN + Adaboost	**14**	145	1000	96.2	92.9	0.828	0.000002
DNM + Bagging	20	144	200	98.3	94.7	0.874	0.002336
DNM + Adaboost	16	144	200	**98.4**	94.8	0.881	0.012258
DDNM + Bagging	31	108	200	98.2	94.8	0.881	0.018897
DDNM + Adaboost	18	108	200	97.8	94.4	0.875	0.000002
RDNM	23	108	200	98.3	**95.0**	**0.888**	0.082213
RF + MDM	31	-	-	98.5	94.3	0.870	0.000002
BPNN + Bagging + MDM	12	145	1000	97.4	93.1	0.841	0.000002
DNM + Bagging + MDM	15	144	200	98.5	94.8	0.882	0.035578
DDNM + Bagging + MDM	24	108	200	**98.6**	94.8	0.887	0.021147
DDNM + Bagging + LID	8	108	200	98.4	94.7	0.880	0.004511
RDNM + MDM	22	108	200	**98.6**	95.1	0.891	0.354711
RDNM + LID	19	108	200	98.4	**95.1**	**0.891**	-

Table VII Experimental results for vertebral column

				Train	Test		
Method	$\bar{H}$	$\bar{W}$	Iterations	$\bar{A}(\%)$	$\bar{A}(\%)$	F1-score	p-Value
RF	23	-	-	98.9	85.1	0.895	0.000002
BPNN + Bagging	60	70	1000	85.9	85.1	0.892	0.000002
BPNN + Adaboost	13	70	1000	85.8	85.1	0.889	0.000002
DNM + Bagging	19	70	200	87.3	85.7	0.894	0.038874
DNM + Adaboost	18	70	200	87.2	85.9	0.891	0.050121
DDNM + Bagging	21	35	140	87.3	85.6	0.892	0.033451
DDNM + Adaboost	8	35	140	87.3	85.4	0.892	0.008771
RDNM	29	35	140	87.8	86.1	0.897	0.106456
RF + MDM	17	-	-	99.1	85.3	0.896	0.000002
BPNN + Bagging + MDM	31	70	1000	86.2	85.4	0.893	0.002458
DNM + Bagging + MDM	21	70	200	87.5	86.0	0.893	0.054789
DDNM + Bagging + MDM	20	35	140	87.5	85.7	0.895	0.041254
DDNM + Bagging + LID	15	35	140	87.6	85.9	0.895	0.031142
RDNM + MDM	21	35	140	87.8	86.1	0.897	0.080578
RDNM + LID	7	35	140	87.8	86.3	0.898	-

Table VIII Experimental results for blood transfusion

				Train	Test		
Method	$\bar{H}$	$\bar{W}$	Iterations	$\bar{A}(\%)$	$\bar{A}(\%)$	F1-score	p-Value
RF	41	-	-	81.4	79.1	0.423	0.000006
BPNN + Bagging	29	49	1200	78.6	78.0	0.239	0.000002
BPNN + Adaboost	21	49	1200	80.4	78.9	0.264	0.000002
DNM + Bagging	36	48	1200	81.5	79.1	0.433	0.000002
DNM + Adaboost	13	48	1200	80.7	79.5	0.448	0.000002
DDNM + Bagging	31	32	1200	80.8	79.6	0.451	0.025542

Table VIII Continued

Method	$\bar{H}$	$\bar{W}$	Iterations	Train $\bar{A}(\%)$	Test $\bar{A}(\%)$	F1-score	p-Value
DDNM + Adaboost	32	32	1200	80.8	**79.8**	**0.459**	0.078021
RDNM	71	32	1200	80.7	79.7	0.447	0.026524
RF + MDM	32	-	-	81.3	79.3	0.450	0.000006
BPNN + Bagging + MDM	11	49	1200	79.6	78.4	0.340	0.000002
DNM + Bagging + MDM	25	48	1200	81.0	79.4	0.461	0.000002
DDNM + Bagging + MDM	39	32	1200	81.2	79.7	0.450	0.036354
DDNM + Bagging + LID	15	32	1200	81.3	**79.8**	0.455	0.057213
RDNM + MDM	39	32	1200	81	79.7	0.444	0.036587
RDNM + LID	**9**	32	1200	**81.6**	**79.8**	**0.462**	-

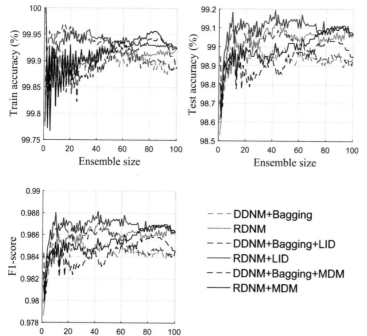

Scan the QR code to see colorful figures

Fig. 5 Influence of ensemble size on test accuracy of Wisconsin Breast-Cancer. Increasing ensemble size does not necessarily lead to better performance. However, LID can significantly improve DDNM ensemble and outperform MDM.

Interpretability Diversity for Decision-Tree-Initialized Dendritic Neuron Model Ensemble

Significant differences are detected. We check if RDNM + LID (the one with the lowest average ranking) is significantly better than others using the Nemenyi test. The results of this test can be visually represented with a simple diagram. Fig. 3 shows the results of the analysis of the data from Table XIII, the lowest (best) rank is RDNM + LID since we perceive the methods on the left side as better. We calculate the critical difference $C_D = 8.106$. The performance of the two ensemble classification methods is significantly different if their horizontal segments do not overlap. From the performance in Fig. 3, RDNM + LID is not significantly better than the others, but it is the best among them. Furthermore, we can confidently conclude that RDNM + LID outperforms almost all of them on these datasets.

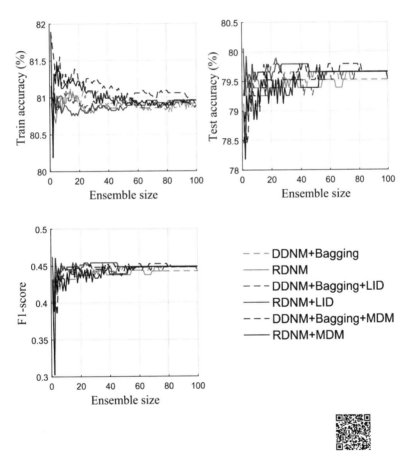

Scan the QR code to see colorful figures

Fig. 6 Influence of ensemble size on test accuracy of Blood Transfusion. Increasing ensemble size does not necessarily lead to better performance. However, LID can significantly improve DDNM ensemble and outperform MDM.

Table IX Experimental results for parkinsons

Method	$\bar{H}$	$\bar{W}$	Iterations	Train $\bar{A}(\%)$	Test $\bar{A}(\%)$	F1-score	p-Value
RF	37	-	-	99.9	90.2	0.936	0.000002
BPNN + Bagging	**10**	419	1000	93.8	98.1	0.925	0.000002
BPNN + Adaboost	15	419	1000	93.5	88.9	0.923	0.000002
DNM + Bagging	100	396	800	24.7	24.7	0	0.000002
DNM + Adaboost	100	396	800	32.7	26.7	0.1	0.000002
DDNM + Bagging	28	198	800	**99.7**	91.7	**0.942**	0.025542
DDNM + Adaboost	25	198	800	98.9	91.6	0.941	0.000211
RDNM	25	198	800	**99.7**	91.4	0.941	0.066587
RF + MDM	22	-	-	99.7	90.5	0.939	0.000002
BPNN + Bagging + MDM	10	419	1000	96.7	90.2	0.935	0.000002
DNM + Bagging + MDM	1	396	800	92.5	86.6	0.911	0.000002
DDNM + Bagging + MDM	23	198	800	99.5	91.7	0.94	0.036654
DDNM + Bagging + LID	19	198	800	99.8	**92.1**	**0.946**	0.344142
RDNM + MDM	23	198	800	99.5	91.4	0.941	0.056587
RDNM + LID	**8**	198	800	**99.9**	91.9	0.944	-

Table X Experimental results for ionosphere

Method	$\bar{H}$	$\bar{W}$	Iterations	Train $\bar{A}(\%)$	Test $\bar{A}(\%)$	F1-Score	p-Value
RF	46	-	-	98.5	92.6	0.945	0.000002
BPNN + Bagging	18	546	1000	98.4	93.1	0.948	0.000002
BPNN + Adaboost	12	546	1000	98.0	92.2	0.941	0.000002
DNM + Bagging	100	544	1000	36.0	36.0	0	0.000002
DNM + Adaboost	100	544	1000	36.0	36.0	0	0.000002

Table X Continued

Method	$\bar{H}$	$\bar{W}$	Iterations	Train $\bar{A}(\%)$	Test $\bar{A}(\%)$	F1-Score	p-Value
DDNM + Bagging	19	272	1000	99.4	93.7	0.951	0.000912
DDNM + Adaboost	**11**	272	1000	98.9	94.1	0.953	0.004817
RDNM	20	272	400	**99.6**	94.4	**0.956**	0.061132
RF + MDM	30	-	-	98.4	93.5	0.951	0.000002
BPNN + Bagging + MDM	16	546	1000	98.5	93.3	0.950	0.000002
DNM + Bagging + MDM	100	544	1000	36.0	36.0	0	0.000002
DDNM + Bagging + MDM	19	272	1000	**99.8**	94.3	0.952	0.009154
DDNM + Bagging + LID	11	272	1000	99.5	94.4	0.954	0.031142
RDNM + MDM	17	272	400	**99.9**	94.3	0.955	0.012148
RDNM + LID	9	272	400	99.8	**94.6**	**0.958**	-

Table XI Experimental results for heart

Method	$\bar{H}$	$\bar{W}$	Iterations	Train $\bar{A}(\%)$	Test $\bar{A}(\%)$	F1-score	p-Value
RF	72	-	-	**94.5**	82.8	0.843	0.000002
BPNN + Bagging	39	392	1000	89.3	83.7	0.851	0.001004
BPNN + Adaboost	46	392	1000	89.5	84.0	0.852	0.044773
DNM + Bagging	**15**	390	200	92.9	83.7	0.846	0.000019
DNM + Adaboost	22	390	200	**93.1**	83.6	0.849	0.000789
DDNM + Bagging	38	196	100	90.0	83.0	0.85	0.000002
DDNM + Adaboost	21	198	100	92.1	83.1	0.847	0.000002
RDNM	49	195	100	90.9	**84.0**	**0.853**	0.053221
RF + MDM	40	-	-	**94.5**	83.2	0.847	0.001123

Table XI Continued

Method	$\bar{H}$	$\bar{W}$	Iterations	Train $\bar{A}(\%)$	Test $\bar{A}(\%)$	F1-score	p-Value
BPNN + Bagging + MDM	21	392	1000	91.2	83.9	0.852	0.021141
DNM + Bagging + MDM	13	390	200	93.0	83.8	0.85	0.001021
DDNM + Bagging + MDM	30	196	100	**91.1**	83.1	0.847	0.000002
DDNM + Bagging + LID	15	196	100	90.8	83.4	0.848	0.000002
RDNM + MDM	17	195	100	91.5	**84.2**	0.853	0.022954
RDNM + LID	12	195	100	91.8	**84.2**	**0.855**	-

Table XII Experimental results for Wisconsin breast-cancer

Method	$\bar{H}$	$\bar{W}$	Iterations	Train $\bar{A}(\%)$	Test $\bar{A}(\%)$	F1-score	p-Value
RF	46	-	-	98.5	92.6	0.945	0.000002
BPNN + Bagging	18	546	1000	98.4	93.1	0.948	0.000002
BPNN + Adaboost	12	546	1000	98.0	92.2	0.941	0.000002
DNM + Bagging	100	544	1000	36.0	36.0	0	0.000002
DNM + Adaboost	100	544	1000	36.0	36.0	0	0.000002
DDNM + Bagging	19	272	1000	99.4	93.7	0.951	0.000912
DDNM + Adaboost	**11**	272	1000	98.9	94.1	0.953	0.004817
RDNM	20	272	400	**99.6**	**94.4**	**0.956**	0.061132
RF + MDM	30	-	-	98.4	93.5	0.951	0.000002
BPNN + Bagging + MDM	16	546	1000	98.5	93.3	0.950	0.000002
DNM + Bagging + MDM	100	544	1000	36.0	36.0	0	0.000002
DDNM + Bagging + MDM	19	272	1000	**99.8**	94.3	0.952	0.009154
DDNM + Bagging + LID	11	272	1000	99.5	94.4	0.954	0.031142
RDNM + MDM	17	272	400	**99.9**	94.3	0.955	0.012148
RDNM + LID	**9**	272	400	99.8	**94.6**	**0.958**	-

Table XIII Average rank comparison obtained by Friedman aligned-rank test with 15 methods for seven different datasets

	Glass	Vertebral Column	Blood Transfusion	Parkinsons	Ionosphere	Heart	Wisconsin Breast-Cancer	average rank
BPNN + Bagging	14	14	15	11	10	7.5	15	12.35
BPNN + Adaboost	15	14	13	12	12	3.5	13	11.78
RF	12	14	11	9.5	11	15	9	11.64
DNM + Bagging	8.5	7.5	12	15	14	7.5	12	10.92
BPNN + Bagging + MDM	13	10.5	14	9.5	9	5	14	10.71
RF + MDM	11	12	10	8	8	11	7.5	9.64
DNM + Adaboost	5.5	5.5	8	14	14	9	11	9.57
DNM + Bagging + MDM	5.5	4	9	13	14	6	10	8.78
DDNM + Adaboost	10	10.5	2	5	6	12.5	7.5	7.64
DDNM + Bagging	5.5	9	7	3.5	7	14	4.5	7.21
DDNM + Bagging + MDM	5.5	7.5	5	3.5	4.5	12.5	4.5	6.14
DDNM + Bagging + LID	8.5	5.5	2	1	2.5	10	2	4.50
RDNM	3	2.5	5	6.5	2.5	3.5	4.5	3.92
RDNM + MDM	1.5	2.5	5	6.5	4.5	1.5	4.5	3.71
RDNM + LID	1.5	1	2	2	1	1.5	1	1.42

5.3 Instance Analysis

We perform two experimental sets to investigate the relationship between diversity and performance on **Glass**. To conduct each set of experiments, we require 100 ensembles, and for each ensemble, we randomly select an ensemble size (H') from 100 DDNMs. In Fig. 4, observing the relationship between ensemble accuracy and diversity, a positive correlation can be found between them. Thus, it is meaningful to consider diversity in terms of interpretability.

To demonstrate the advantages of LID in DDNM ensemble (including RDNM and DDNM + Bagging), we analyzed each ensemble method in two datasets (**Wisconsin Breast-Cancer** and **Blood Transfusion**). As shown in Figs. 5 and 6, the red solid line represents the ensemble curve of RDNM after enhancement by LID. It shows LID only needs a few base learners to reach the highest of all metrics and has obvious advantages over other algorithms. We can observe the following points.

1) With the enhancement of LID, we can reduce the number of DDNM ensemble while improving accuracy and $F1-\text{score}$.
2) In all metrics ($\bar{H}$ testing accuracy and $F1-\text{score}$), LID improves DDNM ensemble better than MDM.

6. Conclusion

As we all know, ensemble learning can achieve better accuracy and lower generalization errors than a single traditional classifier. In this article, we propose the DDNM ensemble and advocate for the inclusion of interpretability diversity in ensemble methods. Specifically, we introduce the LID measure for DNM (to be precise, it refers to DNM in DDNM) and implement ensemble selection based on LID to achieve significant improvements in DDNM ensemble performance, with advantages over MDM.

An RDNM is the outstanding representative of the DDNM ensemble, it is used for the classification and compared with RF, DNM ensemble, BPNN ensemble, and other DDNM ensembles. The results indicate that RDNM + LID performs the best overall in terms of the average accuracy and ensemble size for the seven popularly applied bench-mark datasets, namely, **Glass, Vertebral Column, Blood Transfusion, Parkinsons, Ionosphere, Heart**, and **Wisconsin Breast-Cancer**. Our experiments on selective ensemble validate the necessity of considering interpretability diversity, providing a new perspective on optimizing ensemble learning.

Thus, RDNM + LID has great potential for solving real-world classification problems. It needs to be further explored (e.g., finding better interpretability diversity measures, ensemble selection methods, and combining with other heuristic algorithms to improve ensemble performance) and applied to regression, multivariate classification, and other challenging problems. In future work, interpretability diversity will become an essential direction for our research. SVMs, graph neural networks, etc. are also interpretability machine learning algorithms, and how the interpretability diversity among them can be measured and used also needs to be further studied in depth.

References

[1] DASARATHY B V, SHEELA B V. A composite classifier system design: concepts and methodology[J]. Proc. IEEE, 1979, 67(5): 708-713.
[2] ZHOU Z H. Ensemble learning[J]. Encyclopedia biometrics, 2009, 1: 270-273.
[3] DIETTERICH T G. Machine learning research: four current directions[J]. Artif. intell. mag., 1997, 18(4): 97-136.
[4] QUINLAN J R. Induction on decision tree[J]. Mach. learn, 1986, 1(1): 81-106.
[5] HECHT-NIELSEN R. Theory of the backpropagation neural network[C]//Neural Networks for Perception, Amsterdam. The Netherlands: Elsevier, 1992: 65-93.
[6] KOCH C, POGGIO T, TORRE V. Retinal ganglion cells: a functional interpretation of dendritic morphology[J]. Phil. trans. roy. soc. London, 1982, 298(1090): 227.
[7] TANG Z, TAMURA H, KURATU M, et al. A model of the neuron based on dendrite mechanisms[J]. Electron. commun. jpn. III, fundam. electron. sci., 2001, 84(8): 11-24.
[8] GAO S, ZHOU M, WANG Y, et al. Dendritic neuron model with effective learning algorithms for classification, approximation, and prediction[J]. IEEE trans. neural netw. learn. syst., 2019, 30(2): 601-614.
[9] JI J, GAO S, CHENG J, et al. An approximate logic neuron model with a dendritic structure[J]. Neurocomputing, 2016, 173: 1775-1783.
[10] QIAN X, WANG Y, CAO S, et al. Mr2DNM: a novel mutual information-based dendritic neuron model[J]. Comput. intell. neurosci., 2019: 1-13.
[11] JI J, SONG S, TANG Y, et al. Approximate logic neuron model trained by states of matter search algorithm[J]. Knowl.-based syst., 2019, 163: 120-130.
[12] TODO Y, TANG Z, TODO H, et al. Neurons with multiplicative interactions of nonlinear synapses[J]. Int. J. neural syst., 2019, 29(8): 115-133.
[13] TANG Y, JI J, GAO S, et al. A pruning neural network model in credit classification analysis[J]. Comput. intell. neurosci., 2018: 1-22.
[14] TANG Y, JI J, ZHU Y, et al. A differential evolution-oriented pruning neural network model for bankruptcy prediction[J]. Complexity, 2019: 1-21.
[15] SHA Z, HU L, TODO Y, et al. A breast cancer classifier using a neuron model with dendritic nonlinearity[J]. IEICE trans. inf. syst., 2015, E98.D(7): 1365-1376.
[16] YU Y, SONG S, ZHOU T, et al. Forecasting house price index of China using dendritic neuron model[C]//Int. Conf. Prog. Informat. Comput. (PIC), Dec. 2016, pp. 37-41.

[17] JIANG T, GAO S, WANG D, et al. A neuron model with synaptic nonlinearities in a dendritic tree for liver disorders[J]. IEEJ trans. electr. electron. eng., 2017, 12(1): 105-115.

[18] ZHOU T, GAO S, WANG J, et al. Financial time series prediction using a dendritic neuron model[J]. Knowl.-based syst., 2016, 105: 214-224.

[19] GAO S, et al. Fully complex-valued dendritic neuron model[J]. IEEE trans. neural netw. learn. syst., 2023, 34(4): 2105-2118.

[20] SONG Z, TANG Y, JI J, et al. Evaluating a dendritic neuron model for wind speed forecasting[J]. Knowledge-based syst., 2020, 106052: 201-202.

[21] ZHANG Z, LIU H, ZHOU M, et al. Solving dynamic traveling salesman problems with deep reinforcement learning[J]. IEEE trans. neural netw. learn. syst., 2023, 34(4): 2119-2132.

[22] SONG S, CHEN X, SONG S, et al. A neuron model with dendrite morphology for classification[J]. Electronics, 2021, 10(9): 1062.

[23] ZHANG T, LV C, MA F, et al. A photovoltaic power forecasting model based on dendritic neuron networks with the aid of wavelet transform[J]. Neurocomputing, 2020, 397: 438-446.

[24] YU Y, LEI Z, WANG Y, et al. Improving dendritic neuron model with dynamic scale-free network-based differential evolution[J]. IEEE/CAA J. autom. sinica, 2022, 9(1): 99-110.

[25] WEN X, ZHOU M, LUO X, et al. Novel pruning of dendritic neuron models for improved system implementation and performance[C]//IEEE Int. Conf. Syst., Man, Cybern. (SMC), Oct. 2021, pp. 1559-1564.

[26] LUO X, WEN X, ZHOU M, et al. Decision-tree-initialized dendritic neuron model for fast and accurate data classification[J]. IEEE trans. neural netw. learn. syst., 2022, 33(9): 4173-4183.

[27] GEMAN S, BIENENSTOCK E, DOURSAT R. Neural networks and the Bias/Variance dilemma[J]. Neural comput., 1992, 4(1): 1-58.

[28] KROGH A. VEDELSBY J. Neural network ensembles, cross validation, and active learning[C]//Proc. Int. Conf. Neural Inf. Process. Syst., 1995, pp. 1-8.

[29] BREIMAN L. Random forests[J]. Mach. learn., 2001, 45(1): 5-32.

[30] MELVILLE P, MOONEY R J. Constructing diverse classifier ensembles using artificial training examples[C]//Proc. IJCAI, vol. 3, 2003, pp. 505–510.

[31] YU Y, LI Y F, ZHOU Z H. Diversity regularized machine[C]//Proc. 22 Int. Joint Conf. Artif. Intell., 2011, pp. 1-6.

[32] BREIMAN L. Randomizing outputs to increase prediction accuracy[J]. Mach. learn., 2000, 40(3): 229-242.

[33] ZHOU Z H. Ensemble methods: foundations and algorithms[M]. Boca Raton, FL, USA: Chapman & Hall/CRC, 2012.

[34] CHAKRABORTY S, et al. Interpretability of deep learning models: a survey of results[C]//Proc. IEEE SmartWorld, Ubiquitous Intell. Comput., Adv. Trusted Comput., Scalable Comput. Commun., Cloud Big Data Comput., Internet People Smart City Innov., Aug. 2017: 1-6.

[35] GUIDOTTI R, MONREALE A, RUGGIERI S, et al. PEDRESCHI. a survey of methods for explaining black box models[J]. ACM comput. surv., 2019, 51(5): 1-42.

[36] LUNDBERG S, LEE S I. A unified approach to interpreting model predictions[C]//Proc. Adv. Neural Inf. Process. Syst., 2017, pp. 4765-4774.

[37] ANCONA M, CEOLINI E, ÖZTIRELI C, et al. Towards better understanding of gradient-based attribution methods for deep neural networks[J]. arXiv, 2017: 1711.06104.

[38] MURDOCH W J, SINGH C, KUMBIER K, et al. Interpretable machine learning: definitions, methods, and applications[J]. Neurocomputing, 2019, 116(44): 22071-22080.

[39] WU M, et al. Regional tree regularization for interpretability in black box models[J]. arXiv, 2019: 1908.04494.

[40] TOWNSEND J, CHATON T, MONTEIRO J M. Extracting relational explanations from deep neural networks: a survey from a neural-symbolic perspective[J]. IEEE trans. neural netw. learn. syst., 2020, 31(9): 3456-3470.

[41] HE C, MA M, WANG P. Extract interpretability-accuracy balanced rules from artificial neural networks: a review[J]. Neurocomputing, 2020, 387: 346-358.

[42] P BREZILLON, S ABUHAKIMA. Using knowledge in its context: report on the IJCAI-93 workshop[J]. AI Magazine, 1995, 16(1): 87-91.

[43] SUN T, ZHOU Z H. Structural diversity for decision tree ensemble learning[J]. Frontiers comput. sci., 2018, 12(3): 560-570.

[44] BREIMAN L. Bagging predictors[J]. Mach. learn., 1996, 24, (2): 123-140.

[45] SCHAPIRE R E. The strength of weak learnability[C]//Proc. 30th Annu. Symp. Found. Comput. Sci., 1989: 197-227.

[46] FREUND Y, SCHAPIRE R E. A decision-theoretic generalization of on-line learning and an application to boosting[J]. J. comput. syst. sci., 1997, 55(1): 119-139.

[47] YULE G U. On the association of attributes in statistics: with illustrations from the material of the childhood society, &c[J]. Philos. trans. roy. soc. London A, math. phys. sci., 1900, 194(1900): 257-319.

[48] KUNCHEVA L I, WHITAKER C J. Measures of diversity in classifier ensembles and their relationship with the ensemble accuracy[J]. Mach. learn., 2003, 51(2): 181-207.

[49] SKALAK D B. The sources of increased accuracy for two proposed boosting algorithms[C]//Proc. Amer. Assoc. Artif. Intell., AAAI, Integrating Multiple Learned Models Workshop, vol. 1129, Aug. 1996: 1133.

[50] HO T K. The random subspace method for constructing decision forests[J]. IEEE trans. pattern anal. mach. intell., 1998, 20(8): 832-844.

[51] GIACINTO G, ROLI F. Design of effective neural network ensembles for image classification purposes[J]. Image vis. comput., 2001, 19(9-10): 699-707.

[52] CUNNINGHAM P, CARNEY J. Diversity versus quality in classification ensembles based on feature selection[C]//Proc. Eur. Conf. Mach. Learn. Cham, Switzerland: Springer, 2000: 109-116.

[53] HANSEN L K, SALAMON P. Neural network ensembles[J]. IEEE trans. pattern anal. mach. intell., 1990, 12(10): 993-1001.

[54] KOHAVI R, WOLPERT D H. Bias plus variance decomposition for zero-one loss functions[C]// Proc. Int. Conf. Mach. Learn., vol. 96, 1996: 275-283.

[55] CONOVER W J. Statistical methods for rates and proportions[J]. Techno-metrics, 1974,

16(2): 326-327.

[56] PARTRIDGE D, KRZANOWSKI W. Software diversity: practical statistics for its measurement and exploitation[J]. Inf. softw. technol., 1997, 39(10): 707-717.

[57] BANFIELD R E, HALL L O, BOWYER K W, et al. A new ensemble diversity measure applied to thinning ensembles[C]// Proc. Int. Workshop Multiple Classifier Syst. Cham, Switzerland: Springer, 2003: 306-316.

[58] LIU Y, YAO X. Ensemble learning via negative correlation[J]. Neural netw., 1999, 12(10): 1399-1404.

[59] HIGUCHI T, YAO X, LIU Y. Evolutionary ensembles with negative correlation learning[J]. IEEE trans. evol. comput., 2000, 4(4): 380-387.

[60] CHAN Z S H, KASABOV N. Fast neural network ensemble learning via negative-correlation data correction[J]. IEEE trans. neural netw., 2005, 16(6): 1707-1710.

[61] CHEN H, YAO X. Regularized negative correlation learning for neural network ensembles[J]. IEEE trans. neural netw., 2009, 20(12): 1962-1979.

[62] PERALES-GONZÁLEZ C, FERNÁNDEZNAVARRO F, CARBONERORUZ M, et al. Global negative correlation learning: a unified framework for global optimization of ensemble models[J]. IEEE trans. neural netw. learn. syst., 2022, 33(8): 4031-4042.

[63] ZHOU Z H, WU J, TANG W. Ensembling neural networks: many could be better than all[J]. Artif. intell., 2002, 137(1-2): 239-263.

[64] ZHANG Y, BURER S, STREET W N, et al. Ensemble pruning via semi-definite programming[J]. J. mach. learn. res., 2006, 7(7): 1-24.

[65] MARGINEANTU D D, DIETTERICH T G. Pruning adaptive boosting[C]//Proc. Int. Conf. Mach. Learn., vol. 97, 1997: 211-218.

[66] MARTıNEZMUNOZ G, SUÁREZ A. Aggregation ordering in bagging[C]//Proc. IASTED Int. Conf. Artif. Intell. Appl., 2004: 258-263.

[67] MARTINEZMUNOZ G, HERNANDEZ-LOBATO D, SUAREZ A. An analysis of ensemble pruning techniques based on ordered aggregation[J]. IEEE trans. pattern anal. mach. intell., 2009, 31(2): 245-259.

[68] GUO L, BOUKIR S. Margin-based ordered aggregation for ensemble pruning[J]. Pattern recognit. lett., 2013, 34(6): 603-609.

[69] ZHU X, NI Z, NI L, et al. Improved discrete artificial fish swarm algorithm combined with margin distance minimization for ensemble pruning[J]. Comput. ind. eng., 2019, 128: 32-46.

[70] FUSHIKI T. Estimation of prediction error by using K-fold cross-validation[J]. Statist. comput., 2011, 21(2): 137-146.

[71] BREIMAN L, FRIEDMAN J H, OLSHEN R A, et al. Classification and regression trees[J]. Biometrics, 1984, 40(3): 358.

[72] HOARE C A R. Algorithm 64: quicksort[J]. Commun. ACM, 1961, 4(7): 321.

[73] DEMŠAR J. Statistical comparisons of classifiers over multiple data sets[J]. J. Mach. learn. res., 2006, 7: 1-30.

[74] KIM J H. Estimating classification error rate: repeated cross-validation, repeated hold-out and bootstrap[J]. Comput. statist. data anal., 2009, 53(11): 3735-3745.

[75] RODRIGUEZ J D, PEREZ A, LOZANO J A. Sensitivity analysis of k-fold cross validation in prediction error estimation[J]. IEEE trans. pattern anal. mach. intell., 2010, 32(3): 569-575.

[76] GARCÍA S, MOLINA D, LOZANO M, et al. A study on the use of non-parametric tests for analyzing the evolutionary algorithms' behaviour: a case study on the CEC'2005 special session on real parameter optimization[J]. J. heuristics, 2008, 15(6): 617-644.

[77] HODGES J L, LEHMANN E L. Rank methods for combination of independent experiments in analysis of variance[J]. Ann. math. statist., 1962, 33(2): 482-497.

[78] DEMŠAR J. Statistical comparisons of classifiers over multiple data sets[J]. J. mach. learn. res., 2006, 7: 1-30.

[79] NASH M S. Handbook of parametric and nonparametric statistical procedures[J]. Technometrics, 2001, 43(3): 374.

[80] LUO X D, LONG Y, WEI Z, et al. A novel particle filter based object tracking framework via the combination of state and observation optimization[J]. J. biol. chem., 2013, 30(21): 487-490.

[81] HONG M, LUO Z Q, RAZAVIYAYN M. Convergence analysis of alternating direction method of multipliers for a family of nonconvex problems[J]. SIAM J. Optim., 2016, 26(1): 337-364.

[82] KANG Q, CHEN X, LI S, et al. A noise-filtered under-sampling scheme for imbalanced classification[J]. IEEE trans. cybern, 2017, 47(12): 4263-4274.

[83] LIU H, ZHOU M, LIU Q. An embedded feature selection method for imbalanced data classification[J]. IEEE/CAA J. autom. sinica, 2019, 6(3): 703-715.

[84] ZHANG P, SHU S, ZHOU M. An online fault detection model and strategies based on SVM-grid in clouds[J]. IEEE/CAA J. autom. sinica, 2018, 5(2): 445-456.

[85] DAU H A, et al. The UCR time series archive[J]. IEEE/CAA J. autom. sinica, 2019, 6(6): 1293-1305.

[86] FARD A E, MOHAMMADI M, CHEN Y, et al. Computational rumor detection without non-rumor: a one-class classification approach[J]. IEEE trans. computat. social syst., 2019, 6(5): 830-846.

Pruning of Dendritic Neuron Model with Significance Constraints for Classification*

1. Introduction

There are two chief categories of weight pruning, structured and non-structured. General non-structured pruning can prune arbitrary weight in DNN. Even though it provides high pruning rates (weight reduction), the sparse weight matrix storage and associated indices limit its actual hardware implementation. In contrast, structured pruning can directly reduce the size of weight matrix and maintain the form of an entire matrix. Therefore, it is adaptable with hardware acceleration and has become a research focus in recent years. Structured pruning has various schemes, e.g., filter pruning, channel pruning, and column pruning for convolutional layers of DNN, as summarized in [1], [5], [6], [8]. A systematic solution framework has recently been developed based on a robust optimization tool, namely alternating direction methods of multipliers (ADMM). It applies to schemes of structured and non-structured pruning and has achieved the-state-of-the-art results.

Ullrich et al. presented a simple regularization method based on soft weight sharing, which comprises quantization and pruning in a simple pre-training process. Louizos et al. proposed a practical approach for L0 normalization of neural networks: pruning the network by encouraging the weights to be precisely zero during the training process. In [13], a new sequential learning algorithm for radial basis function (RBF) networks was proposed referred to as a generalized growing and pruning algorithm for RBF networks. Hassibi and Dtork investigated the utilization of information from all second-order derivatives of the error function for network pruning (that is, removing insignificant weights from a trained network) to enhance generalization, simplify networks, reduce hardware and storage requirements, increase the speed of further training, and in some cases achieve rules extraction. Their proposed method called optimal brain surgeon (OBS) is superior to a magnitude-based method and optimal brain damage method, which often removes incorrect

---

* The paper was originally published in *IJCNN*, 2023, and has since been revised with new information. It was co-authored by Xudong Luo, Long Yea, Xiaolan Liu, Xiaohao Wen, and Qin Zhang.

weights. OBS can prune more weights than other methods under the same training error, thus producing better generalization on test data. Han et al. exploited "deep compression" to address the limitation of insufficient hardware resources by combining pruning, trained quantization, and Hoffman coding.

Unlike traditional neural networks, DNM is a single neuronal model consisting of synaptic, dendrite, membrane, and soma layer. Based on the Koch-Poggio-Torre model of the dendritic mechanism, Zhang et al. applied the multiplicative method to the proposed dendritic neuron model (DNM). It has a robust ability to solve nonlinear problems. It has an excellent effect on classification, approximation and prediction, as demonstrated in [18]-[34]. Ji et al. proposed a DNM pruning method based on synaptic connection status in [20]. Although effective, this method has an uncontrollable extent of pruning, leading to excessive pruning and a significant reduction in accuracy. To address this, Wen et al. proposed a pruning method based on the significance of dendrite layers (PBSD), which retains the training and testing accuracy of the original model. However, while PBSD is more accurate than the method in [20], more of the dendrite layer is retained. To further compress the model size while maintaining accuracy, a constraint term is added to reduce the impact of pruning on model accuracy. The constraint term clusters the significance of dendrite layers on a small number of layers, ensuring that the model's accuracy is preserved. The main contributions of this paper can be summarized into the following four aspects:

1) We present a novel pruning method for DNM based on the significance constraint of dendrite layers;
2) We derive an analytical formula for calculating the significance of each dendrite layer based on probability theory;
3) We design an objective function with a penalty term to enforce the significance constraint and use back-propagation algorithm to optimize it; and
4) We extensively evaluate our method on six UCI datasets and provide empirical evidence of its effectiveness in comparison to PBSD.

The rest of this article is organized as follows. The second part introduces the structure and function of a novel DNM and the pruning method. The third section shows the experimental results as well as the analysis. The fifth section draws the conclusion and the prospect.

2. Dendritic Neuron Model with Significance Constraints

2.1 Dendritic Neuron Model

Following the characteristics of biomimetic neurons, as shown in Fig. 1, DNM is composed of four layers: synaptic, dendrite, membrane, and soma. The synaptic layer converts the external signal into neuronal signals using the sigmoid function. The interactions among synaptic signals occur at each dendritic layer, which performs a multiplicative function on the output of the synaptic layer. The membrane layer sums over all dendrite layers' output. The soma layer uses another sigmoid function for the output of the membrane layer and generates its result or transmits it to the next neuron. Its detailed structure is described as follows.

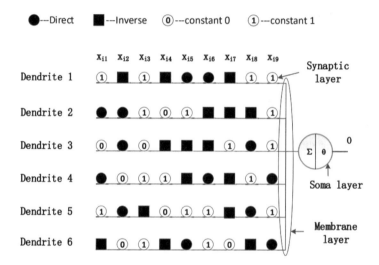

Fig. 1 A six-dendrite and nine-input dendritic neuron model in which the axons of presynaptic neurons (input x_i) connect to a dendrite layer through synaptic (four connection states); and the membrane layer aggregates dendrites to activate and transfer them to the soma layer.

1) Synaptic layer: The synaptic layer plays an important role in regulating and controlling neuronal activity. Synapses include excitatory and inhibitory ones. When receptors receive an ion, and the accumulated ions exceed a threshold, their potential is modified. This process determines if the connecting synapse is excitatory or inhibitory. Assuming we use a sigmoid function to express the connection states, the i th $(i = 1, 2, ..., N)$ synaptic input is connected to the j th $(j = 1, 2, ..., M)$ dendrite layer as follows:

$$\varphi_{ij}(x_i) = \frac{1}{1+e^{-k(w_{ij}x_i - \theta_{ij})}} \qquad (1)$$

where φ_{ij} represents the output from the i th synaptic input to the j th dendrite layer, φ is a nonlinear activation function. The input of DNM with M synaptic layers is a vector $x = (x_1, \ldots, x_i, \ldots, x_N)^T \in X \subseteq [0,1]^N$, where N is the dimension of the input observation space $X \subseteq [0,1]^N$. $x_i \in [0,1]$ is the input of a synapse. k is a positive constant whose value is usually an integer between 1 and 10. Weights w_{ij} and thresholds θ_{ij} are the connection parameters to be trained.

2) Dendrite layer: A dendrite contains multiplicative functions on various synaptic connections in a branch. It represents a typical nonlinear interaction of synaptic signals on each dendrite branch, which can be achieved by classic multiplication. The input and output values of the dendrites correspond to 1 or 0. The synaptic interactions on a dendritic branch are essentially logical *AND* operations, i.e.,

$$\phi_j(x) = \prod_{i=1}^{N} \varphi_{ij}(x_i) \qquad (2)$$

where $\phi_j(x)$ is the j th dendrite's response to the input vector x.

3) Membrane layer: A membrane layer collects the signals from all dendrites. The inputs received from each branch are summarized, much like a logical OR operation. Then, the resultant output is transmitted to the soma layer. This layer's output $f(x)$ is expressed as:

$$f(x) = \sum_{j=1}^{M} \phi_j(x) \qquad (3)$$

4) Soma layer: The soma layer functions to produce the final result. The neuron fires when the output from the membrane layer exceeds the threshold. A sigmoid function is applied, which has the following mathematical description:

$$O(x) = \frac{1}{1+e^{-k_s(f(x)-\theta_s)}} \qquad (4)$$

where $\theta_s \in [0,1]$ is the parameter of the soma layer. k_s is a positive constant, and its value is usually an integer between 1 and 10.

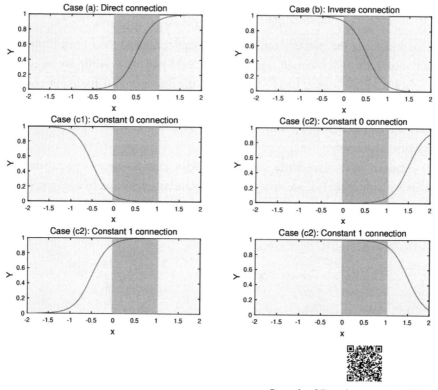

Scan the QR code to see colorful figures

Fig. 2 Six cases correspond to the four connection states in the synaptic layer.

2.2 Pruning Method

According to the various values of w_{ij} and θ_{ij} in DNM, four connection states are defined corresponding to six conditions, as illustrated in Fig. 2. The horizontal axis represents input x_{ij} and vertical one indicates the output of φ_{ij}. The four connection states are:

1) Direct connection (when $0 < \theta_{ij} < \omega_{ij}$) where the output is positively correlated with the input no matter when the input transforms from 0 to 1.

2) Inverse connection (when $\omega_{ij} < \theta_{ij} < 0$) where the output is inversely related to an input no matter when the input is converted from 0 to 1.

3) constant 0 connection (when $\omega_{ij} < 0 < \theta_{ij}$ or $0 < \omega_{ij} < \theta_{ij}$) where output φ_{ij} is approximately 0 whenever an input transforms from 0 to 1.

4) constant 1 connection (when $\theta_{ij} < \omega_{ij} < 0$ or $\theta_{ij} < 0 < \omega_{ij}$) where φ_{ij} is approximately 1 whenever an input transforms from 0 to 1.

Once the synaptic layer is a constant 0 connection, its output is 0, since any value multiplied by 0 results in 0. The multiplication operation sets the whole dendrite to 0 regardless of any other synaptic signals in the dendrite. Since such dendrites do not affect the membrane layer, they should be removed.

1) Significance of Dendrite: This concept of significance of a dendrite relies on its statistical average contribution to all inputs. In sequential learning, a series of training samples are randomly drawn, and learned by the network one by one. Let a series of training samples (x_i, y_i) be drawn randomly from a range X with a sampling density function of $p(x)$. $p(x)$ is defined as:

$$\int \cdots \int_X p(x) dx = 1 \tag{5}$$

In this paper, $\int \cdots \int_X$ is replaced by $\int_X$ for brevity, and the range of integration is represented by X. Suppose a DNM with M dendrites has been obtained, the output of membrane layer for the input x_i is given as:

$$f_1(\omega_1, \theta_1, ..., \omega_M, \theta_M, x_i) = f(x_i) = \sum_{j=1}^{M} \phi_j(x_i) \tag{6}$$

If the dendrite τ is removed, the output of DNM with the remaining $M-1$ dendrites is:

$$f_2(\omega_1, \theta_1, ..., \omega_{\tau-1}, \theta_{\tau-1}, \omega_{\tau+1}, \theta_{\tau+1}, ..., \omega_M, \theta_M, x_i) = \sum_{j=1}^{\tau-1} \phi_j(x_i) + \sum_{j=\tau+1}^{M} \phi_j(x_i) \tag{7}$$

Therefore, for an observation x_i, the error resulting from the removal of dendrite τ is:

$$\varepsilon(\tau, i) = f_1(\omega_1, \theta_1, ..., \omega_M, \theta_M, x_i)$$
$$- f_2(\omega_1, \theta_1, ..., \omega_{\tau-1}, \theta_{\tau-1}, \omega_{\tau+1}, \theta_{\tau+1}, ..., \omega_M, \theta_M, x_i) = \phi_j(x_i) \tag{8}$$

Theoretically, the average error of all n sequentially learned observations caused by removing dendrite τ is:

$$\varepsilon(\tau) = \left(\frac{\sum_{i=1}^{n} \varepsilon(\tau, i)}{n} \right) = \left(\frac{\sum_{i=1}^{n} \phi_\tau(x_i)}{n} \right) \tag{9}$$

However, the computational complexity of $\varepsilon(\tau)$ would be very high if the number of input observations n is very large. There must be some simpler and better way to calculate $\varepsilon(\tau)$ as stated in (9) without the prior knowledge of each specified observation (x_i, y_i).

Assume that the observations (x_i, y_i), $i \in \{1, 2, ..., n\}$ are sampled from a sampling range X with a sampling density function $p(x)$. Divide the sampling range X into n small spaces Δ_j, $j \in \{1, 2, ..., n\}$. The size of Δ_j is represented by $S(\Delta_j)$. As the sampling density function is $p(x)$, there are approximately $n \cdot p(x) \cdot S(\Delta_j)$ samples in each Δ_j, where x_j is a point chosen in Δ_j. From (9), we have

$$\varepsilon(\tau) \approx \left(\frac{\sum_{j=1}^{n} \phi_\tau(x_j) \cdot np(x_j) \cdot S(\Delta_j)}{n} \right) \qquad (10)$$

$$= \left(\sum_{j=1}^{n} \phi_\tau(x_j) p(x_j) S(\Delta_j) \right)$$

When the number of input observations $n \to \infty$ and $\Delta_j \to 0$, we have

$$\lim_{n \to +\infty} \varepsilon(\tau) \approx \lim_{n \to +\infty} \left(\sum_{j=1}^{n} \phi_\tau(x_j) p(x_j) S(\Delta_j) \right) = \left(\int_x \phi_\tau(x) p(x) dx \right)$$

$$= \left(\int_x \prod_{i=1}^{N} \varphi_{i\tau}(x_i) p(x) dx \right) \qquad (11)$$

This is the statistical contribution of dendrite τ to the overall output of DNM, we define this as the significance of the dendrite τ, i.e.,

$$E(\tau) = \lim_{n \to +\infty} \varepsilon(\tau) \qquad (12)$$

In addition, if the N attributes $(x_1, ..., x_i, ..., x_N)^T$ of x are independent of each other, then the density function $p(x)$ can be written as: $p(x) = \prod_{i=1}^{N} p_i(x_i)$, where $p_i(x_i)$ is the density function of the i th attribute x_i. Hence, in this case, significance (12) can be computed as:

$$E(\tau) = \left(\int_x \prod_{i=1}^{N} \varphi_{i\tau}(x_i) p(x) dx \right)$$

$$= \prod_{i=1}^{N} \left(\int_0^1 \varphi_{i\tau}(x_i) p_i(x_i) dx_i \right) \qquad (13)$$

The above equation relates to an integration of the probability density function $p(x)$ in the sampling range X. We can do this analytically for some simple but popularly used $p(x)$ functions in, e.g., uniform and normal distributions.

To facilitate estimation, we assume that the input samples are uniformly drawn from a range $X \subseteq [0,1]^N$, i.e., $p(x) = 1/S(X)$, where $S(X)$ is the size of the range $X \subseteq [0,1]^N$ given by $S(X) = \int_X 1 dx = 1$. Hence, we have:

$$E(\tau) = \left(\int_X \prod_{i=1}^{N} \varphi_{i\tau}(x_i) \frac{1}{S(X)} dx \right) = \left(\int_X \prod_{i=1}^{N} \varphi_{i\tau}(x_i) dx \right)$$

$$= \left(\int_X \prod_{i=1}^{N} \varphi_{i\tau}(x_i) dx \right) = \prod_{i=1}^{N} \left(\int_0^1 \frac{1}{1+e^{-(\omega_{i\tau} x_i - \theta_{i\tau})}} dx \right) \qquad (14)$$

$$= \prod_{i=1}^{N} \left(\frac{1}{\omega_{i\tau}} \ln\left(1+e^{-(\omega_{i\tau} x_i - \theta_{i\tau})}\right) \bigg|_0^1 \right) = \prod_{i=1}^{N} \frac{1}{\omega_{i\tau}} \ln\left(\frac{1+e^{\omega_{i\tau} - \theta_{i\tau}}}{1+e^{-\theta_{i\tau}}} \right)$$

2) Pruning Criterion: To facilitate the establishment of a uniform significance criterion, $E(\tau)$ can be normalized to obtain:

$$\tilde{E}(\tau) = \frac{E(\tau)}{\|E\|_1} = \frac{E(\tau)}{\sum_{t=1}^{M} E(t)}, (E(\cdot) \geq 0) \qquad (15)$$

Therefore, the significance of M dendrite layers in DNM can be represented by the vector $E = [\tilde{E}(1), ..., \tilde{E}(\tau), ..., \tilde{E}(M)]$. If the significance $\tilde{E}(\tau)$ is less than the significance threshold $E_{\min}$, dendrite τ is insignificant and should be removed, otherwise, it is retained. Given the significance threshold $E_{\min}$, it is pruned if

$$\tilde{E}(\tau) < E_{\min} \qquad (16)$$

The above condition implies that after learning each observation, the significance for all dendrites should be computed and checked for possible pruning.

3) Learning with Significance Constraint: Since DNM is a feed-forward model and all functions in it are differentiable, the error back-propagation algorithm (BP) can be effectively used as a learning algorithm. BP continuously tunes the values of w_{ij} and θ_{ij} through derivative and learning rate to decrease the difference between actual output O and desired output $\hat{O}$. The Squared Error (SE) between O and $\hat{O}$ is defined as:

$$l = \frac{1}{2}(\hat{O} - O)^2 \qquad (17)$$

In order to prune as many dendrite layers of DNM as possible, **E** must contain as many values less than $E_{\min}$ as possible. This can be achieved by increasing the variance of **E**. To constrain the variance of **E**, a penalty term C is added to the objective function L

$$C = \frac{1}{2}\|E\|_2^2 = \frac{1}{2}\sum_{\tau=1}^{M}\left(\frac{E(\tau)}{\sum_{t=1}^{M} E(t)}\right)^2 \qquad (18)$$

$$L = l - \lambda C \qquad (19)$$

3. Experimental Results

3.1 Datasets

In this section, to validate the advantages of the proposed pruning method, three datasets from University of California at Irvine (UCI) Machine Learning Repository are selected for experiments, including **Glass, Heart**, and **Wisconsin Breast-Cancer** from University of

California at Irvine Machine Learning Repository. Table I describes these datasets.

Table I Dataset description

Dataset	Samples	Attributes	Classes
Glass	214	9	2
Heart	270	13	2
Wisconsin Breast-Cancer	699	9	2

3.2 Experimental Condition and Evaluation Method

We evaluate the pruning effect proposed in this paper by comparing it with PBSD method proposed by Wen et al. Furthermore, for a fair comparison, the main parameters' settings and descriptions according to the characteristics of each method are shown in Table II. There are three key hyperparameters in DNM (i.e., the synaptic hyperparameter (k_c) in the connection sigmoid function and two soma hyperparameters (k_s and θ_s) in the output sigmoid function). In evaluating the performance of a method, testing is an important step. A K-fold cross-validation is a classical approach to evaluate it. In this validation, the original samples are randomly and averagely divided into K subsamples. Each subsample becomes the testing set once, and the remaining K-1 subsamples are used as the training set. In general, K=5. To ensure the independence and validity of the experiments, we randomly repeat different partitions six times. Hence, the experiments are performed for $6K$ times with these subsamples. The average performance of the experiment is accepted as its final performance. The advantage of this approach is that all observations are used for training and testing, thus overfitting and lack of learning are avoided to produce convincing results.

Table II The Main Parameters

Method	Parameter	Value
DNM	Activation function	Sigmoid
	Synaptic layer parameter k	5
	Soma layer parameter k_s	5
	Threshold θ_s	0.5
	The number of dendrites M	10, 20, 50, 100
	Penalty coefficient λ	0.01

The Figs 3-5 show the relationship between Emin and M. For a certain M, a smaller Emin represents less accuracy lost after pruning. The comparison of the two curves in the lower left region leads to the following two conclusions:

1) For the same E_{min}, our algorithm can trim off more dendrite layers and thus retain a smaller model; and
2) For the same M, our pruning method can have a smaller E_{min} value and thus reduce the accuracy loss caused by model compression.

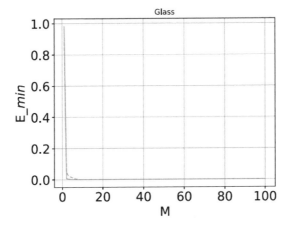

Fig. 3 For Glass, the relationship curves of E_{min} and M for the two pruning methods. The dashed line represents PBSD method and the solid line represents our method.

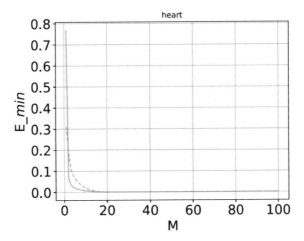

Fig. 4 For Heart, the relationship curves of E_{min} and M for the two pruning methods. The dashed line represents PBSD method and the solid line represents our method.

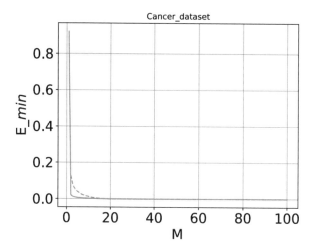

Fig. 5 For Wisconsin Breast-Cancer, the relationship curves of E_{min} and M for the two pruning methods. The dashed line represents PBSD method and the solid line represents our method.

To demonstrate the performance advantage, we compare the accuracy of both. Statistical test is implemented for measuring and assessing whether there are significant differences between our pruning method and that of PBSD. Wilcoxon signed ranks test is used to detect if there are significant differences between them. We have also recorded the accuracy results for the proposed and original methods in terms of $R+$ and $R-$ and $p-value$. $R+$ denotes the sum of ranks for the problems in which the proposed method outperforms the original one, and $R-$ is the opposite. A smaller $p-value$ indicates that the proposed method performs significantly better than the original one. Average accuracy ($\bar{A}$) is an important criterion to evaluate performance of a model. Δ is used to represent the difference in accuracy between our method and PBSD.

The purpose of pruning is to compress a given model. The more excellent pruning means that its pruned model achieves the comparable result as the original one while having the fewer dendrite layers in DNM. However, the accuracy of pruning depends on the number of remaining layers and is usually reduced after pruning. Therefore, the criterion for comparing the two pruning methods is that the method has higher accuracy while preserving the same number of dendrite layers. In our records, Δ represents the difference between our accuracy and that of PBSD, and a positive number means we have a higher accuracy.

In general, from Tables III—V, the accuracy of the model used in our method often is higher than that of PBSD while retaining the same dendrite layers. These indicate that information lose is mild with the proposed pruning method and the compressed model is closer to the original one.

Table III Experimental results for glass

M	Acc			R+	R-	p − value
	Our	PBSD	Δ			
1	**77.9**	76.6	1.3	28.0	0.0	0.00887796
2	**81.6**	80.8	0.7	110.0	61.0	0.14192864
3	**84.8**	84.1	0.8	147.5	83.5	0.13261879
4	**88.4**	87.4	1.0	123.5	66.5	0.12525446
5	**90.4**	89.8	0.6	107.0	83.0	0.31416330
6	91.7	**92.0**	-0.3	102.5	128.5	0.67520294
7	92.1	**92.5**	-0.4	97.5	133.5	0.73626433
8	92.6	**92.8**	-0.2	99.5	110.5	0.58167309
9	93.0	93.0	0.0	77.5	75.5	0.48095591
10	**93.2**	93.2	0.0	67.0	69.0	0.52073068
11	**93.4**	93.2	0.2	61.5	43.5	0.28465509
12	**93.6**	93.5	0.1	45.0	33.0	0.31776011
13	**93.7**	93.7	0.1	48.0	43.0	0.43027185
14	**93.8**	93.7	0.1	56.0	49.0	0.41244934
15	**94.0**	93.9	0.1	51.0	40.0	0.34911938
16	**94.1**	93.9	0.2	50.5	27.5	0.18116462
40	**94.2**	93.9	0.3	50.5	27.5	0.18116462
100	**94.2**	93.9	0.3	58.5	32.5	0.17995751

Table IV Experimental results for heart

M	Acc			R+	R-	p − value
	Our	PBSD	Δ			
1	**49.3**	49.2	0.1	84.0	106.0	0.67110916
2	54.8	**55.4**	-0.6	207.5	227.5	0.58564338
3	**61.1**	60.4	0.7	226.5	208.5	0.42280058

Table IV Continued

M	Acc			R+	R-	p-value
	Our	PBSD	Δ			
4	**66.8**	64.5	2.3	240.0	138.0	0.11011091
5	**69.9**	67.5	2.4	248.0	158.0	0.15267851
6	**72.3**	69.0	3.3	255.0	96.0	0.02166451
7	**74.4**	71.1	3.3	329.0	77.0	0.00204488
8	**76.8**	72.1	4.6	359.5	75.5	0.00106339
9	**77.4**	73.4	3.9	250.0	50.0	0.00212658
10	**78.5**	75.2	3.2	288.5	62.5	0.00202934
11	**79.8**	76.8	3.0	312.0	66.0	0.00154574
12	**80.1**	77.8	2.3	291.0	87.0	0.00707771
13	**80.2**	79.4	0.8	236.0	142.0	0.12898184
14	**80.7**	79.3	1.4	218.5	81.5	0.02499526
15	**80.5**	79.5	1.0	196.5	103.5	0.09153540
16	**80.5**	79.5	1.0	214.5	110.5	0.08037897
17	**80.5**	79.5	1.0	157.0	74.0	0.07429690
18	**80.7**	79.6	1.0	184.0	92.0	0.08022523
19	**80.8**	79.7	1.0	202.5	97.5	0.06611243
20	**80.7**	79.9	0.9	194.5	105.5	0.10090929
21	**80.7**	79.9	0.9	194.5	105.5	0.10090929
22	**80.8**	79.9	0.9	195.0	105.0	0.09832977
100	**80.8**	79.9	0.9	195.0	105.0	0.09832977

Table V Experimental results for Wisconsin breast-cancer

M	Acc			R+	R-	p-value
	Our	PBSD	Δ			
1	**94.1**	88.7	5.4	296.5	3.5	0.00001417
2	**95.8**	93.4	2.4	282.0	43.0	0.00064348

Table V Continued

M	Acc			R+	R-	p – value
	Our	PBSD	Δ			
3	**96.2**	95.3	0.9	201.0	99.0	0.07218941
4	**96.4**	95.6	0.9	182.0	94.0	0.09004021
5	**96.6**	95.8	0.8	141.0	90.0	0.18721663
6	**96.7**	96.0	0.7	119.0	112.0	0.45133445
7	**96.7**	96.1	0.7	124.0	129.0	0.53252004
13	**96.7**	96.2	0.5	120.5	132.5	0.57753618
22	**96.7**	96.3	0.4	120.5	132.5	0.57753618
24	**96.7**	96.7	0.0	120.5	132.5	0.57753618
100	**96.7**	96.7	0.0	120.5	132.5	0.57753618

4. Conclusion

This paper presents a new pruning method for dendritic neuron model that can compress the model size while maintaining accuracy. The method clusters the significance of dendrite layers on a few layers using a penalty term and removes the rest of the low-significance layers. The experiments show that the method outperforms the PBSD method in terms of compression and performance.

We also discuss some insights and implications of the new pruning method. We demonstrate that the significance of each dendrite layer can be computed analytically using probability theory, which eliminates the need for empirical or approximate methods. We also illustrate that the penalty term can enforce the significance constraint and skew the significance distribution, which enables pruning more dendrite layers.

5. Future Works

There are some possible directions for future work. One direction is to validate the effectiveness of the proposed pruning method on more datasets and more complex tasks, such as image classification, natural language processing, or reinforcement learning. Another direction is to explore other ways to measure or constrain the significance of dendrite layers or synaptic connections, such as using information theory or regularization techniques. A third direction is to combine pruning with other compression techniques, such as quantization or coding, and optimize them jointly to achieve higher compression rates and better performance.

Acknowledgement

This work was supported in part by the National Natural Science Foundation of China under Grant No. 61971383 and No. 62263002, and in part by the Guangxi Vocational Education Teaching Reform Research Project in 2020 under Grant GXGZJG2020B101.

References

[1] LUO J H, WU J. An entropy-based pruning method for cnn compression[J]. arXiv preprint arXiv, 2017: 1706.05791.

[2] GUO Y, YAO A, CHEN Y. Dynamic network surgery for efficient dnns[C]//Advances in Neural Information Processing Systems, December 5-10, 2016, Barcelona, Spain. 2016: 1379-1387.

[3] HAN S, POOL J, TRAN J, et al. Learning both weights and connections for efficient neural network[C]//Advances in Neural Information Processing Systems, December 7-12, 2015, Montreal, Quebec, Canada.2015: 1135-1143.

[4] ZHANG T, YE S, ZHANG K, et al. A systematic dnn weight pruning framework using alternating direction method of multipliers[C]//The European Conference on Computer Vision, September 8-14, 2018, Munich, Germany. Springer, c2018: 184-199.

[5] WEN W, WU C, WANG Y, et al. Learning structured sparsity in deep neural networks[C]// Advances in Neural Information Processing Systems, December 5-10, 2016, Barcelona, Spain. 2016: 2074-2082.

[6] HE Y, ZHANG X, SUN J. Channel pruning for accelerating very deep neural networks[C]// The IEEE International Conference on Computer Vision, October 22-29, 2017, Venice, Italy. IEEE, c2017: 1389-1397.

[7] MIN C, WANG A, CHEN Y, et al. 2pfpce: two-phase filter pruning based on conditional entropy[J]. arXiv preprint arXiv, 2018: 1809.02220.

[8] ZHANG T, ZHANG K, YE S, et al. Adam-admm: a unified, systematic framework of structured weight pruning for dnns[J]. arXiv preprint arXiv, 2018: 1807.11091.

[9] BOYD S, PARIKH N, CHU E, et al. Distributed optimization and statistical learning via the alternating direction method of multipliers[J]. Foundations and trends in machine learning, 2011, 3(1): 1-122.

[10] SUZUKI T. Dual averaging and proximal gradient descent for online alternating direction multiplier method[C]//International Conference on Machine Learning, June 16-21, 2013,

Atlanta, GA, USA. JMLR.org, c2013: 392-400.

[11] ULLRICH K, MEEDS E, WELLING M. Soft weight-sharing for neural network compression[J]. arXiv preprint arXiv, 2017: 1702.04008.

[12] LOUIZOS C, WELLING M, KINGMA D P. Learning sparse neural networks through l 0 regularization[J]. arXiv preprint arXiv, 2017: 1712.01312.

[13] HUANG G B, SARATCHANDRAN P, SUNDARARAJAN N. A generalized growing and pruning RBF (GGAP-RBF) neural network for function approximation[J]. IEEE transactions on neural networks, 2005, 16(1): 57-67.

[14] HASSIBI B, STORK D G. Second order derivatives for network pruning: optimal brain surgeon[J]. Advances in neural information processing systems, 1993, 5: 164-171.

[15] CUN Y L, DENKER J S, SOLLA S A. Optimal brain damage[C]//Advances in Neural Information Processing Systems, November 27-30, 1989, Denver, Colorado, USA. Morgan Kaufmann, c1989: 598-605.

[16] HAN S, MAO H Z, DALLY W J. Deep compression: compressing deep neural networks with pruning, trained quantization and huffman coding[J]. Fiber, 2015, 56(4): 3-7.

[17] KOCH C, POGGIO T, TORRE V. Retinal ganglion cells: a functional interpretation of dendritic morphology[J]. Philosophical transactions of the royal society of London, 1982, 298(1090): 227.

[18] TANG Z, TAMURA H, KURATU M, et al. A model of the neuron based on dendrite mechanisms[J]. Electronics and communications in Japan, 2010, 84(8): 11-24.

[19] GAO S, ZHOU M, WANG Y, et al. Dendritic neuron model with effective learning algorithms for classification, approximation, and prediction[J]. IEEE transactions on neural networks and learning systems, 2018, 30(2): 601-614.

[20] JI J, GAO S, CHENG J, et al. An approximate logic neuron model with a dendritic structure[J]. Neurocomputing, 2016, 173: 1775-1783.

[21] TODO Y, TANG Z, TODO H, et al. Neurons with multiplicative interactions of nonlinear synapses[J]. International journal of neural systems, 2019, 29(08): 115-133.

[22] TANG Y J, JI J K, GAO S C, et al. A pruning neural network model in credit classification analysis[J]. Computational intelligence and neuroscience, 2018, 2018(4): 1-22.

[23] TANG Y, JI J, ZHU Y, et al. A differential evolution-oriented pruning neural network model for bankruptcy prediction[J]. Complexity, 2019(1): 1-21.

[24] SHA Z J, LIN H U, TODO Y, et al. A breast cancer classifier using a neuron model with dendritic nonlinearity[J]. IEICE transactions on information and systems, 2015, E98.D(7): 1365-1376.

[25] JIANG T, GAO S, WANG D, et al. A neuron model with synaptic nonlinearities in a dendritic tree for liver disorders[J]. IEEJ transactions on electrical and electronic engineering, 2017, 12: 105-115.

[26] ZHOU T L, GAO S C, WANG J H, et al. Financial time series prediction using a dendritic neuron model[J]. Knowledge-based systems, 2016, 105(C): 214-224.

[27] GAO S, ZHOU M, WANG Z, et al. Fully complex-valued dendritic neuron model[J]. IEEE transactions on neural networks, 2021: 1-14.

[28] SONG Z, TANG Y, JI J, et al. Evaluating a dendritic neuron model for wind speed

forecasting[J]. Knowledge based systems, 2020, 106052: 201-202.

[29] ZHANG Z, LIU H, ZHOU M, et al. Solving dynamic traveling salesman problems with deep reinforcement learning[J]. IEEE transactions on neural networks and learning systems, 2021, 34(4): 2119-2132.

[30] SONG S, CHEN X, SONG S, et al. A neuron model with dendrite morphology for classification[J]. Electronics, 2021, 10(9): 1062.

[31] ZHANG T, LV C, MA F, et al. A photovoltaic power forecasting model based on dendritic neuron networks with the aid of wavelet transform[J]. Neurocomputing, 2020, 397: 438-446.

[32] YU Y, LEI Z, WANG Y, et al. Improving dendritic neuron model with dynamic scale-free network-based differential evolution[J]. IEEE/CAA journal of automatica sinica, 2022, 9(1): 99-110.

[33] WEN X, ZHOU M, LUO X, et al. Novel pruning of dendritic neuron models for improved system implementation and performance[C]//IEEE International Conference on Systems, Man, and Cybernetics (SMC), 2021, 1559-1564.

[34] LUO X, WEN X, ZHOU M, et al. Decision-tree-initialized dendritic neuron model for fast and accurate data classification[J]. IEEE transactions on neural networks and learning systems, 2022, 33(9): 4173-4183.

[35] DEMŠAR J. Statistical comparisons of classifiers over multiple data sets[J]. Journal of machine Learning research, 2006, 7(01): 1-30.

[36] KIM J H. Estimating classification error rate: repeated cross-validation, repeated hold-out and bootstrap[J]. Computational statistics & data analysis, 2009, 53(11): 3735-3745.

[37] RODRIGUEZ J D, PEREZ A, LOZANO J A. Sensitivity analysis of k-fold cross validation in prediction error estimation[J]. IEEE transactions on pattern analysis and machine intelligence, 2009, 32(3): 569-575.

[38] GARCÍA S, MOLINA D, LOZANO M, et al. A study on the use of non-parametric tests for analyzing the evolutionary algorithms' behaviour: a case study on the cec'2005 special session on real parameter optimization[J]. Journal of heuristics, 2008, 15: 617.

A General Paradigm of Knowledge-Driven and Data-Driven Fusion*

1. Introduction

While the term artificial intelligence (AI) was first coined by John McCarthy in 1956, AI has made huge progress thanks to better algorithms, more data, and faster and stronger computers and technology. Artificial intelligence technology has recently been integrated into all aspects of people's lives and production. Machine learning, especially deep learning, has effectively expanded the boundaries of problems that artificial intelligence can solve. The current popular deep learning mainly belongs to data-driven artificial intelligence technology. Although deep learning has achieved inevitable success, it has not yet constructed true intelligence. The current artificial intelligence systems based on deep learning are not mature enough, and the application scenarios are usually stringent. For example, the scenarios always need reasonable certainty, obtain complete information, a relatively static working environment, a relatively single work task, and so on. If the application scenario does not meet these conditions, the performance of the artificial intelligence system based on deep learning does not work well or even becomes unusable.

The current deep learning has a distinctive feature that is hardly unexplainable and incomprehensible, mainly manifested in two aspects. On the one hand, people cannot understand and explain the decision-making process of deep learning. On the other hand, it is not easy to accurately understand people's intentions using deep learning methods. It is precisely because of this inexplicability and incomprehensibility that deep learning has gained the title of "alchemy".

AI projects used to depend on human experts, who gave their knowledge to knowledge engineers. They were clear and easy for people to design and understand. For example, expert systems, often based on decision trees, are good models of how humans make

---

* The paper was originally published in *2023 15th International Conference on Advanced Computational Intelligence (ICACI)*, 2023, and has since been revised with new information. It was co–authored by Fei Hu, Long Ye, Qin Zhang, Wei Zhong, and Danting Duan.

decisions and are easy for both developers and end-users to understand. But in the last decade, the main AI method changed to machine learning systems based on Deep Neural Networks (DNN), which made understanding harder. The current systems are like "black boxes", hard for humans to understand but very good at learning new things and doing things that worry us. As long as they have Big Data and Huge Compute, they don't need any human knowledge to do better than humans. Because of their new abilities, DNN-based AI systems are used to make decisions in investing, security, and many other important areas. Since many of these areas have legal rules, it is good that these systems can explain how they made their decisions, mainly to show that they are not biased.

Due to the inexplicability and incomprehensibility, people are usually required to make further decisions based on deep learning results. A production line product defect detection system using deep learning can detect suspected defective products but ultimately requires people to determine whether it is a genuine defective product. A communication anti-fraud platform using deep learning can detect suspected fraudulent text messages, but ultimately Humans are also needed to determine whether it is a fraudulent text message. Self-driving cars can drive under normal road conditions, but human participation is still needed to deal with emergencies and so on.

Cooperation between man and machine is an actual demand in the future. The inexplicability and incomprehensibility of machine learning affect the ability of humans and machines to cooperate. The key to human-machine dialogue is the mutual understanding between machines and humans, while machines based on deep learning understand things based on probability and statistics. This method cannot achieve a deep understanding of human intentions. With the development of machine learning-based speech recognition and natural language processing technologies, the ability of machines to understand natural human language will become stronger and stronger; that is, machines will understand humans more and more.

Nevertheless, on the other hand, people still cannot understand the working principle of deep learning, and it is not easy to understand the machine decision process and processing results based on deep learning. Failure to understand leads to distrust, which causes enormous problems for man-machine cooperation. The anti-interference ability of deep learning is also weak, the dependence on data samples is too high, and the certainty is insufficient. Some samples cannot be obtained in the production process, making it challenging to learn various behaviors or features that need to be detected through data training. For example, it is difficult for autonomous driving based on deep learning to respond to unexpected road conditions, and it is difficult for network security detection systems based on trial learning to discover new types of network attack events, and so on. These sudden traffic incidents and new cyber-attacks are challenging to learn in advance.

The success of deep learning comes from its practical analysis and utilization of big data, and its shortcomings also come from its excessive reliance on big data. This shortcoming of deep learning limits its application and development space.

The strength of human intelligence lies in the processing of knowledge and the use of knowledge to find the essence of things. Human knowledge includes prior models, logical rules, representation learning, statistical constraints, etc. The machine's strength lies in processing data and in finding characteristics of things from the data. The ability of machines to apply data is more potent than that of humans, and that of humans to apply knowledge is more potent than that of machines. If the two can be organically combined, artificial intelligence can serve the development of the industry well. Human intelligence cannot be learned from purely data-driven artificial intelligence systems. The specialty of the data-driven artificial intelligence system lies in the grasp of the local characteristics of things, and the specialty of the knowledge-driven artificial intelligence system lies in the grasp of the global attributes of things. Data-driven and knowledge-driven fusion will be able to organically combine the local and global, the machine's implicit intuition, and people's common-sense judgment. Introducing knowledge into a data-driven artificial intelligence system can enable the artificial intelligence system to have specific reasoning and decision-making capabilities and acquire the ability to deal with emergencies, reducing the dependence on large data sets. However, the existing fusion models depend on specific tasks. This paper expects to mine the existing fusion models and build a general fusion paradigm for fusion tasks.

2. Knowledge-Driven and Data-Driven Fusion Paradigm

The role of knowledge in the fusion model can be reflected in the following three levels: input, rule and structure. We summarize the fusion paradigm as

$$\min \sum_d L\left(F\left(x_d + f_k, x_k, \theta + \theta_k\right), y_d, \delta + \delta_k\right)$$

while L is the object function, $(x_d, y_d)_d$ denotes the dataset, f_k denotes the preprocess for input data, x_k denotes the knowledge tensor, θ denotes the parameters of the data model F, θ_k denotes the knowledge rule of the model, δ and δ_k denotes the parameters and the knowledge rule of the objective function.

According to the expression form and the application of knowledge in the fusion model, the basic fusion mode can be divided into the following three modes:

- **Knowledge as Input** When knowledge can be directly represented as a standard tensor, the tensor can be used as a complement to the raw input of data-driven models. This mode can be represented as follows:

$$\min \sum_d L\left(F(x_d, x_k, \theta), y_d, \delta\right)$$

- **Knowledge as Rules** Knowledge is not used directly as input, but rather a constraint on the data-driven model. This mode can be represented as follows:

$$\min \sum_d L\left(F(x_d + f_k, \theta), y_d, \delta + \delta_k\right)$$

- **Knowledge an Implicit Modal** If knowledge can guide the construction of data-driven models. This mode can be represented as follows:

$$\min \sum_d L\left(F(x_d, x_k, \theta + \theta_k), y_d, \delta\right)$$

These three modes are basic modes, which can be used separately or mixed in actual modeling. Some of the knowledge fusion methods presented in this article are based on existing methods, while others represent novel methods or variants of an existing method. The current deep learning network is the most fruitful data-driven method, so in this article, the network will generally refer to data-driven models.

3. Basic Modes

3.1 Knowledge as Input

1) Knowledge Represented by Tensor: In this case, knowledge is data-related and can be expressed in structured or unstructured variables. In that case, we can directly process and calculate the variables through the network model to obtain the corresponding knowledge representation, directly used as auxiliary information for a single training improvement. The vectorization of the sentence can well represent the knowledge expressed in the language as a vector. The knowledge vector can be directly used as an auxiliary input.

2) Knowledge Graph: In many cases, knowledge cannot be directly expressed by vectors. However, we can display these complex knowledge fields through data mining, information processing, knowledge measurement, and graph drawing, reveal the law of dynamic development in the field of knowledge, and provide practical and valuable references for subject research. This knowledge can be represented as a knowledge graph. The knowledge graph's entities, concepts, and relationships are expressed in discrete and explicit symbolic representations. However, these discrete symbolic representations are difficult to apply directly to neural networks based on numerical representations. The representation of the knowledge graph can obtain the real-valued vectorized representation of the knowledge graph's constituent elements (nodes and edges). These continuous vectorized representations can be used as the input of the neural network so that the neural network model can fully use a large amount of prior knowledge in the knowledge graph.

For example, the range of images generated by GAN is too extensive and difficult to control. The scene we need can be entered into the network as a knowledge map to generate images of specified content. On the one hand, the image content is more accurate, and on the other hand, meaninglessness is reduced.

3.2 Knowledge as Rules

In this case, knowledge is a rule-based restriction on a data- driven approach and cannot be used directly as input. We summarize this type of mode into the following types:

1) Artificial prior: Artificial prior refers to manually setting rules or defining problems based on prior knowledge. Knowledge-based loss function such as regularization loss and multi-loss belongs to artificial prior. We take a classification task of the image data set CIFAR-10 as an example to explain the power of knowledge-based loss. We construct a simplified convolutional neural network that maintains the same number of layers as AlexNet as a baseline. The baseline model has trained 101 epochs, the optimization algorithm is Adam, the learning rate is set to 0.001, and the result is present in Fig. 1.

Considering the sparsity of the prediction result, a 0.1-fold L1 limitation is added to the loss function. The result is shown in Fig. 2, which is a further increase of 4% compared to the baseline.

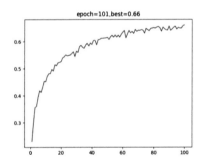

Fig. 1 The result of the baseline model.

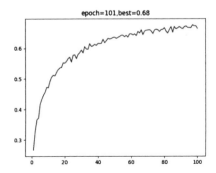

Fig. 2 The result with knowledge-based loss.

2) Knowledge-based preprocessing: Data preprocessing can be divided into data-based and knowledge-based. Data-based preprocessing involves transforming raw data into well-formed data sets with statistical methods so that data mining analytics can be applied. Knowledge-based preprocessing preprocess input data or labels according to existing knowledge. We still use the classification experiment of CIFAR-10 to illustrate the knowledge-based preprocessing.

To reduce redundancy and improve data integrity, we count the distribution of the dataset and then normalize the dataset. As shown in Fig. 3, the model with knowledge-based preprocessing is six percentage points higher than the baseline model. The knowledge-based preprocess dramatically improves the performance of the model.

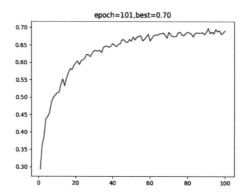

Fig. 3 The result with knowledge-based preprocessing.

3) Artificial posterior: Artificial posterior refers to knowledge based on the posterior, and using the knowledge to optimize the solution or directly prediction. Posterior knowledge can be learning from samples or artificially summarizing from posterior fact. If the rules can be represented in a definite form, then the rules learned from the training data can be used to discriminate unseen examples. Rule learning or knowledge graph generation both belong to this category. Taking an association rule discovery task as example. Association rule learning finds patterns among the datapoints in a large dataset. It uses data mining to discover useful business connections or regularities among the variables. It is currently used in the sales industry to predict if a person will buy item A based on their previous purchase B. It is widely used by businesses to find out how customers buy items together. It uses if/then statements. It mainly reveals how datapoints or entities are related when they are used together. A set of transactions is given below, from which we can get association rules and predict if an item will appear based on other items. The rule item3 → item4 specifies that people who buy item3 also tend to buy item4.

(a) item1, item3, item4

(b) item2, item5
(c) item1, item2, item3, item4
(d) item2, item3, item4, item6
(e) item1, item2, item5

3.3 Knowledge as Implicit Modal

In some cases, knowledge is at a higher level, which can be used to guide the design of data-driven models. We summarize this mode into the following types:

1) Knowledge Transfer: Knowledge Transfer is to transfer the "knowledge" of a trained model to another network, or to learn the "knowledge" of the trained model or imitate the behavior of the model through the network. To transfer knowledge of other modalities, corresponding modules or additional tasks can be added according to the semantic information of specific tasks to realize cross-modal knowledge transfer. There are two common ways to achieve knowledge transfer:

- Transfer learning. Transfer learning is when elements of a pre-trained model are reused in a new machine learning model. Transfer learning is not a distinct type of machine learning algorithm, instead it's a technique or method used whilst training models. The knowledge developed from previous training is recycled to help perform a new task. The new task will be related in some way to the previously trained task, which could be to categorise objects in a specific file type. The original trained model usually requires a high level of generalisation to adapt to the new unseen data. Transfer learning means that training won't need to be restarted from scratch for every new task. Training new machine learning models can be resource-intensive, so transfer learning saves both resources and time. It is becoming an important part of the evolution of machine learning and is increasingly used as a technique within the development process.

- Knowledge distillation. Knowledge distillation is a way of making a big and complex model or group of models smaller and simpler so that they can work better in real situations. It is a kind of model compression that was first shown to work by Bucilua and others in 2006. Since knowledge distillation is often used for neural network models that have many layers and parameters, it has become more popular in the last ten years with the rise of deep learning. Deep learning has been successful in many fields like speech recognition, image recognition, and natural language processing. Knowledge distillation helps make deep learning models more practical for real applications.

2) Relevant Experience Transfer: Relevant Experience Transfer differs from direct Experience Transfer. In many cases, integrating new information too quickly can

create catastrophic disruptions, and previously acquired knowledge is suddenly lost. Complementary Learning Systems Theory (CLST) suggests that new memories can be gradually integrated into the neocortex by interweaving them with existing knowledge. Many studies have now successfully used interleaved replay to achieve lifelong learning in neural networks. Rajat Saxena et al. also showed that training with similarity-weighted interleaving of old items with new ones allows deep networks to learn new items rapidly without forgetting while using substantially less data.

3) Knowledge Refinement Task: Knowledge Refinement Task means refine tasks through knowledge, refine in stages, refine modules, and increase the interpretability of the entire learning step. Task refinement and feature decoupling both belong to this category. Taking nonlinear distortion cancellation with neural network as an example. For a system S on space R with an unknown nonlinear distortion function $F(x) = x + x^3$, the sampling frequency is f_s, $\min_G \max_{x_i \in R} \sum_{i=1}^{f_s} \|F(G(x_i)) - x_i\|^2$ where G is a neural network. $x = \sum_i^m \sin\left(2\pi * f_i * \dfrac{j}{f_s}\right), 0 \leq j \leq 10k$

The goal is to explore the ability of neural networks to perform nonlinear distortion cancellation. Nonlinear distortion cancellation is a classic and difficult problem. Due to the high requirements of high-precision prediction for the task, we decompose the task into three steps:

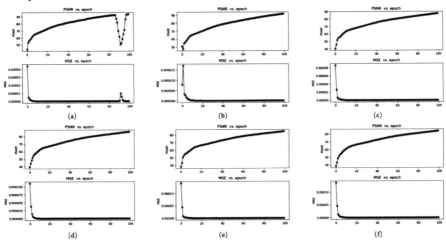

Fig.4 Network solution with different overlaps on a single frequency $f_i = 10\text{Hz}$ (a)-(f) overlap=50; 40; 20; 10; 5; 0.

A General Paradigm of Knowledge–Driven and Data–Driven Fusion

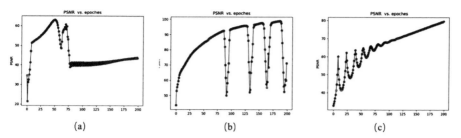

Fig.5 Network solution on different frequencies at a single frequency. (a)-(c)
$$f_i = 1\text{Hz}, 10\text{Hz}, 100\text{Hz}$$

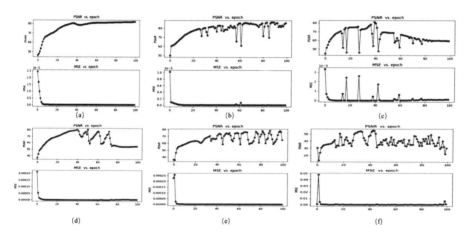

Fig. 6 Network solution at multiple frequencies. (a)-(f) m=4,5,6,8,10,18.

Step1. Network solution with different overlaps on a single frequency. As shown in Fig. 4, the optimal PSNR gradually decreases as the overlap length decreases. In the case of overlap=50, there is an unstable situation at the end of training, and other overlapping situations are stable.

Step2. Network solution on different frequencies at a single frequency. As shown in Fig. 5, when the frequency is too low, the network will be more difficult to converge.

Step3. Network solution at multiple frequencies. As shown in Fig. 6, the stability of the network fit decreases as the number of frequencies increases. Under a certain number of frequencies, the fitting ability of the network is relatively stable. Excessive frequencies still require further study.

4. Conclusion

This paper proposes a general framework of the fusion paradigm. Under this paradigm, we introduce and show the three basic models: Knowledge as Input, Knowledge as Rules, and Knowledge as Implicit Modal. The techniques of knowledge fusion have essential applications in building, updating, and applying data-driven models. There are still many problems to be solved in related research, such as how to improve the quality and confidence of the knowledge base and so on. As the data scale and the number of knowledge base methods continue to increase, so will the demand for knowledge fusion. Based on the analysis of this paper, we hope it can provide some helpful guidance and inspiration for the research of knowledge fusion of future networks.

References

[1] KAHN K, WINTERS N. Constructionism and ai: a history and possible futures[J]. Br. J. educ. technol., 2021, 52: 1130-1142.

[2] ZHANG J, TAO D. Empowering things with intelligence: a survey of the progress, challenges, and opportunities in artificial intelligence of things[J]. IEEE internet of things journal, 2021, 8: 7789-7817.

[3] SHI F, WANG J, SHI J, et al. Review of artificial intelligence techniques in imaging data acquisition, segmentation, and diagnosis for covid-19[J]. IEEE reviews in biomedical engineering, 2021, 14: 4-15.

[4] GURGITANO M, ANGILERI S A, RODA G M`, et al. Interventional radiology ex-machina: impact of artificial intelligence on practice[J]. La Radiologia Medica, 2021, 126: 998-1006.

[5] WERBOS P J. Elastic fuzzy logic: a better fit to neurocontrol and true intelligence[J]. J. intell. fuzzy syst., 1993, 1: 365-377.

[6] GABRIEL I. Artificial intelligence, values and alignment[J]. Minds and machines, 2020, 30(3): 411-437.

[7] HUTSON M. Has artificial intelligence become alchemy?[J]. Science, 2018, 360(6388): 478.

[8] YAMPOLSKIY R V. Leakproofing the singularity artificial intelligence confinement problem[J]. Journal of consciousness studies, 19(1-2): 194-214.

[9] ARMSTRONG S, YAMPOLSKIY R. Security solutions for intelligent and complex systems[M]//Security solutions for hyperconnectivity and the internet of things. Pennsylvania, United States: IGI Global, 2020: 37-88.

[10] SILVER D, SCHRITTWIESER J, SIMONYAN K, et al. Mastering the game of go without human knowledge[J]. Nature, 2017, 550: 354-359.

[11] ROBERT C P. Superintelligence: paths, dangers, strategies[J]. Chance, 2017, 30: 42-43.

[12] TRIPPI R R, TURBAN E. Neural networks in finance and investing: using artificial

intelligence to improve real-world performance[M]. New York: McGraw-Hill, Inc., Professional Book Group, 1995.

[13] NOVIKOV D V, YAMPOLSKIY R, REZNIK L. Artificial intelligence approaches for intrusion detection[C]//IEEE Long Island Systems, Applications and Technology Conference, 2006: 1-8.

[14] NOVIKOV D, YAMPOLSKIY R V, REZNIK L. Anomaly detection based intrusion detection [C]//Third International Conference on Information Technology: New Generations (ITNG'06), 2006: 420-425.

[15] WANG H, WANG N, YEUNG D Y. Collaborative deep learning for recommender systems [C]//Proceedings of the 21th ACM SIGKDD International Conference on Knowledge Discovery and Data Mining, 2015.

[16] KOREN Y, BELL R M, VOLINSKY C. Matrix factorization techniques for recommender systems[J]. Computer, 2009, 42(8): 30-37.

[17] GLIKSON E, WOOLLEY A W. Human trust in artificial intelligence: review of empirical research[J]. The academy of management annals, 2020, 14: 627-660.

[18] DETTMERS T, MINERVINI P, STENETORP P, et al. Convolutional 2d knowledge graph embeddings[C]//The Thirty-Second AAAI Conference on Artificial Intelligence, February 2-7, 2018, New Orleans, Louisiana, USA. AAAI Press, c2018.

[19] LIU W, ZHOU P, ZHAO Z, et al. K-bert: enabling language representation with knowledge graph[C]//The Thirty-Fourth AAAI Conference on Artificial Intelligence, February 7-12, 2020, New York, NY, USA. AAAI Press, c2020.

[20] WANG Z, ZHANG J, FENG J, et al. Knowledge graph embedding by translating on hyperplanes[C]//The Twenty-Eighth AAAI Conference on Artificial Intelligence, July 27 -31, 2014, Québec City, Québec, Canada. AAAI Press, c2014.

[21] WANG M, QIU L, WANG X. A survey on knowledge graph embeddings for link prediction[J]. Symmetry, 2021, 13: 485.

[22] ILIEVSKI F, SZEKELY P A, ZHANG B. Cskg: the commonsense knowledge graph[C]// European Semantic Web Conference (ESWC), June 6-10, 2021, Virtual Event. Springer, c2021.

[23] CHAMI I, WOLF A, JUAN D C, et al. Low-dimensional hyperbolic knowledge graph embeddings[C]//The 58th Annual Meeting of the Association for Computational Linguistics, July 5-10, 2020, Virtual Event. ACL, c2020.

[24] KIPF T, WELLING M. Semi-supervised classification with graph convolutional networks[J]. ArXiv, 2017, 1609. 02907: 1-14.

[25] LOSHCHILOV I, HUTTER F. Decoupled weight decay regularization[C]//The 7th International Conference on Learning Representations, May 6-9, 2019, New Orleans, LA, USA. OpenReview.net, c2019.

[26] ZHU Z, JIANG X, ZHENG F, et al. Viewpoint-aware loss with angular regularization for person re-identification[C]//The Thirty-Fourth AAAI Conference on Artificial Intelligence, February 7-12, 2020, New York, NY, USA. AAAI Press, c2020.

[27] LETCHER A. On the impossibility of global convergence in multi-loss optimization[J]. ArXiv, 2005.12649, 2021.

[28] DU X, YU Z, ZHU B, et al. Bytecover: cover song identification via multi-loss training[C]// IEEE International Conference on Acoustics, Speech and Signal Processing, June 6-11, 2021, Toronto, ON, Canada. IEEE, c2021: 551-555.

[29] FAN C, CHEN M, WANG X, et al. A review on data preprocessing techniques toward efficient and reliable knowledge discovery from building operational data[J]. Frontiers in energy research, 2021, 9: 652801.

[30] GNANAPRIYA S, SUGANYA R, DEVI G, et al. Data mining concepts and techniques[J]. Data mining and knowledge engineering, 2010, 2: 256-263.

[31] CACHIN C, WIESMANN H J. Pd recognition with knowledge-based preprocessing and neural networks[J]. IEEE transactions on dielectrics and electrical insulation, 1995, 2: 578-589.

[32] TAN P N, STEINBACH M S, KUMAR V. Introduction to data mining[M]//Data mining and machine learning applications. Beverly, MA: Scrivener Publishing, 2022: 1-19.

[33] SALAM A, SCHWITTER R, ORGUN M A. Probabilistic rule learning systems[J]. ACM computing surveys (CSUR), 2021, 54: 1-16.

[34] ZHOU Z H. Rule learning[M]//Machine Learning. Singapore: Springer, 2021: 373-398.

[35] MEILICKE C, CHEKOL M W, RUFFINELLI D, et al. Anytime bottom-up rule learning for knowledge graph completion[C]//International Joint Conference on Artificial Intelligence, IJCAI, August 10-16, 2019, Macao, China. Morgan Kaufmann, c2019.

[36] CHEN Y, LI H, QI G, et al. Outlining and filling: hierarchical query graph generation for answering complex questions over knowledge graph[J]. ArXiv, 2111.00732, 2021.

[37] KIM T, YUN Y, KIM N. Deep learning-based knowledge graph generation for covid-19[J]. Sustainability, 2021, 13: 2276.

[38] HEYVAERT P, MEESTER B D, DIMOU A, et al. Rule-driven inconsistency resolution for knowledge graph generation rules[J]. Semantic web, 2019, 10: 1071-1086.

[39] KOTSIANTIS S B, KANELLOPOULOS D N. Association rules mining: a recent overview[J]. GESTS international transactions on computer science and engineering, 2006, 32: 71-82.

[40] ABDEL-BASSET M, MOHAMED M, SMARANDACHE F, et al. Neutrosophic association rule mining algorithm for big data analysis[J]. Symmetry, 2018, 10: 106.

[41] TANC, SUN F, KONG T, et al. A survey on deep transfer learning[J]. ArXiv, 1808.01974, 2018.

[42] PAN S J, YANG Q. A survey on transfer learning[J]. IEEE transactions on knowledge and data engineering, 2010, 22: 1345-1359.

[43] ZHUANG F, QI Z, DUAN K, et al. A comprehensive survey on transfer learning[J]. Proceedings of the IEEE, 2021, 109: 43-76.

[44] HINTON G E, VINYALS O, DEAN J. Distilling the knowledge in a neural network[J]. ArXiv, 1503.02531, 2015.

[45] CHEN P, LIU S, ZHAO H, et al. Distilling knowledge via knowledge review[C]//IEEE/CVF Conference on Computer Vision and Pattern Recognition, June 19-25, 2021, Virtual Event. IEEE, c2021: 5006-5015.

[46] GOU J, YU B, MAYBANK S J, et al. Knowledge distillation: a survey[J]. ArXiv, 2006.05525, 2021.

[47] BUCILA C, CARUANA R, NICULESCU-MIZIL A. Model compression[C]//The Twelfth ACM SIGKDD International Conference on Knowledge Discovery and Data Mining, Philadelphia, August 20-23, 2006, PA, USA. ACM, c2006.

[48] GEPPERTH A R T, KARAOGUZ C. A bio-inspired incremental learning architecture for applied perceptual problems[J]. Cognitive computation, 2016, 8: 924-934.

[49] KEMKER R, KANAN C. Fearnet: brain-inspired model for incremental learning[J]. ArXiv, 1711.10563, 2018.

[50] SAXENA R, SHOBE J L, MCNAUGHTON B L. Learning in deep neural networks and brains with similarity-weighted interleaved learning[J]. Proceedings of the national academy of sciences of the United States of America, 2022, 119(27): e2115229119.

[51] ZISER Y, REICHART R. Task refinement learning for improved accuracy and stability of unsupervised domain adaptation[C]//The 57th Annual Meeting of the Association for Computational Linguistics, July 28-August 2, 2019, Florence, Italy. ACL, c2019.

[52] WAHLSTEN D. Task refinement and standardization[M]//Mouse behavioral testing, London: Elsevier, 2011: 215-233.

[53] SAHARIA C, HO J, CHAN W, et al. Image super-resolution via iterative refinement[J]. ArXiv, 2104.07636, 2021.

[54] FENG Z, XU C, TAO D. Self-supervised representation learning by rotation feature decoupling[C]//IEEE/CVF Conference on Computer Vision and Pattern Recognition, June 15-20, Long Beach, CA, USA. IEEE, c2019: 10356-10366.

[55] ZHOU F, HANG R, LIU Q. Class-guided feature decoupling network for airborne image segmentation[J]. IEEE transactions on geoscience and remote sensing, 2021, 59: 2245-2255.

[56] CHEN W, WANG Q, HUANG S, et al. Dfdm: a deep feature decoupling module for lung nodule segmentation[C]//IEEE International Conference on Acoustics, Speech and Signal Processing, June 6-11, 2021, Toronto, ON, Canada. IEEE, c2021: 1120-1124, 2021.